Lecture Notes in Physics

For information about Vols. 1–110, please contact your bookseller or Springer-Verlag.

Vol. 111: H.-J. Schmidt, Axiomatic Characterization of Physical Geometry. V, 163 pages. 1979.

Vol. 112: Imaging Processes and Coherence in Physics. Proceedings, 1979. Edited by M. Schlenker et al. XIX, 577 pages. 1980.

Vol. 113: Recent Advances in the Quantum Theory of Polymers. Proceedings 1979. Edited by J.-M. André et al. V, 306 pages. 1980.

Vol. 114: Stellar Turbulence. Proceedings, 1979. Edited by D. F. Gray and J. L. Linsky. IX, 308 pages. 1980.

Vol. 115: Modern Trends in the Theory of Condensed Matter. Proceedings, 1979. Edited by A. Pekalski and J. A. Przystawa. IX, 597 pages. 1980.

Vol. 116: Mathematical Problems in Theoretical Physics. Proceedings, 1979. Edited by K. Osterwalder. VIII, 412 pages. 1980.

Vol. 117: Deep-Inelastic and Fusion Reactions with Heavy Ions. Proceedings, 1979. Edited by W. von Oertzen. XIII, 394 pages. 1980.

Vol. 118: Quantum Chromodynamics. Proceedings, 1979. Edited by J. L. Alonso and R. Tarrach. IX, 424 pages. 1980.

Vol. 119: Nuclear Spectroscopy. Proceedings, 1979. Edited by G. F. Bertsch and D. Kurath. VII, 250 pages. 1980.

Vol. 120: Nonlinear Evolution Equations and Dynamical Systems. Proceedings, 1979. Edited by M. Boiti, F. Pempinelli and G. Soliani. VI, 368 pages. 1980.

Vol. 121: F. W. Wiegel, Fluid Flow Through Porous Macromolecular Systems. V, 102 pages. 1980.

Vol. 122: New Developments in Semiconductor Physics. Proceedings, 1979. Edited by F. Beleznay et al. V, 276 pages. 1980.

Vol. 123: D. H. Mayer, The Ruelle-Araki Transfer Operator in Classical Statistical Mechanics. VIII, 154 pages. 1980.

Vol. 124: Gravitational Radiation, Collapsed Objects and Exact Solutions. Proceedings, 1979. Edited by C. Edwards. VI, 487 pages. 1980.

Vol. 125: Nonradial and Nonlinear Stellar Pulsation. Proceedings, 1980. Edited by H. A. Hill and W. A. Dziembowski. VIII, 497 pages. 1980.

Vol. 126: Complex Analysis, Microlocal Calculus and Relativistic Quantum Theory. Proceedings, 1979. Edited by D. Iagolnitzer. VIII, 502 pages. 1980.

Vol. 127: E. Sanchez-Palencia, Non-Homogeneous Media and Vibration Theory. IX, 398 pages. 1980.

Vol. 128: Neutron Spin Echo. Proceedings, 1979. Edited by F. Mezei. VI, 253 pages. 1980.

Vol. 129: Geometrical and Topological Methods in Gauge Theories. Proceedings, 1979. Edited by J. Harnad and S. Shnider. VIII, 155 pages. 1980.

Vol. 130: Mathematical Methods and Applications of Scattering Theory. Proceedings, 1979. Edited by J. A. DeSanto, A. W. Sáenz and W. W. Zachary. XIII, 331 pages. 1980.

Vol. 131: H. C. Fogedby, Theoretical Aspects of Mainly Low Dimensional Magnetic Systems. XI, 163 pages. 1980.

Vol. 132: Systems Far from Equilibrium. Proceedings, 1980. Edited by L. Garrido. XV, 403 pages. 1980.

Vol. 133: Narrow Gap Semiconductors Physics and Applications. Proceedings, 1979. Edited by W. Zawadzki. X, 572 pages. 1980.

Vol. 134: $\gamma\gamma$ Collisions. Proceedings, 1980. Edited by G. Cochard and P. Kessler. XIII, 400 pages. 1980.

Vol. 135: Group Theoretical Methods in Physics. Proceedings, 1980. Edited by K. B. Wolf. XXVI, 629 pages. 1980.

Vol. 136: The Role of Coherent Structures in Modelling Turbulence and Mixing. Proceedings 1980. Edited by J. Jimenez. XIII, 393 pages. 1981.

Vol. 137: From Collective States to Quarks in Nuclei. Edited by H. Arenhövel and A. M. Saruis. VII, 414 pages. 1981.

Vol. 138: The Many-Body Problem. Proceedings 1980. Edited by R. Guardiola and J. Ros. V, 374 pages. 1981.

Vol. 139: H. D. Doebner, Differential Geometric Methods in Mathematical Physics. Proceedings 1981. VII, 329 pages. 1981.

Vol. 140: P. Kramer, M. Saraceno, Geometry of the Time-Dependent Variational Principle in Quantum Mechanics. IV, 98 pages. 1981.

Vol. 141: Seventh International Conference on Numerical Methods in Fluid Dynamics. Proceedings. Edited by W. C. Reynolds and R. W. MacCormack. VIII, 485 pages. 1981.

Vol. 142: Recent Progress in Many-Body Theories. Proceedings. Edited by J. G. Zabolitzky, M. de Llano, M. Fortes and J. W. Clark. VIII, 479 pages. 1981.

Vol. 143: Present Status and Aims of Quantum Electrodynamics. Proceedings, 1980. Edited by G. Gräff, E. Klempt and G. Werth. VI, 302 pages. 1981.

Vol. 144: Topics in Nuclear Physics I. A Comprehensive Review of Recent Developments. Edited by T.T.S. Kuo and S.S.M. Wong. XX, 567 pages. 1981.

Vol. 145: Topics in Nuclear Physics II. A Comprehensive Review of Recent Developments. Proceedings 1980/81. Edited by T. T. S. Kuo and S. S. M. Wong. VIII, 571-1.082 pages. 1981.

Vol. 146: B. J. West, On the Simpler Aspects of Nonlinear Fluctuating. Deep Gravity Waves. VI, 341 pages. 1981.

Vol. 147: J. Messer, Temperature Dependent Thomas-Fermi Theory. IX, 131 pages. 1981.

Vol. 148: Advances in Fluid Mechanics. Proceedings, 1980. Edited by E. Krause. VII, 361 pages. 1981.

Vol. 149: Disordered Systems and Localization. Proceedings, 1981. Edited by C. Castellani, C. Castro, and L. Peliti. XII, 308 pages. 1981.

Vol. 150: N. Straumann, Allgemeine Relativitätstheorie und relativistische Astrophysik. VII, 418 Seiten. 1981.

Vol. 151: Integrable Quantum Field Theory. Proceedings, 1981. Edited by J. Hietarinta and C. Montonen. V, 251 pages. 1982.

Lecture Notes in Physics

Edited by H. Araki, Kyoto, J. Ehlers, München, K. Hepp, Zürich
R. Kippenhahn, München, H. A. Weidenmüller, Heidelberg
and J. Zittartz, Köln

192

Heidelberg Colloquium on Spin Glasses

Proceedings of a Colloquium
held at the University of Heidelberg
30 May – 3 June, 1983

Edited by J. L. van Hemmen and I. Morgenstern

Springer-Verlag
Berlin Heidelberg GmbH 1983

Editors

J. L. van Hemmen
Sonderforschungsbereich 123
Universität Heidelberg
Im Neuenheimer Feld 294
D-6900 Heidelberg 1

I. Morgenstern
Institut für Theoretische Physik
Universität Heidelberg
Philosophenweg 19
D-6900 Heidelberg 1

ISBN 978-3-540-12872-4 ISBN 978-3-540-38761-9 (eBook)
DOI 10.1007/978-3-540-38761-9

Originally published by Springer-Verlag Berlin Heidelberg New York Tokyo in 1983

2153/3140-543210

PREFACE

These are the proceedings of the Heidelberg Colloquium on Spin Glasses which was held at the University of Heidelberg from 30 May until 3 June, 1983. Since Les Houches 1978 (Ill-Condensed Matter), the developments in spin-glass theory and experiment have been extensive and this volume aims at presenting at least a substantial part of the newly gained insights and results.

There seems to be increasing evidence that metallic (long-range) and insulating (short-range) spin glasses are in different universality classes, perhaps even with different values of the lower critical dimensionality. We placed this issue before the conference participants and there now appears to be some agreement that metallic spin glasses indeed have lower critical dimensionalities. The reader may consult, for instance, the papers by Maletta, Omari et al., and Young.

"And what is the use of a book, without pictures or conversations" (Lewis Carroll, Alice in Wonderland). We, therefore, are very grateful to two individuals for providing us with appropriate illustrations: J. Souletie who, apparently, always discovers relevant news in newspapers (p.1), and W.F. Wolff who expertly summarized the conclusion (p.347).

The colloquium was organized with financial support from the Deutsche Forschungsgemeinschaft via the Sonderforschungsbereich 123 at the University of Heidelberg. In preparing the colloquium we greatly profited from the helpful suggestions of John Mydosh and Richard Palmer. We are also grateful to Jochen Canisius, John Chalker, and Aernout van Enter for their assistance during and after the conference. Finally, it is a pleasure to thank Christine Pendl (Springer-Verlag) for her expert help and advice.

Heidelberg, 28.X.1983

Leo van Hemmen
Ingo Morgenstern

CONTENTS

FRUSTRATION AND DISORDER
NEW PROBLEMS IN STATISTICAL MECHANICS
SPIN GLASSES IN A HISTORICAL PERSPECTIVE

Gérard Toulouse

Laboratoire de Physique de l'Ecole Normale Supérieure
24 rue Lhomond, 75231 Paris Cedex 05

This title was assigned to me by Leo and Ingo. It is a translation from a review paper, in French, written two years ago (1981) [1]. This present paper contains a revised history (of the period before 1975), with great help coming from the recent historical account of B.R. Coles : "The origin and influences of the spin-glass problem" [2]. Following is an updated survey of developments and spin-offs, since 1981, and specially since the Orsay meeting (January 1983). Though it was feared at some moments that these two meetings might be too close in time, I see at least three exciting lines of development which were not yet discussed in Orsay :

i) on the theoretical side, the physical interpretation of the order function q(x), a so far rather mysterious output of replica symmetry breaking in the mean field theory of spin glasses [3,4],

ii) on the numerical side, the exploration of the notion of relative defects for Heisenberg spin glasses in dimension three [5,6],

iii) on the experimental side, a beginning bridge on the gap between the physics well below T_f and the physics around T_f, with the Orsay torque experiments [7].

1. A revised history

In any history, one should read between the lines. Looking between the lines of my 1981 sketch-history of spin glasses, here are some of the things one can see :

- 1931, a study by J.W. Shih [8] of dilute alloys of iron in gold,
- 1951, the first observation of a spin glass property : resistivity maximum at low temperatures [9],
- 1959, observation of the difference between the magnetisation profiles, as obtained by either field cooling or cooling in a field opposite to the measurement field [10],
- 1968, observation of the susceptibility cusp and of its low field rounding, in LaGd [11].

JUDO

CHAMPIONNATS D'EUROPE

Sur les traces de Parisi

Paris. — Les judokas français n'ignorent pas qu'ils ont donné de mauvaises habitudes au public parisien, en ne ratant jamais un grand rendez-vous dans le vieux stade de Coubertin. Le dernier en date se soldant au mois de décembre dernier par quatre médailles d'or pour les jeunes filles aux championnats du monde féminins.

Ce public peut donc légitimement espérer la poursuite de cette série de succès, d'aujourd'hui jusqu'à dimanche, pour les championnats d'Europe messieurs.

Un retour en arrière pourrait cependant contribuer à tempérer quelque peu les illusions des plus optimistes. Il y a tout juste un an, une bien belle sélection se rendait à Rostock (R.D.A.), dans le but d'y maintenir sa place, la première à égalité avec l'U.R.S.S., sur l'échiquier européen. Angelo Parisi, Thierry Rey, Bernard Tchoullouyan, Roger Vachon échouaient tour à tour. Depuis 1978 à Helsinski, jamais les Français n'avaient subi pareil revers.

Une telle mésaventure n'est pas à exclure. Il suffirait que Parisi chute dès le premier jour, face, par exemple, au Soviétique Valeri Devesenko qu'il n'a jamais vu, que, dans le même temps, le revenant belge Robert Van De Walle déborde Roger Vachon pour que le doute s'installe.

Il est cependant plus raisonnable d'imaginer une équipe conquérante, dont six des sept éléments déjà titularisés — le « toutes catégories » ne sera désigné que samedi — reconnaissent bien volontiers avoir subi une préparation quasi idéale.

Seul le jeune marseillais Richard Melillo risque de ne pas être prêt, psychologiquement, et cela s'explique. Ce n'est que mardi qu'il a appris le forfait de Marcel Pietri. Surprenant champion de France au mois de mars, Melillo n'a guère eu l'occasion de se frotter aux meilleurs combattants de sa catégorie des légers. Il est vrai qu'il était pratiquement inconnu avant de remporter le titre national.

Rey n'a pas oublié l'uchi mata l'ayant projeté au sol l'année dernière. Son auteur avait pour nom Reissmann, membre d'une formation est-allemande portée par son public, accaparant six médailles dont deux d'or. Ce dernier, peut-être diminué puisqu'il a subi une opération au tendon de la main droite il y a un mois et demi, relèvera le défi.

Roger Vachon rêve pour sa part de retrouver sa suprématie, l'Autricien Kostenberger n'ayant, selon lui, assuré qu'un intérim d'un an...

Guy Lebaupin se souvient de 1977. Cette année là, l'Orléanais, encore junior, ne succombait qu'en finale, sur blessure, face au Soviétique Pogorelov. Physiquement, mais aussi moralement, il eut beaucoup de mal à supporter ce coup du sort, à vivre l'ascencion de Thierry Rey dans la plus petite catégorie, à repousser enfin les assauts de Guy Delvingt.

Enfin, le Normand Fabien Canu ne se pose guère de questions. Succéder à Bernard Tchoullouyan ne l'émeut guère. Il est vrai qu'il a atteint le plus haut niveau. Et il sait qu'avec un peu de chance...

Un succès d'Angelo Parisi, très probable dans la mesure où il semble le mieux armé sur le vieux continent, entraînerait vraisemblablement dans son sillage six combattants avides de gloire.

BOULISME

Toulouse seul en tête

Le nouveau classement établi par la F.F.S.B. pour le trophée des Clubs 51 à la date du 5 mai (c'est-à-dire après le 1er championnat de France en individuel), voit, en 1re division, Toulouse rester seul en tête :

1. Toulouse (Cannizzo) 320 pts ; 2. Ste-Foy-lès-Lyon (Cheviet) 260 ; 3. Clos Vert (A. Boursier) 215 ; 4. St-Jean-en-Royans (Cuzet) 200 ; 5. St-Etienne-de-Crossey (Brochier) et C.R.O. (Bouvet) 195 ; 7. Cusset (Evandiloff) 190 ; 8. Bourg (Gourbeyre) et Carriers (Grimaldi) 180 ; 10. Côte-Saint-André (Gondrand) 160 ; 11. Valence (Cotte) 150 ; 12. Marseille (San Martin) 145 ; 13. Grenoble (Giroud) 135 ; 14. ASCUL (Lapierre) 150 ; 15. Lagnieu (Carre) et Aubigny (Jullien) 100.

Valencin (Faure) 2e en 2e division

En 2e division, la situation est la suivante :

1. Ris-Orangis (Meyssonnier) 270 pts ; 2. Valencin (Faure) 200 ; 3. Bourgoin (Lanza), Romans (Izérable) 190 ; 5. Clermont (Leroy) 180 ; 6. St-Claude (Cavalli) 160 ; 7. St-Claude (Delcourt) 150 ; 8. Eybens (Coing), Portes-lès-Valence (Planta), Aix (Sbalchiero) 140 ; 11. Valence (Charlaix), Miribel (Drujon), Dortan (Berrodier), Argenteuil (Chocet) 130 ; 15. C.R.O. (Plasse) 120.

Pour la coupe de France 1 et 2, Toulouse, Ste-Foy, Clos Vert, occupent les trois premières places. Valencin se classe 4e, Bourgoin 5e devant St-Etienne-de-Crossey, Cusset, Ris-Orangis, St-Jean-en-Royans, la C.R.O., les Carriers et Romans.

Les jeunes du Dauphiné-Savoie à Millau

L'omnium interrégional des jeunes débute ce week-end, avec deux rencontres éliminatoires des groupes 1 et 2, les autres (groupes 3 et 4, avec le Lyonnais et la Bourgogne) se disputant les 28 et 29 mai.

Le Dauphiné-Savoie va retrouver, à Millau, les mêmes adversaires que l'an dernier, en espoirs comme en cadets, c'est-à-dire la Provence-Côte d'Azur (qui l'avait précédé en cadets), les Pyrénées (vainqueurs en espoirs), le Languedoc.

Ce sera donc avec un réel esprit de revanche que les deux sélections, que managera M. Pioppo, vont accomplir, cette fin de semaine, le voyage de Millau.

Les sélections ont été constituées ainsi :

Espoirs. — Combet, Genthial (Drôme), Augustin (Ardèche), Tirard, Parmilleux, Chaffurin (Isère).

Cadets. — Miller, Teyssier (Ardèche), Patrice Didier (Drôme), Barret, Philpa, Jaccenin (Isère).

Il faut avouer que la sélection espoirs apparaît très séduisante. Elle l'était aussi l'an dernier, donnée même comme favorite, et elle avait échoué d'un cheveu à cette même éliminatoire. Elle va donc être, on le pense, doublement attentive.

R.P.

This illustrates how cautious one should be in trying and giving a historical account. Further revisions are to be expected in the future. For good reasons, I shall therefore make no attempt at covering the explosive period post-1975. It is to be noted though that the debate on the existence of a phase transition in spin glass materials is still raging. I have never understood why so many people are so prejudiced about it, with sometimes dramatic reversals from one side of the fence to the other. To those sitting on the fence, it appears worth the effort to explore how sharp the transition can be found, in various and possibly new ways, and to analyze the universal features and also the differences between various classes of materials. From this vantage point, one great merit of the mean field theory lies in its heuristic power, as a guide for suggesting new experiments.

2. Some experimental developments (since 1981)

There is still a growing extension of the class of materials which are found to display spin glass properties but simultaneously one observes also a refocusing on the archetypal spin glasses : CuMn, AuFe and alike.

Some of the standing questions are :

i) Is the phase diagram of insulating EuSrS (or CdMnTe) qualitatively different from the one of metallic AuFe, as far as reentrant ferromagnetism and the existence of mixed phases are concerned ?

ii) Uniform anisotropies : can one find good representative materials for Ising, planar and Heisenberg spins and can one detect significant differences between those ?

iii) Random and dipolar anisotropies : what effect on the transition temperature ?

A division of the field has sharpened during the last three years : the physics far below T_f versus the physics around T_f.

i) Physics far below T_f [12]

Thanks to NMR, ESR, hysteresis and torque measurements, there has been great progress in the development of the triad theory and in the understanding of the role of Dzyaloshinski-Moriya interactions in metallic spin glasses.

ii) Physics around T_f [13]

This includes the study of the critical properties of the non-linear susceptibility and of the corresponding equation of state [14], the determination of the phase diagram and the exploration of the mixed phases, the remarkable magnetocaloric effect and consequent developments [15], new evidence from muon spin relaxation data [16].

There has been much mention recently of the very accurate measurements on the relaxation of the field-cooled magnetisation by the Uppsala group. This group claims that "clearly, experimental field-cooled curves cannot be characterized as equilibrium or reversible curves" [17] and this statement has often been construed as an argument in favor of the absence of a phase transition. It is therefore worth noting that the observed relaxations are a minute fraction of the total magnetisation ($\sim 10^{-3}$) occurring on a rather short time scale ($\sim 10^2$ sec), but it is specially important stressing that the relaxation is toward a smaller magnetisation and not toward a larger one, as a paramagnetic Curie-like behaviour would imply. Of course, it is not excluded that at longer times the sign of the relaxation might reverse but, to my knowledge, such an upturn has never been observed experimentally. This is a challenge for experimentalists.

3. Some developments in computer simulation

Besides the increase of sample sizes and computer times in Monte-Carlo simulations [18], one has seen during the last three years the development of exact enumeration methods for somewhat smaller samples [19,20]. The main lesson drawn from these improved studies is that it is simply not possible anymore to say that computer spin glasses with shortrange interactions are insensitive to space dimensionality ($2 \leqslant d \leqslant 5$). In d = 2, for equal probability of positive and negative bonds, there is a large consensus in favor of no transition [18]. G. Paladin has recently explored the domain of existence of 1/f noise in the magnetisation fluctuations due to overcoming of finite energy barriers [21]. Close to the ferromagnetic region in the phase diagram, the possible existence of a random antiphase region [20] remains open, specially in view of the enigmatic significance of the Nishimori line [22]. In d = 3, for equal probability of positive and negative bonds, the Monte-Carlo data are compatible with the existence of a phase transition [18] and this has led to a reassessment of the high-temperature series.

An important progress, mentioned at the beginning, has been made in the detailed analysis of low-energy configurations for Heisenberg spins in dimension three. This has allowed C.L. Henley [5] to test which relative defects, among the different predictions produced by the topological classification of triad theory : walls, lines, textures [23], occur most frequently and an edge has been found in favor of walls. An amazing fit with the Bray-Moore [24] estimate for the mean size of clusters ($\sim$ 100 spins), as derived for the SK model, is also worth noting on the list of successes of mean field theory.

Finally, it should be mentioned that the main spinoff of spin glass physics, so far, has been in the transfer of Monte-Carlo annealing techniques to optimization problems [25]. A less wellknown stream exists in the opposite direction with the use of Edmonds' algorithm to find various exact ground state properties of spin glasses [20] and this stream may well grow in the future (valley projection function, random field problem, etc).

4. Brief report on frustration

Frustration means competing interactions and therefore unsatisfactory states, degeneracy, metastability, sensitivity to external parameters.

The three basic concepts are those of gauge invariance, frustration function, curvature [26]. In two dimensions, the pairing of frustrated plaquettes, which should be minimal in a ground state, leads to a conflict between local optimization and a global constraint : if one would start pairing each plaquette to its nearest neighbour, one would end up doing very badly.

This provides contact with many optimization problems, such as the traveling salesman problem [25]. Another generalization of the frustration concepts has been toward amorphous packing and random networks : the odd lines of N. Rivier [27] come from a conservation of odd faces (on the surface of a polyhedron) which is the strict analogue of the conservation of frustration in three dimensions [26] (clue : put (-1) interactions on the bonds).

Hard discs can be perfectly tiled on a plane (but not on a curved surface). Tetrahedra cannot be perfectly packed in three dimensions : this impossibility can be viewed as a frustration or curvature effect (J.F. Sadoc, M. Kleman, D.R. Nelson). In two dimensions, a homogeneous fine tuning of frustration may be mimicked by varying the curvature of the space (hard discs on a curved surface).

Coming back to magnetic systems, the most conspicuous effect of frustration is that it leads to canting of vector spins. This is essential for the notions of irreducible defects and relative defects [5,23].

The properties of diversity and stability, in ergodicity-broken systems, as due to frustration, have been brought to bear on biological problems : neural networks and content-addressable memories [28], the origin of life [29].

A wealth of important results have been obtained in the study of periodic frustrated Ising models [22]. For fully frustrated models in three dimensions, much less is known and I list here two pending questions :

i) What is the nature of the transition (universality class) for Ising spins on a fully frustrated simple cubic lattice ?

ii) Is there a phase transition for Heisenberg spins on a fully frustrated face centered cubic lattice ?

Another way of introducing a homogeneous fine tuning of frustration is by applying a magnetic field on a two-dimensional wire network. References for this problem which has interesting new experimental applications can be found in [30].

After these general considerations, we turn our attention to the present status of the mean field theory of spin glasses.

5. The infinite range SK model for Ising spins

During the last two years, a series of assaults have been launched against the Parisi solution. Meanwhile, this solution has passed successfully the stability tests [31]. This is not enough : there are also interpretation tests.

i) Is the linear response susceptibility χ_{LR} given by

$$\chi_{LR} = \frac{1 - q(1)}{T} \qquad ? \tag{1}$$

ii) Is the equilibrium susceptibility χ_{eq} given by

$$\chi_{eq} = \frac{1 - \int_0^1 q(x)dx}{T} \quad ? \tag{2}$$

Remember the Golden Rule : we are interested, firstly and ultimately, in the thermodynamic limit ($N \to \infty$) and in the response to a uniform field.

iii) Is the shape of the Parisi order function q(x), with its plateaux, physically significant ? Or is it defined up to some "gauge transform" ?

There are even simpler questions which are not fully settled. Such is the nature of the phase transition in the presence of a field. The PaT hypothesis would have predicted a cusp in the magnetisation (second order transition in the Ehrenfest sense). It was recognized later that PaT could not hold rigorously because it was not compatible with the fact that the spin glass free energy must be higher than the continuation of the paramagnetic free energy below the transition [32]. The suggestion was then that the transition is third order, with a cusp in the susceptibility and not in the magnetisation. However a different prediction has been put forward recently [33].

6. Valley projection analysis

Already several years ago, the experts in numerical simulation (K. Binder, D. Stauffer, L.R. Walker, R.E. Walstedt,...) had found it cogent to compare two spin configurations 1 and 2 by computing the mutual projection defined by :

$$q_{12} = \frac{1}{N} \sum_i S_i^1 S_i^2 \tag{3}$$

(we shall mostly consider Ising spins ; the generalization to m-vector spins is obvious with the scalar q becoming the trace of an m×m matrix). This number can be viewed as a scalar product in phase space : it gives a measure of the distance between two spin configurations in this space.

For a system which has a multiplicity of equilibrium configurations, it is natural to condense the information on the various valleys (or attractor basins) into a probability function P(q) :

$$P(q) = \sum_{s,s'} W_s W_{s'} \, \delta(q_{ss'} - q) \quad , \tag{4}$$

where W_s is a weight associated with valley s (in thermodynamics, this will be the Boltzmann factor $e^{-\beta f_s}/Z$, where f_s is the free energy of valley s, Z the total partition function, $\sum_s W_s = 1$) and $q_{ss'}$ is the overlap of valleys s and s' ; for spin problems, following (3), this will be :

$$q_{ss'} = \frac{1}{N} \sum_i \langle S_i \rangle_s \langle S_i \rangle_{s'} \quad . \tag{5}$$

Obviously, this distribution P(q) is potentially of interest in many other fields where multistate systems are considered (hydrodynamics, cellular automata, memory, amorphous structures, etc). One has to define adequate weights and distances. It is in some sense complementary to the entropy as defined in information theory. Before getting some physical intuition for the meaning of this function, let us consider the infinite range model for which it was first defined [3,4], and where

$$P(q) = \frac{dx}{dq} \quad , \tag{6}$$

i.e. P(q) is the derivative of the inverse function of q(x).

Three years ago already, it was recognized that such a definition was a neat way of getting rid of the dummy variable x :

i) normalization of the probability distribution function P(q) is automatically satisfied :

$$\int_0^1 P(q)\,dq = \int_0^1 dx = 1 \quad ,$$

ii) positivity of P(q) derives from monotonicity of q(x),

iii) physical quantities, such as the equilibrium susceptibility (2) and the internal energy, are expressed in terms of integrals over x, which are just the two first moments of P(q) :

$$\int_0^1 q(x)\,dx = \int_0^1 P(q)\,q\,dq \quad , \quad \int_0^1 q^2(x)\,dx = \int_0^1 P(q)\,q^2\,dq \quad .$$

iv) moreover, it can be shown easily that

$$\int_0^1 q^2(x)\,dx = \frac{1}{N^2} \sum_{ij} \langle S_i S_j \rangle_T^2 \tag{7}$$

where $\langle \ldots \rangle_T$ means thermodynamic average ; this led C. Itzykson, in a discussion during a Saclay seminar, to formulate the guess :

$$\int_0^1 q^k(x)\,dx = \frac{1}{N^k} \overline{\sum_{i_1,\ldots i_k}} \langle S_{i_1} \ldots S_{i_k} \rangle_T^2 \tag{8}$$

as a natural generalization of the expressions for the first two moments. Note that such a guess implied that the whole function q(x) was physically meaningful.

Now, if we plug (4) and (6) into the left hand side of (8), we find that the equality does hold. This is not difficult to check.

A particularly neat way of reexpressing (8) has been given by Parisi [3]. It involves the characteristic function of the distribution P(q), which can be written as :

$$\int_0^1 e^{yq} P(q) dq = \langle e^{y\frac{1}{N}\sum_i S_i\sigma_i} \rangle = g(y) \tag{9}$$

where S and σ denote two real replicas of the same system. The average is taken over the statistical ensemble corresponding to two uncoupled replicas. That such a formulation involving two real replicas might be useful had been intuitively foreseen by the late André Blandin [34] [35]. His idea was that, for a phase transition with no simple symmetry breaking in the Landau sense, the role of the coupling to a symmetry breaking field (which selects among the degenerate equilibrium states) had to be taken over by coupling to another copy of the same system.

In order to convey some of the richnesses and potentialities of this projection analysis, we shall work it out on some of the simplest models of statistical mechanics. For this purpose, we need an operational definition of P(q), which contains definition (4) in the adequate limit, but which should be much more straightforward to work with. The natural recipe that suggests itself is to use formula (9) as the definition of P(q), because the right hand side is a well defined thermodynamic average, very similar to the standard definition of a partition function.

Thus we define the characteristic function $\phi(u)$ as

$$\phi(u) = \langle e^{u\sum_{i=1}^{N} S_i\sigma_i} \rangle \quad , \tag{10}$$

$\phi(0) = 1$ by definition. The projection function P(q) is then defined and obtained from :

$$\phi(u) = \int_0^1 P(q)\ e^{Nuq}\ dq \quad , \tag{11}$$

where N is the number of spins. The logarithm of $\phi(u)$ is expected to be an extensive quantity, and it is therefore natural to define

$$\psi(u) = \lim_{N\to\infty} \frac{1}{N} \ln \phi(u) \quad , \tag{12}$$

which is somewhat analogous to a free energy per spin.

In a disordered (paramagnetic) phase, $\psi(u)$ is analytic in u at u = 0. A phase transition is signaled by breaking of this analyticity. This criterion is a very general one, encompassing the case of spin glass transitions as well as ordinary transitions. Let us see how it works for the Ising chain with nearest-neighbour interactions.

7. The nearest-neighbour Ising chain

The hamiltonian for two uncoupled replicas is

$$\mathcal{H} = - J \sum_i \left(S_i S_{i+1} + \sigma_i \sigma_{i+1}\right) - H \sum_i \left(S_i + \sigma_i\right) \quad . \tag{13}$$

In order to compute $\phi(u)$ via formula (10), we introduce a coupling term of the form :

$$- T\, u \sum_i S_i\, \sigma_i \quad , \tag{14}$$

and we look for the eigenvalues of the transfer matrix. There are four eigenvalues $\lambda_1, \ldots, \lambda_4$ from which $\phi(u)$ is obtained as :

$$\phi(u) = \frac{\lambda_1^N(u) + \lambda_2^N(u) + \lambda_3^N(u) + \lambda_4^N(u)}{\lambda_1^N(0) + \lambda_2^N(0) + \lambda_3^N(0) + \lambda_4^N(0)} \quad , \tag{15}$$

for periodic boundary conditions (other boundary conditions, e.g. frustrating ones, may be also of interest).

Note that the denominator is just the square of the partition function for one replica. For $u = 0$, the four eigenvalues are just products of the two eigenvalues obtained for one replica.

Setting $K = \beta J$, $\beta = T^{-1}$, the eigenvalues in zero field ($H = 0$) are easily calculated :

$$\lambda_1(u) = 2e^{-u} \text{ sh } 2K \quad , \tag{16}$$

$$\lambda_2(u) = 2e^{u} \text{ sh } 2K \quad , \tag{17}$$

$$\lambda_{3,4}(u) = 2 \text{ ch } 2K \text{ ch } u \pm 2\sqrt{\text{ch}^2 2K \text{ch}^2 u - \text{sh}^2 2K} \quad . \tag{18}$$

For K finite, $\lambda_3(u)$ is the largest eigenvalue and therefore

$$\psi(u) = \lim_{N\to\infty} \frac{1}{N}\, \phi(u) = \ell n \frac{\lambda_3(u)}{\lambda_3(0)} \quad .$$

It is of interest to consider the behaviour of $\psi(u)$ for u small and large. Note that $\psi(u)$ is even in u (this is a zero-field property).

$$\psi(u) \sim \frac{u^2}{2} \text{ ch } 2K + O(u^4) \quad , \text{ u small} \quad , \tag{19}$$

$$\psi(u) \sim |u| + \ell n \left(\frac{\text{ch } 2K}{1 + \text{ch } 2K}\right) + O(e^{-u}) \quad , \text{ u large} \quad . \tag{20}$$

We observe that $\psi(u)$ is analytic in u as it should. We know that a phase transition takes place at $T = 0$, ferromagnetic for $K \to +\infty$, antiferromagnetic for $K \to -\infty$. Indeed, the coefficient of u^2 diverges in (19) when $K \to \infty$. Its physical significance in terms of P(q) is obtained by expanding (11) in powers of u (a legitimate procedure if $\psi(u)$ is analytic) :

$$\ell n\ \phi(u) = uN\langle q\rangle + \frac{u^2}{2} N^2\langle q^2\rangle_c + \ldots + \frac{u^p}{p!} N^p\langle q^p\rangle_c + \ldots \quad , \tag{21}$$

where $\langle q^p\rangle_c$ is the cumulant of order p of the distribution P(q). Therefore (19) implies that :

$$\langle q\rangle = 0 \quad , \quad \langle q^2\rangle = \frac{\text{ch } 2K}{N} \quad . \tag{22}$$

and in the thermodynamic limit, P(q) is a delta-function at the origin. More generally, anywhere in the paramagnetic phase, $\psi(u)$ is analytic in u and P(q) is a delta-function, as a consequence of (21).

Let us consider now the limit $K \to +\infty$. The eigenvalues become

$$e^{2K+u+2v} \quad , \quad e^{2K+u-2v} \quad , \quad e^{2K-u} \text{ (twice)} \quad , \tag{23}$$

where $v = \beta H$. Thus $\phi(u)$ is

$$\phi(u) = \text{ch } Nu + \text{th}^2 Nv.\text{sh } Nu \quad , \tag{24}$$

and

$$\psi(u) = |u| \quad ,$$

as could be foreseen from (20). CLearly, $\psi(u)$ is nonanalytic in u at u = 0 and we also observe that the information on P(q) is lost. To retrieve it, we have to consider the function g(y) as defined in (9), which unfolds the singularity at u = 0 by using the variable y = Nu.

Thus

$$\ell(y) = \ell n\ g(y) = y\langle q\rangle + \frac{y^2}{2}\langle q^2\rangle_c + \ldots + \frac{y^p}{p!}\langle q^p\rangle_c + \ldots \quad . \tag{26}$$

In the case at hand, (24) gives

$$g(y) = \text{ch } y + \text{th}^2 Nv.\text{sh } y \quad , \tag{27}$$

and P(q) may be written as :

$$P(q) = \left(\frac{1+m^2}{2}\right) \delta(q-1) + \left(\frac{1-m^2}{2}\right) \delta(q+1) \quad , \tag{28}$$

where m = th Nv is the magnetisation per spin, for N rigidly coupled spins.

As a conclusion for this section, one observes that for ferromagnetic couplings, the behavior of $\psi(u)$ is by and large rather similar to f(H), free energy as a function of uniform magnetic field. One virtue of the coupling between replicas however is its "gauge invariance" : the antiferromagnetic case yields exactly the same properties for $\psi(u)$, $g(y)$, $P(q)$, without need for any knowledge of the nature of the ordering, in contrast with the use of an external field coupled to the order parameter. We have seen also that a transition is signaled by a break of analyticity in $\psi(u)$ at $u = 0$ and that the information on $P(q)$ in the low temperature phase is then contained in $g(y)$. On the other hand, in the high temperature phase, $\psi(u)$ contains information on the approach to the thermodynamic limit and on the approach to the transition, which is lost in $g(y)$.

8. Assembly of fine grains

We consider now an assembly of non interacting magnetic grains of various sizes. This is essentially the Néel model for spin glasses and it is natural to explore its predictions for the projection function. Let $f(t)dt$ be the number of spins which belong to a grain of size between t and $t+dt$. By definition,

$$\int_0^\infty f(t)dt = N \tag{29}$$

and $\int_0^\infty f(t) \frac{dt}{t}$ is the total number of grains.

By the additivity property of $\psi(u)$

$$\psi(u) = \frac{1}{N}\int_0^\infty f(t) \frac{dt}{t} \ell n\left(ch\ ut + m^2(t)\ sh\ ut\right) \quad , \tag{30}$$

where $m(t) = th\left(\beta H t m_0(t)\right)$ and $m_0(t)$ is the saturation magnetisation, per spin, of the grains (which are supposed to be rigidly coupled but which may be ferromagnetic, ferrimagnetic, antiferromagnetic,...).

Of course, if $f(t)$ is temperature-independent (original Néel model), no phase transition can occur. However, by introducing a suitable temperature and field dependence of $f(t)$, it is possible to mimick a transition. In the spin-glass phase, one has to consider :

$$\begin{aligned} \ell(y) &= \int_0^\infty f(t) \frac{dt}{t} \ell n\left(ch \frac{yt}{N} + m^2(t)\ sh \frac{yt}{N}\right) \\ &= \ell n \left(\int_{-1}^{+1} P(q)\ e^{yq}\ dq\right) \end{aligned} \tag{31}$$

in order to determine $P(q)$. We will not explore further here this track which should meet somewhere the phenomenological two-level state approach of J. Souletie [36].

9. Other examples

A simple way of getting a large ground-state degeneracy is achieved by considering a spin density wave on a chain. Let us suppose that the magnetic cell consists of M spins up and M spins down : we call this magnetic structure (M,M). There are M distinct ground states obtained by translating the structure along the chain.

For a (1,1) structure

$$P(q) = \frac{1}{2}\left(\delta(q-1) + \delta(q+1)\right) \quad , \tag{32}$$

this is a special case of Section 7 (limit $K \to -\infty$).

For a (2,2) structure

$$P(q) = \frac{1}{4}\left(\delta(q-1) + 2\delta(q) + \delta(q+1)\right) \quad , \tag{33}$$

and in the limit $M \to \infty$ (which is equivalent to a ferromagnet with antiperiodic boundary conditions), P(q) tends toward a constant distribution on the interval $[-1,+1]$.

We turn now our attention to classical vector spins. For simplicity, we consider N ridigly coupled spins. Writing

$$q = \cos\theta \quad , \tag{34}$$

the distribution P(q) is given by

$$P(q)dq = \frac{\Sigma^{(m-1)}}{\Sigma^{(m)}} (\sin\theta)^{m-2} \, d\theta \quad , \tag{35}$$

where m is the number of spin components and

$$\Sigma^{(m)} = \frac{2\pi^{m/2}}{\Gamma(\frac{m}{2})} \tag{36}$$

is the surface of the unit bowl in a space of dimension m. Hence

$$P(q) = \frac{1}{\sqrt{\pi}} \frac{\Gamma(\frac{m}{2})}{\Gamma(\frac{m-1}{2})} (1-q^2)^{\frac{m-3}{2}} \quad . \tag{37}$$

For planar spins, P(q) diverges at q = 1. For Heisenberg spins (m = 3), it is a constant distribution. For large spins, P(q) tends toward a δ-function at the origin, $\langle q^2 \rangle = \frac{1}{m}$.

Note that P(q) is non-trivial despite the fact that all ground states lie, in some sense, within the same "valley". There is no ambiguity here because we chose definition (11), rather than the less precise definition (4). (However, in the case of the SK model for vector spin glasses for instance, the distance between states should be the minimal one up to global symmetry operations [5], or so it seems, if one wants to keep to relation (6) with the function q(x), as usually defined). Thus here the function $\psi(u)$ is obtained as :

$$\psi(u) = \ln \left\{ \Gamma(\tfrac{m}{2}) (\tfrac{u}{2})^{1-\frac{m}{2}} I_{\frac{m}{2}-1}(u) \right\} \quad , \tag{38}$$

and not surprisingly, involves modified Bessel functions [37].

For u small :

$$\psi(u) = \frac{u^2}{2m} + O(u^4) \quad . \tag{39}$$

For u large :

$$\psi(u) = |u| - \frac{(1-m)}{2} \ln|u| + O(1) \quad . \tag{40}$$

(39) and (40) agree with the Ising (m = 1) expressions (19), (20), (25) in the appropriate limits.

It would be interesting to compute $\phi(u)$ for a chain of vector spins, specially since one knows that a finite temperature transition occurs for $m < 1$ [37]. Another extension is the study of the effect of a magnetic field. Here is the expression for the distribution P(q) for one planar spin in a magnetic field H :

$$P(q) = \frac{1}{\pi(1-q^2)^{1/2}} \cdot \frac{I_0\left(2\beta H(\frac{1+q}{2})^{1/2}\right)}{I_0^2(\beta H)} \quad , \tag{41}$$

which shows that the distribution becomes biased toward q positive but keeps its divergences at $q = \pm 1$.

10. Spin glasses again

After this somewhat lengthy digression meant to provide some intuitive feeling for the distribution P(q) and for the two-replicas approach, we return to more general considerations and then to the infinite-range model of spin glasses.

From (10) and (21), we note that

$$\langle q^2\rangle_c = \frac{1}{N} \sum_{ij} \left(\langle S_i S_j\rangle^2 - \langle S_i\rangle^2 \langle S_j\rangle^2\right) = \frac{1}{N} \mathrm{Tr}(G^2) \quad , \tag{42}$$

where the matrix element G_{ij} of the correlation matrix G is

$$G_{ij} = \langle (S_i - \langle S_i\rangle)(S_j - \langle S_j\rangle)\rangle \quad . \tag{43}$$

Therefore, a very general criterion for the existence of a phase transition (in spin systems) is obtained by the divergence of $Tr(G^2)$, which signals the appearance of a non trivial P(q). In this context, the Almeida-Thouless line of the Ising SK model loses some of its exotic character.

One point should be clarified. In Sections 5 and 7, the P(q) derived from q(x) was non zero only for $0 \leqslant q \leqslant 1$, whereas on general grounds P(q) should be even in q, in zero field. Strictly speaking, this P(q) should be called $P_+(q)$ with :

$$P(q) = \frac{1}{2}\left(P_+(q) + P_+(-q)\right) \quad . \tag{44}$$

In each system where time reversal invariance is spontaneously broken (and in this respect spin glasses share an analogy with ferromagnets), it is useful to consider the distribution $P_+(q)$ in zero field (in particular, its odd moments do not vanish trivially and contain useful information).

In order to get some insight for the distribution $P_+(q)$ of the SK model, we shall examine the predictions of [38], which have the merit of being open to numerical confirmation or rejection, and the only ones on the market for this purpose.

$P_+(q)$ comprises two contributions. A δ-function at $q = \overline{q}(T)$, the Edwards-Anderson order parameter whose weight is A(T) and a continuous function TV(q), which contributes for $0 < q < \overline{q}$.

$$P_+(q) = TV(q) + A\delta(q-\overline{q}) \quad . \tag{45}$$

Furthermore, it is predicted that :

$$V(q) = 2 - 12q + O(q^2) \quad , \text{ for q small,} \tag{46}$$

implying a cusp in P(q) at q = 0, and

$$V(q) \sim \frac{\sqrt{\alpha}}{4} \frac{1}{(1-q)^{3/2}} \quad , \text{ for } q \sim 1, \tag{47}$$

where the value of α is 3/2 in the context of the PaT hypothesis [39]. The weight A(T) is predicted to vary as

$$A = 1 - 2\tau + 6\tau^2 + O(\tau^3) \quad , \text{ for } \tau = 1 - T \text{ small,} \tag{48}$$

and to tend toward the finite value A = 1/2, at low temperatures.

With such a distribution P(q), the function $\psi(u)$ as defined by (11) and (12) is bound to be non analytic in u at u = 0, because for u positive it is dominated by the δ-function component (high q) whereas for u negative it is governed by the continuous component (low q). In finite field H, another δ-function component appears at $q = \hat{q}(H)$

[38] and analyticity is recovered when the two merge, squeezing the continuous component.

It remains somewhat of a mystery to understand this coexistence (if it is real) of a continuous and a delta component. Somehow, it seems to suggest that all valleys are not equivalent and that one couple of time-reversed states is deeper (in free energy) than the others by an amount of order $\ell n N$ (using expression (4)).It is possibly useful to mimick such a distribution by a suitable choice of f(t) in Section 8.

11. The envelope picture

A rather neat picture of ergodicity breaking in the spin glass phase and of the consequent existence of two susceptibilities, a linear response one and an equilibrium one, can be visualized by considering plots of the free energy versus the magnetisation [40]. Each valley contributes a curve in the (F,M) diagram, and the equilibrium free energy is the convex envelope of this family of valleys.

The zero field susceptibility is the radius of curvature at the bottom of the free energy curves. Since the radius of curvature of the envelope has to be larger than the radius of curvature of the individual valleys, this implies

$$\chi_{eq} > \chi_{LR} \quad , \tag{49}$$

the difference being equal to the radius of curvature of the locus of minima.

This result can be derived geometrically, assuming parabolic valleys and parabolic locus of minima :

$$F = \frac{(M-m^s)^2}{2\chi^s} + f \quad , \qquad f = \frac{a}{2}(m^s)^2 \quad . \tag{50}$$

The envelope of this family of curves, parametrized by m^s is

$$F_{eq} \sim M^2 \frac{a}{2(1+a\chi^s)} \quad , \tag{51}$$

and therefore $\chi_{eq} = \chi^s + \frac{1}{a}$, where χ^s is the linear response, intravalley, susceptibility.

An alternate, thermodynamic, derivation is obtained by writing the free energy, as a function of the field H, G(H)

$$G(H) = -T\ \ell n \sum_s e^{-G^s/T} \quad , \tag{52}$$

where s is again a valley index, as in Section 6 [41], and

$$\frac{G^s}{N} = f^s - Hm^s - \chi^s \frac{H^2}{2} \quad . \tag{53}$$

Double differentiation with respect to H yields :

$$\chi_{eq} = \frac{1}{Z}\sum_{s} e^{-f^s/T}\left(\chi^s + \frac{N(m^s)^2}{T}\right) \tag{54}$$

$$= \chi^s + \frac{1}{T}\,\frac{\int e^{-\frac{am^2}{2T}} m^2 dm}{\int e^{-\frac{am^2}{2T}} dm} = \chi^s + \frac{1}{a} \quad ,$$

as above (51).

However, there seems to be more in the SK model than in this simple parabolic envelope picture, namely the singularities in zero field. To complete the envelope picture is therefore still another cnallenge for the theoreticians.

12. Conclusion

At the end of this introductory review, I propose three rules for a successful meeting on spin glasses.

Golden Rule (generalized) : we should know what it is that we are talking about.

A lot of confusion has arisen from misunderstandings on local and global susceptibilities, nature of averaging, order of limits, etc.

Silver Rule : we need not have prejudices, only working hypotheses are useful.

As a working hypothesis, the mean field theory of spin glasses, it seems fair to say, has considerably fertilized the topic but tension coming from unbelievers is also healthy.

Bronze Rule : experimentalists should always beware of theoreticians, and of theories of any color, in order to be worthy of trust and to allow for useful comparisons of the materials they study.

Acknowledgements

The author expresses his gratitude to many colleagues, too numerous to be listed here, for fruitful discussions. Special thanks are due to R. Rammal and J. Vannimenus, for useful remarks on the manuscript. And I wish to express my respect for the memory of my thesis supervisor, André Blandin.

REFERENCES

1. G. Toulouse, Congrès de la Société Française de Physique, Clermont-Ferrand 1981 (Les éditions de physique, 1982) ; see also in Lecture Notes in Physics, Vol.149 (Springer 1981).
2. B.R. Coles, Proc. of the Geilo Institute on Multicritical Phenomena, 1983, to be published (Plenum).
3. G. Parisi, Phys. Rev. Letters 50 (1983) 1946.
4. C. De Dominicis, A.P. Young, J. Phys. A 16 (1983) 2063.
5. C.L. Henley, to appear in Phys. Rev. B.
6. R.E. Walstedt, this conference.
7. I.A. Campbell, D. Arvanitis, A. Fert, Phys. Rev. Letters 51 (1983) 57.
8. J.W. Shih, Phys. Rev. 38 (1931) 2051.
9. A.N. Gerritsen, J.O. Linde, Physica 17 (1951) 573.
10. R. Street, J.H. Smith, J. Phys. Radium 20 (1959) 82.
11. D.K. Finnemore, L.J. Williams, F.H. Spedding, D.C. Hopkins, Phys. Rev. 176 (1968) 712.
12. H. Alloul, this conference.
13. J.A. Mydosh, this conference.
14. J. Souletie, this conference.
15. A. Berton, J. Chaussy, J. Odin, R. Rammal, R. Tournier, J. Physique Lett. 43 (1982) L153 ; J. Odin, Thèse (Grenoble, 1982).
16. D.E. Mc Laughlin, this conference.
17. L. Lundgren, P. Svedlindh,O. Beckman, Phys. Rev. B 26 (1982) 3990.
18. A.P. Young, this conference.
19. K. Binder, this conference.
20. R. Maynard, R. Rammal, J. Physique Lett. 43 (1982) L347.
21. G. Paladin, private communication ; E. Marinari, G. Paladin, G. Parisi, A. Vulpiani, to be published.
22. W. Wolff, A. Sütő, this conference.
23. G. Toulouse, Phys. Rep. 49 (1979) 267.
24. A.J. Bray, M.A. Moore, J. Phys. C 14 (1981) 2629.
25. S. Kirkpatrick, C.D. Gelatt Jr., M.P. Vecchi, Science 220 (1983) 671.
26. G. Toulouse, Commun. Phys. 2 (1977) 115.
27. N. Rivier, Phil. Mag. 40 (1979) 859.
28. J.H. Hopfield, Proc. Natl. Acad. Sci. USA, 79 (1982) 2554.
29. P.W. Anderson, Proc. Natl. Acad. Sci. USA, 80 (1983) 3386.
30. R. Rammal, G. Toulouse, M.T. Jackel, B.I. Halperin, Phys. Rev. B 27 (1983) 5142 ; G. Toulouse, Proc. of the 1983 Geilo Institute, to be published (Plenum).
31. C. De Dominicis, I. Kondor, Phys. Rev. B 27 (1983) 606.
32. G. Toulouse, M. Gabay, T.C. Lubensky, J. Vannimenus, J. Physique Lett. 43 (1982) L109.
33. D. Elderfield, this conference.
34. A. Blandin, M. Gabay, A.T. Garel, J. Phys. C 13 (1980) 403.
35. G. Parisi, in Lecture Notes in Physics, Vol.149 (Springer 1981).
36. J.J. Préjean, J. Souletie, J. Physique 41 (1980) 1335 ; and following work.
37. R. Balian, G. Toulouse, Ann. of Phys. 83 (1974) 28.
38. J. Vannimenus, G. Toulouse, G. Parisi, J. Physique 42 (1981) 565.
39. G. Parisi, G. Toulouse, J. Physique Lett. 41 (1980) L 361.
40. G. Toulouse, Springer Series in Solid-State Sciences, Vol.39 (1982) 218.
41. R.G. Palmer, Adv. in Phys. 31 (1982) 669.

EXPERIMENTAL STUDIES OF THE LOW-TEMPERATURE PROPERTIES OF SPIN GLASSES

H.ALLOUL

Laboratoire de Physique des Solides,* Université de Paris Sud
91405 Orsay, France

ABSTRACT

The recent experimental advances in the characterization of the low T properties of spin glasses are reviewed. A macroscopic anisotropy with a triadic character is found to characterize the spin glass state at low T, whatever its remanence σ, and can be modified at will in RKKY spin glasses by addition of non magnetic impurities. The irreversible properties of spin glasses might be explained by successive jumps in phase space over energy barriers with a flat distribution of heights which are associated with the exchange couplings rather than the anisotropic interactions. Other marked differences with the case of fine magnetic grains, such as a single crossover, form non-ergodic to paramagnetic behavior and the probable occurrence of an equilibrium field cooled state, are stressed. The low-T specific heat and the correlated resistivity data in metallic spin glasses indicate the existence of a large density of magnon-like excitations, which have been simulated numerically. From zero field NMR experiments it can be seen that the local spin correlation functions do not markedly differ throughout the sample. The limited NMR and μSR relaxation data available indicate that these local correlation functions have long time tails although it cannot be decided yet whether they correspond at low T to a $1/\omega$ or $\omega^{\nu-1}$ noise spectrum, with $\nu<1$.

I. INTRODUCTION

Experimental spin glasses display non ergodic phenomena at low T, although it is not yet clear whether breaking of ergodicity can be associated here with a phase transition. Therefore below T_g experimentalists have to face a myriad of phenomena depending on the experimental timescales, the thermomagnetic history of the sample, etc..[(1)]. Experiments with various probes have been aimed at performing systematic studies of the frequency dependence of the magnetic properties around the temperature of the susceptibility cusp in order to decide between a sharp onset of spin freezing and a progressive extension of non ergodic behaviour throughout the sample when T is lowered [(2)].

However two novel experimental approaches have been developed in parallel these last few years. Although irreversible behaviour seems to set in below T_g, experiments until recently did point out that field cooling (FC) or zero field cooling (ZFC) a spin glass through T_g could be considered as a thermal equilibrium process. Therefore many attempts to characterize the criticality of the "phase transition" have been done with this assumption as a working hypothesis (3). A second kind of experimental approach has resulted from the evidence that the time decay of the remanence in a spin glass slows down when the sample has been cooled down to T = O. In such a case it can be assumed that the spin glass is trapped in a non ergodic energy well during the experimental time. Interestingly enough, data taken on short time in such a quasi-equilibrium state give information on the ZFC state as well, as has been found for the study of the macroscopic anisotropy of RKKY spin glasses (4).

I have been asked to survey the experimental properties of spin glasses at low T and shall therefore not consider the experiments performed around T_g. I shall further restrict my review to the specific case of Heisenberg spin glasses (for which crystal fields are of no importance), with no average ferromagnetic interaction between spins, i.e. I shall not discuss the eventual occurrence of a transition from ferromagnetism to "spin glass" transverse freezing (5).

In Section II, I shall briefly recall the results on the macroscopic anisotropy of RKKY spin glasses and its main manifestations when the random anisotropic couplings are much weaker than the random exchange couplings. I shall then review, in Sect. III, the main information gained from systematic studies of the non equilibrium properties, such as the remanence, its time decay as well as the time decay of the energy of a non equilibrium state. As these properties bear real analogies with the magnetic aftereffects for a collection of fine magnetic grains, I shall stress the differences with the Néel model (6) as well, which will allow me to point out that the nature of the non-ergodicity in spin glasses is quite different from that of fine magnetic grains. Finally, in Section IV, I shall consider those few experiments which directly bring some insight into the low-T excitations of the spin-glass state both from the macroscopic and the microscopic point of view.

II. MACROSCOPIC ANISOTROPY: A QUASI EQUILIBRIUM PROPERTY

As for any real experimental magnetic system the interactions between spins involve scalar (exchange) couplings as well as anisotropic (dipole-dipole at least) couplings. The very existence of a remanence $\vec{\sigma}$ which keeps the direction of the cooling field is a direct evidence of the presence of anisotropic couplings between local moments. It was well known that sharp hysteresis loops displaced with respect to zero field do occur at low T in CuMn alloys, which points out the existence of some unidirectional anisotropy, i.e. the spin system keeps some memory of the cooling field direction (7). These properties were assumed to be linked with an inhomogeneous description of the spin system, subdivided into domains with more or less ferromagnetic order. However various experiments have confirmed recently that the anisotropy has a macroscopic character (4).

1/ Evidences for a macroscopic anisotropy in spin glasses

For $T \ll T_g$, when blocked in one of its possible configurations, the spin system undergoes a rigid rotation if submitted to a small external field, as initially emphasized by NMR experiments[8]. This has led to develop a simple magnetostatic model in which the macroscopic free energy of a spin glass is split into three terms

$$F = (\vec{M} - \vec{\sigma})^2/2\chi_0 - \vec{M}.\vec{H} + E_a(\theta). \quad (1)$$

The first two terms involve the remanence $\vec{\sigma}$ and the isotropic magnetization $\chi_0\vec{H}$ induced by an applied field whatever $\vec{\sigma}$ (here $\vec{M}$ is the total magnetization). The anisotropy energy $E_a(\theta)$ allows to stabilize $\vec{\sigma}$ near the cooling field direction $\vec{H}_c//z$, θ being the angle $(\vec{\sigma},z)$. The validity of such a free energy has been probed by many experiments for which a small-angle approximation $E_a(\theta) = K\theta^2/2$ can be used. For instance the transverse susceptibility $\chi_\perp$ taken along y(see fig 1) is found[9] to follow accurately

$$\chi_\perp - \chi_0 = \sigma(H_z + K/\sigma)^{-1}, \quad (2)$$

H_z being a small field applied in the direction of $\vec{H}_c$. Similarly in the limit $\sigma \gg \chi_{iso}H_z$ an ESR mode is detected[10] at

$$\omega = \gamma(H_z + K/\sigma), \quad (3)$$

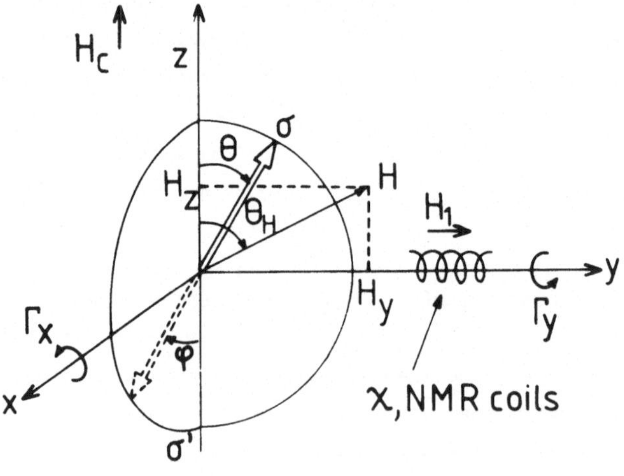

Fig.1: Geometry of the various experiments. H_1 is the ac field in $\chi_\perp$ experiments and the rf field in NMR or ESR experiments. During torque measurements the field $\underset{\sim}{H} > K/\sigma$ is rotated in the yz plane, which rotates $\vec{\sigma}$, with $\theta=\pi$ for $\theta_H=\pi$. Then a rotation in the xz plane yields a φ rotation of $\vec{\sigma}$. The respective torques acting on the sample during these two rotations are Γ_x and Γ_y.

and the initial torque $\Gamma_x \equiv -dE_a/d\theta$ which acts on the sample when its magnetization is rotated through an angle θ by a field applied in the yz plane, is directly given by[11]

$$\Gamma_x = K\,\theta\ , \qquad (4)$$

a unique value for K being derived from these experiments for a given sample at a given temperature[9].

2/ <u>Triad character of the anisotropy</u>

This macroscopic anisotropy energy presented above, which behaves as in a ferromagnetic material has been shown[9] to be independent of the magnitude of the remanence σ (fig.2) at a given temperature T. It is defined as well in the ZFC state, as a shifted ESR line associated with the isotropic magnetization $\chi_o H$ has been detected in ZFC samples[12] in small fields (for which $\sigma \ll \chi_o H$). This collective ESR mode with a limiting zero field frequency $\omega = \gamma(K/\chi_o)^{1/2}$ rather corresponds to an antiferromagnetic like resonance.

The fact that K and therefore $E_a(\theta)$ is an intrinsic property of the spin glass state is easy to understand. The ZFC configuration taken by the spins is such that it minimizes the total free energy involving both exchange and anisotropic couplings. *Then any rigid rotation of the spin system with respect to the lattice costs some anisotropy energy*. However, due to the overall isotropy of the state, this energy $E_a(\theta)$ will be the same *whatever the axis of the rotation*. As experiments tell us

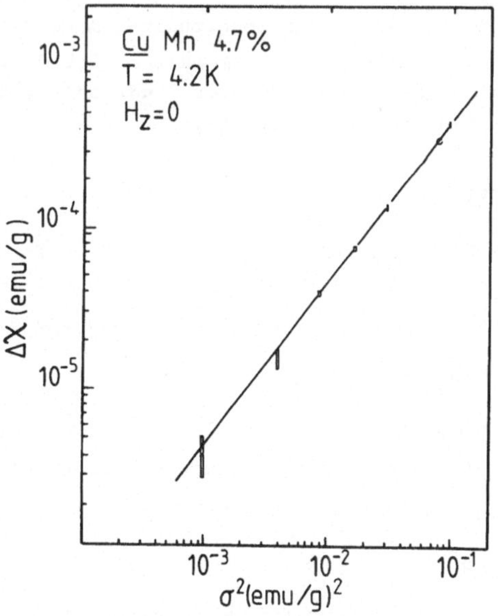

Fig.2: $\Delta\chi = \chi_\perp - \chi_o$ as measured for $H_z = 0$ plotted versus σ^2 for various values of the remanence. The straight line with slope unity perfectly fits the data, which allows to conclude from Eq(2) that K is independent of the remanence (ref.9).

that the modification of the spin configuration leading to the existence of a remanence does not change K, the same property is likely to be true in the field cooled spin glass. Of course rotations of the spin system cannot be induced experimentally for the ZFC spin glass, but the presence of $\vec{\sigma}$ makes it possible, by applying an external field, to rotate the spin system around any axis perpendicular to $\vec{\sigma}$, while rotation around $\vec{\sigma}$ cannot be controlled. Therefore in Eq(1) θ has to represent *the total rotation of the spin system*, its state being characterized by a *triad* rather than an axis [13].

This triadic character of the anisotropy has been directly detected first by torque experiments [11], as for instance after a rotation of π around x induced by rotating an applied field from z to -z in the yz plane (fig.3), the spin system does not have the same responses in the yz and xz planes as would be the case for a vector anisotropy (Fig.4). Indeed the torques which oppose to further rotations of σ are found quite different in these two planes. Especially Γ_y is found nearly zero for a further rotation of φ in the xz plane, as the total rotation undertaken from the initial state is still π (fig.4). This anisotropy of the magnetic response has been recently confirmed by $\chi_\perp$ measurements [14]. The consequence of this triadic character on the ESR modes has been studied by Henley et al [15] and Leggett[16] who produced a full solution for the resonance modes, introducing equations of motion for $\vec{M}$ and $\vec{\theta}$, the vector which specifies the rotation of the spin system. For $\vec{H}//z$ they found two transverse modes and one longitudinal mode at $\omega_L = \gamma(K/\chi_0)^{1/2}$. These three modes are degenerate for $\vec{H} = 0$ and $\vec{\sigma} = 0$. The two transverse modes do correspond to those already detected [10],[12] while the longitudinal mode, which results from the rotation of the spin system induced by a longitudinal field, is more difficult to detect because it is field independent. However, indirect evidence for the existence of these three modes has been obtained recently [17].

3/ Origin of the anisotropy

In the particular case of CuMn [11][14][18] or ternary CuT Mn[9][18] alloys, where T is a non magnetic impurity, the anisotropy is found to be of purely (DM) Dzyaloshinski-Moriya type, i.e. the microscopic couplings are given by

$$H_{ij} = \vec{D}_{ij}(\vec{S}_i \times \vec{S}_j) , \tag{5}$$

where $\vec{D}_{ij}$ is a vector linked to the lattice. Therefore this microscopic anisotropy energy has the symmetry of a vector with respect to rotations in 3d space, which yields a simple large angle relation for the macroscopic anisotropy

$$E_a(\theta) = - K \cos\theta , \tag{6}$$

as has been probed by $\chi_\perp$ [14][18] and torque experiments[11]. This microscopic anisotropy results from an indirect interaction between spins i and j mediated by the conduction electrons scattered on a third center (either Mn, or a ternary impurity) with spin orbit scattering λ [20]. The macroscopic anisotropy per Mn atom is found

Fig.3: Rotation of the anisotropy triad during the torque experiment of Fig.4. A rigid rotation of the spin system is induced from θ = 0 to θ=π when H is rotated from θ_H = 0 to θ_H=π (see fig.1). When a subsequent rotation φ is induced in the xz plane, the total rotation angle of the triad is still π with respect to the θ= 0 starting position, the axis of the rotation being the bisector of angle ($\vec{\sigma}$,Z) and E_a remains maximum (E_a = K). However, a non-controlled irreversible π rotation of the anisotropy triad around $\vec{\sigma}$ brings back the rotation angle to π-φ and has been observed in ref (11).

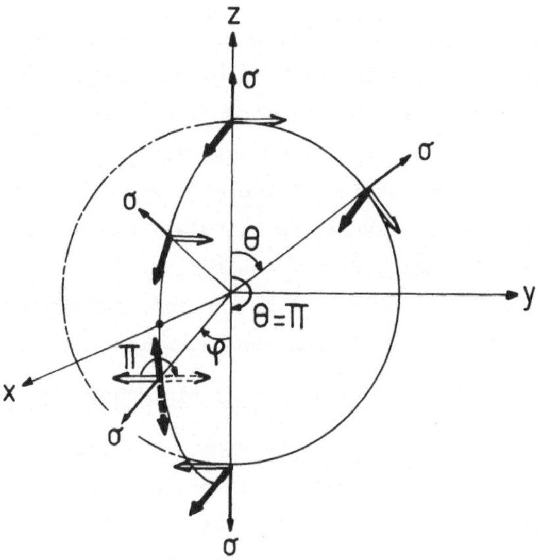

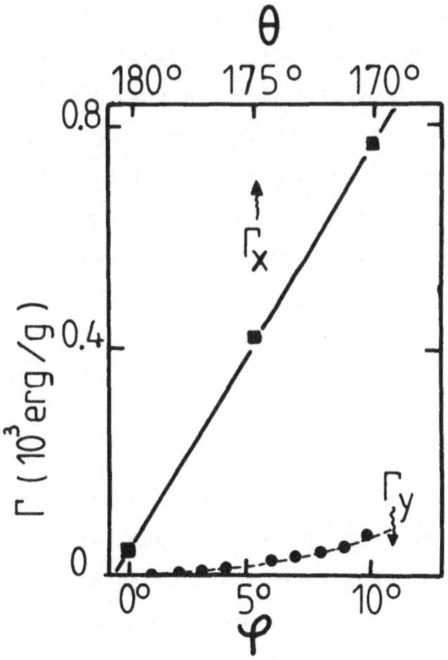

Fig.4: Torque measurements in a CuMn 20% sample at 1.5°K. The torque Γ_x and Γ_y measured after a π rotation of $\vec{\sigma}$ in the yz plane (see fig.3) are reported. These two responses are quite different and point out the triad character of the anisotropy. Γ_y is found to be very weak for small φ as the total anisotropy energy is then φ independent (from ref 11).

to increase linearly with the concentration of the ternary impurities [9,19]. Introduction of impurities with large spin orbit coupling provides then a very simple means for increasing the anisotropic couplings in RKKY spin glasses without significantly modifying the RKKY couplings. Such a method will be seen to be quite useful to separate the properties linked with anisotropy from those associated with exchange interactions. In insulating materials the anisotropy should be of quite a different origin. Although very few cases have been studied, a similar triad character of the anisotropy is suggested from the independence of K upon the magnitude of σ observed with $\chi_\perp$ measurements in an amorphous insulating spin glass [21].

4/ Some limitations of the rigid body approximation

Some of the effects depicted above correspond to the idealized case where a large angle rigid rotation of the spin system can be produced reversibly within the experimental time. They can only be observed for $T \ll T_g$, and in a limited angular range, even in pure CuMn systems for which the weakest anisotropy energy has been found up to now. They could certainly not be generalized as such to any spin glass system as the field $H \sim K/\sigma$ needed to produce a sizeable rotation of $\vec{\sigma}$ will very often induce irreversible changes of the magnetic state of the sample. Irreversibilities during the rotation of $\vec{\sigma}$ were even detected in the CuMn case [11] [18], as will be discussed in next section, and pure rigid rotation is very often limited to small θ rotations. This also explains why the interpretation of the hysteresis cycles [22] assuming a rigid rotation of the spin system have been somewhat controversial. It is now clear [11] [14] that, in a magnetization reversal along the hysteresis cycle loop the sample might be subdivided into domains in which the anisotropy triad follows different paths. A rigid behaviour of the spin system is not either expected to occur on large timescales when T is increased towards T_g, and its validity on short timescales as well as the T dependence of K might only be checked with high frequency probes.

III. IRREVERSIBLE PROPERTIES - REMANENCE ETC....

The existence of magnetic remanence and after effects in spin glasses is well known, and these features have been recognized for long to be quite similar to those of fine magnetic grains [1]. This analogy with the magnetism of rocks is at the origin of attempts to describe spin glasses as purely inhomogeneous systems which could be split into magnetically independent entities [23]. It will be shown hereafter that this simple mapping of one problem onto the other is not supported by various experimental evidences. However, the physics of irreversible processes is not an easy matter and the analogies between spin glasses and fine magnetic grains justify an attempt to mimic the data on the irreversible properties [24] [25] with a phenomenology derived from the Néel model [6]. For fine ferromagnetic grains, the creation of the remanence, its log t time decay, are associated with successive reversals of the magnetization of individual grains along their anisotropy axis through activated jumps over the energy barrier associated with the energy anisotropy of the grain (Fig.5). A spin glass prepared in a non-equilibrium state below T_g certainly undergoes also transitions between different energy wells above energy barriers. If τ_o is the attempt frequency in a given energy well, separated by an energy barrier W from the next, a jump will occur in a time

$$\tau = \tau_o \exp W/kT . \quad (6)$$

If τ_o is assumed to be negligibly dependent of the energy well, and if one assumes a flat distribution of energy barriers W surrounding a given well, then the only barriers which can be crossed in an experimental time t at a temperature T are those for which

$$W < k\,T \,\mathrm{Log}\, t/\tau_o \; . \quad (7)$$

This shows that the physical properties as the energy of the system, its magnetization should be a function of this cut-off parameter only. Such a universal behavior in T Log t/τ_o has been found for the saturated remanence [24] (Fig.6) as well as for the corresponding time decay of the energy [26]. These experiments are convincing evidences for the existence of such activated jumps and for a flat distribution of energy barriers in spin glasses. However, no matter how strong are the analogies, the differences with fine magnetic particles are also striking and it might be somewhat premature to take advantage of the analogies of irreversible properties

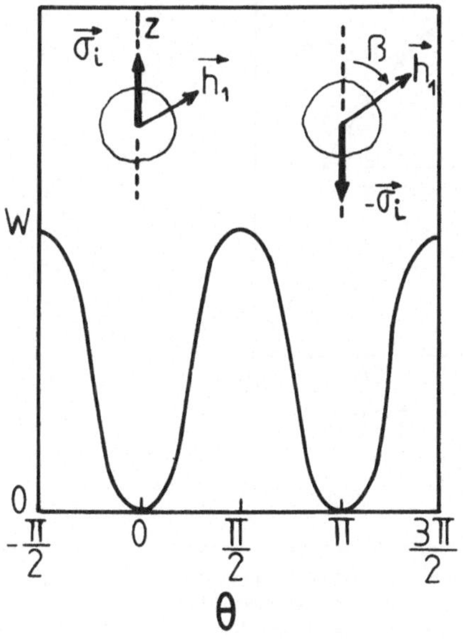

Fig.5: Anisotropy energy of a fine magnetic grain $E = W\sin^2\theta$, where θ is the angle of $\vec{\sigma}_i$ with z, the direction of the easy axis of magnetization. The susceptibility of the fine grain associated with the rotation of $\vec{\sigma}_i$ is $\chi = |\sigma|^2 \sin\beta/W$ in the direction of $\vec{h}_1$, whether σ_i is in the up or down position. For a collection of grains with random orientations of easy axes χ is isotropic and independent of the magnetization.

to justify microscopic models or to conclude for the absence of a phase transition. I intend therefore to discuss the qualitative differences of physical properties of spin glasses and fine magnetic grains which have been put forward so forth.

1/ Spin glasses and magnetic grains

For a collection of fine independent magnetic grains a contribution of the remanence to the susceptibility is not expected. Indeed for a given grain the susceptibility is independent of the orientation of the grain magnetization $\vec{\sigma}_i$ along the anisotropy axis. Therefore as the variations of the total magnetization are only associated with individual reversals of $\vec{\sigma}_i$, χ should be quite independent of remanence, and always isotropic if the anisotropy axes are oriented at random. The inhomogeneous description in terms of independent fine magnetic grains is therefore obviously impossible to reconcile with the coherence needed to describe the collective responses described in section II [8]. Similarly magnetoresistance data points out that the remanence is distributed homogeneously on the scale of the electronic mean free path in RKKY spin glass samples[27].

Even if one attempts to introduce small interparticle interactions between fine grains [28] to account for a collective behaviour at low T, a strong qualitative difference remains. For ferro or antiferromagnetic grains the coherence of the low T state is associated with the large intraparticle exchange interactions which allows for a ferromagnetic (or antiferro) order at T_c (or T_N). The energy barriers W are de-

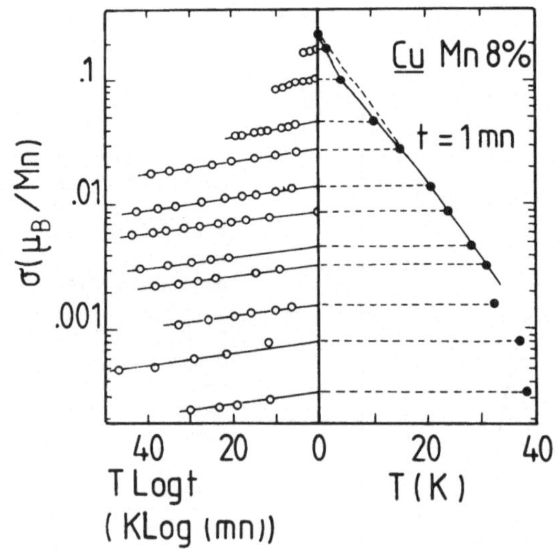

Fig.6: The log of the saturated remanence σ of a CuMn 8% sample is shown at different temperatures versus T log t in the left hand side of the figure. The initial points for t = 1 mn are reported versus T in the right. It can be seen that $\text{Log}\,\sigma = \text{Log}\,\sigma_o - \alpha T - \beta\, T\, \text{Log}\, t$ *which can be reduced in the form* $\text{Log}/\sigma/\sigma_o = -\beta T\, \text{Log}(t/\tau_o)$, *and therefore σ is a unique function of* $T \log(t/\tau_o)$ *(ref 24).*

termined by the much smaller anisotropy energies of the individual particles. These two energy scales therefore correspond to two temperatures scales T_c and $T_B(W)$, the blocking temperature at a given measurement time τ given by eq (6). In RKKY spin glasses the data of Section II have learned us that the dominant microscopic coupling in spin glasses is the DM interaction. An important result obtained from magnetization measurements [19] and $\chi_\perp$ data [9], is that the value of σ, as well as its time variation are not significantly modified if the strength of the DM coupling is increased by a factor 30 by introducing 1% Au in a $\underline{Cu}Mn_{0.01}$ alloy. (In such a case K is still two orders of magnitude smaller than k_BT_g). Therefore *the remanent properties of spin glasses are completely independent on the magnitude of the anisotropic interactions*. At variance with fine magnetic grains, the same exchange couplings are responsible for the low T coherence of the spin glass state and for the barrier heights. This is indeed coherent with the experimental fact that remanent properties follow scaling laws in RKKY spin glasses [1].

In order to transform a fine magnetic grain system into a spin glass the random interactions between particles should be larger than the individual grain anisotropy energies, so that freezing should be associated with spin glass like ordering rather than particle blocking. Such a situation, as well as intermediate cases, might indeed occur among the myriad of systems which are considered as spin glasses because they show up remanence and a susceptibility cusp.

2/ Remanence and susceptibility cusp

Although the microscopic analogy between spin glasses and fine magnetic grains is not supported by experiment it seems nevertheless worthwhile considering whether the simple assessment of a flat distribution of energy barriers, needed to explain the decay of the remanence at low T, would explain the properties of the susceptibility cusp which is a non equilibrium response. It is immediately clear that a flat T independent distribution of W would yield a log ν frequency dependence of the temperature at which the isotropic susceptibility departs from a Curie law, as well as a flat distribution of characteristic ("blocking") temperatures, and would never yield any sharp feature at a given temperature [29]. Experimentally it is well known that the frequency dependence of the cusp, at least in RKKY spin glasses is rather small and at most compatible with a Fulcher law [30]

$$\tau = \tau_o \exp\{ W'/k_B(T_o-T)\}, \qquad (8)$$

where W' is no more equivalent to an energy barrier. One is therefore forced to conclude that the decay of the remanence and the dynamics of the spin system near T_g are not associated with the same energy processes, or in other words *that the energy barriers abruptly disappear at a given temperature* ($\sim T_o \sim T_g$). All these results are coherent with a crossover from a paramagnetic to a non ergodic state at low T. Similarly there are growing evidences that the crossover is field dependent [3].

3/ Energy decay of the non-equilibrium states

It has been found that after field cooling a spin glass does not release energy to a thermal bath as long as the magnetic field has not been modified[26][31]. However, it releases energy to the bath when the field has been increased or decreased below T_g (while the magnetization increases or decreases[24]). These experimental observations are the best evidences that a FC spin glass is not far from equilibrium. This is again a qualitative difference with fine magnetic grains which, being mere paramagnets with long relaxation times, should continuously release energy to (absorb from) the bath when H is decreased (or increased). Similarly fine magnetic particles should cool down slowly by adiabatic demagnetization as the *entropy* is progressively transferred to the lattice, while a spin glass heats[32] as its internal energy decreases when its magnetization decreases. Such a difference is to be emphasized as it implies that in spin glasses the energy decay associated with the decay of σ is linked in priority with spin rearrangements and a subsequent *decrease of the macroscopic exchange energy* rather than with simple free cluster reversals which would only contribute to *entropic* terms. Then, while a collection of fine magnetic grains cooled in a given field does not reach thermal equilibrium and has a magnetization which increases in time towards the Curie value, this is far from being the case in spin glasses. Although recent challenging experiments[33] might indicate that a FC spin glass is not in thermal equilibrium either, the only observation up to now is that after field cooling in a very small external field

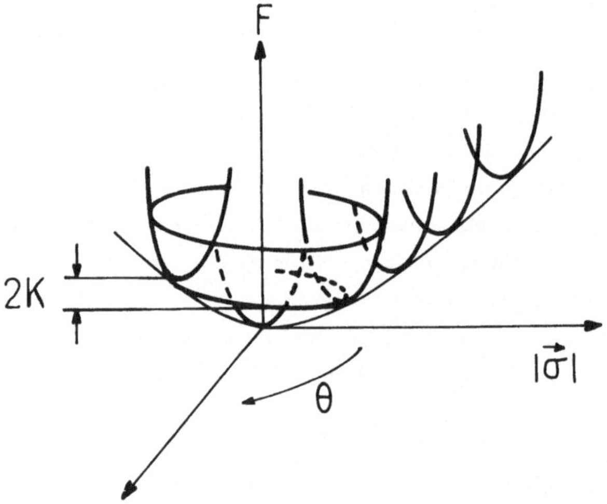

Fig.7: Quasi equilibrium free energy minima versus $\vec{\sigma}$. The rigid rotation of sec.II corresponds to $\vec{\sigma}$ being driven in the bottom of the "mexican hat". Non-rigid rotation of the spin system with constant $|\vec{\sigma}|$ might correspond to the existence of small energy barriers and wells along the rim of the hat.

just below T_g, a minute *decrease* of the magnetization ($\Delta\sigma/\sigma \sim 10^{-2}$) is observed, which is still quite far from the expected *increase* towards the Curie magnetization expected for fine grains or from numerical simulations for Ising spins and short range interactions [34]. Some of these differences between fine grains and spin glasses have been put forward by Préjean and Souletie [24], who were led to propose a phenomenological model for the irreversible properties, assuming a distribution of ill defined asymmetric two level systems (TLS) which might explain in details the non-equilibrium magnetization curves at low T, as soon as the equilibrium curve is taken as the field cooled one. This phenomenological model is obviously also bound to assume that the energy barriers vanish in order to explain the abrupt onset of reversible paramagnetic behaviour at high T (or high field). It is further hereagain clear that *no fixed spatial significance* should be given to these TLS, as there is no ensemble of spins which is bound in a spin glass to switch between only two configurations, as is shown clearly by numerical simulations. Any ensemble of spins rather has a large variety of possible spin configurations in the various intermediate states reached during a given time evolution [35].

4/ Non Equilibrium phenomena linked with the anisotropy

In this series of discussions I have considered only the experiments which monitor the time evolution of spin glasses with the energy or the magnitude of the magnetization as physical observables. This corresponds to considering evolutions in the 2d projection of phase space onto the $E, |\vec{\sigma}|$ plane. However, the studies on anisotropy have allowed to consider a third observable, the orientation of the spin system. In the quasi-equilibrium limit it has been seen that rigid body rotation of the spin system can be performed. If one uses polar angles to describe the rotation of $\vec{\sigma}$, the free energy in a given well transforms into a "mexican hat" picture (Fig.7). In the case of DM couplings the "mexican hat" is tilted by 2K, the state with $\vec{\sigma}$ along the cooling field H_c being more stable than that with $\vec{\sigma}$ antiparallel to H_c. However, experiments in which large angle rotations have been attempted at intermediate temperatures, have demonstrated that irreversible rearrangements do occur, leading to non rigid behaviour of the spin system or to an irreversible rotation of the anisotropy triad [11] [18]. These effects are detected while the magnitude of $|\sigma|$ is stable. Therefore these rearrangements which might be linked with the magnitude of the anisotropy energy correspond to a special viscosity of the spin glass state which involves other degrees of freedom in phase space. They can be described by a deformation of the rim of the "mexican hat" of fig.7, with appearance of weak energy barriers, and an eventual stabilization of the magnetization in the $-\sigma$ direction. Although these effects have been qualitatively evidenced, further more detailed work is still needed to characterize these type of irreversible processes.

IV. LOW T EXCITATIONS OF THE SPIN GLASS STATE

Apart for the already mentioned studies of the macroscopic anisotropy, very few efforts have been aimed these last few years at clarifying the equilibrium or quasi-equilibrium properties of the low T spin glass state. I shall recall first the well known results on the thermal and transport properties and will then discuss the experiments which probe the local dynamics such as some NMR experiments started in my group recently.

1/ Specific heat, resistivity and initial susceptibility

The occurrence of a large T-linear contribution to $C_v(T)$ well below T_g has been known for long [1][36] and, in the early molecular field theories, was attributed to those spins submitted to small molecular fields [1]. This possibility was recognized to be wrong for Heisenberg spins, and recent numerical simulations by Walker and Walstedt[37] for classical spins and RKKY interactions demonstrate that this low T specific heat can for a large part be accounted for by magnon-like excitations of the spin system around either of its equilibrium configurations. It seems also natural to consider that these excitations will contribute to the scattering of conduction electrons. Assuming a Bose statistics and a constant interaction strength, Campbell[38] derived a simple relationship between $C_v(T)$ and the T-dependent part of the impurity resistivity $\Delta\rho(T)$

$$C_v(T) = N\, k_B\, d(T\Delta\rho(T)/\rho_\infty)/dT \qquad (9)$$

This relationship is perfectly obeyed experimentally (Fig.8), and it can be then considered that both $C_v(T)$ and $\Delta\rho(T)$ probe the same low-energy excitations. The magnitude of the low T reversible contribution to the susceptibility which is also found quite insensitive to the presence of a remanence $\vec{\sigma}$, is also correctly explained by the same numerical simulations [37]. However for both $C_v(T)$ and $\Delta\rho(T)$ the universality of the low T behaviour is not completely settled down as for $T \lesssim T_g/10$ C_v is found proportional to T^α with $1<\alpha<2$ depending upon the system [40][41]. The low T behaviour cannot either be predicted reliably by numerical simulations as the accuracy for the lowest energy excitations is restricted by sample size limitations. It is not clear whether TLS are still required[42], as in real glasses, to describe the thermal properties at the lowest temperatures.

2/ Neutrons, NMR, μSR

All the efforts to detect directly spin waves from inelastic neutron scattering experiments have not been successful yet. We ignore here of course the various experiments in the reentrant phase for which ferromagnetic magnons are detected down to the lowest temperatures[43], with a T dependent spin wave stiffness constant [44]. Similarly the resolution of the Neutron Spin Echo experiments is not sufficient to yield any insight on the shape of the local correlation function for

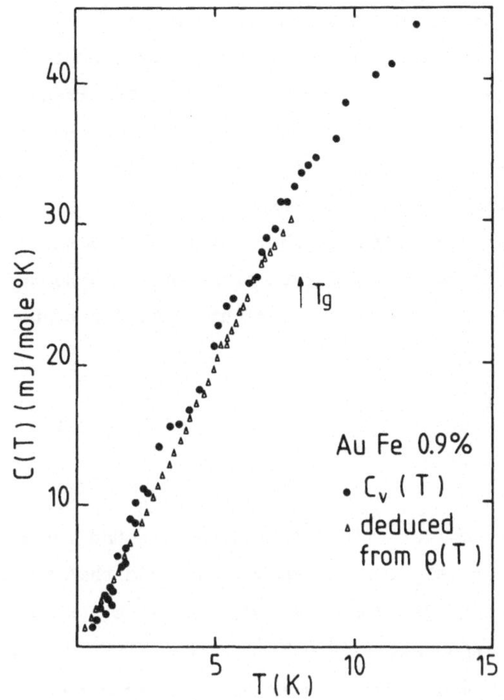

Fig.8: The specific heat data in a AuFe *0.9% sample(T_g=8K) are plotted together with the values obtained from the alloy resistivity data using Eq(9). Here a normalization factor ρ_∞= 0.5 μΩcm has been taken. The perfect agreement between the two sets of data is an illustration of the validity of the assumption done by Campbell[38].*

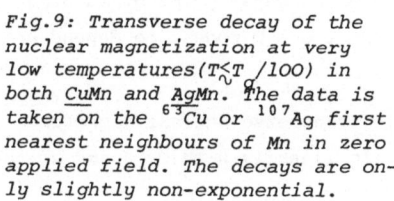

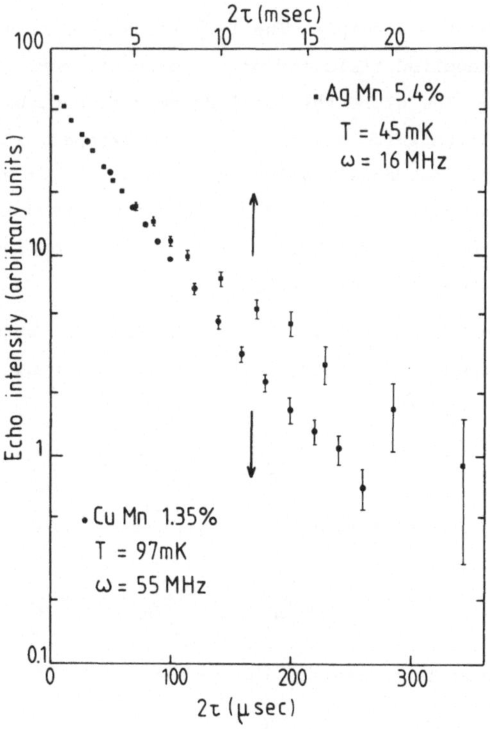

Fig.9: Transverse decay of the nuclear magnetization at very low temperatures($T \lesssim T_g/100$) in both CuMn *and* AgMn*. The data is taken on the ^{63}Cu or ^{107}Ag first nearest neighbours of Mn in zero applied field. The decays are only slightly non-exponential.*

for $T<0.5\ T_g$ [45]. Taking advantage of the NMR enhancement in spin glasses in presence of remanence [8], nuclear spin relaxation data has been taken in zero applied field in CuMn as well as AgMn spin glasses at very low temperature [46]. The transverse spin relaxation rate, being strongly T-dependent, is dominated by local moment fluctuations. A given nuclear spin $\vec{I}$ is exchange coupled ($\mathcal{H} = A(r)\vec{I}.\vec{S}$) with the neighbouring Mn local spin at distance r. The ^{63}Cu or ^{107}Ag NMR frequency $\omega_n = A(r)S/\hbar\,\gamma_n$ depends upon the considered near-neighbour shell of the Mn atoms. NMR relaxation of the nuclei of a given shell is then a direct probe of the dynamic susceptibility of the neighbouring Mn local spin S with a contribution of the transverse fluctuations given by

$$\frac{1}{T_\perp} = \frac{A^2(r)}{2} \int_{-\infty}^{+\infty} <\Delta S^+(t)\ \Delta S^-(o)> e^{i\omega_n t}\, dt \tag{10}$$

$$\simeq \omega_n^2\, J(\omega_n, T)\ .$$

Here $J(\omega_n, T)$ is the Fourier transform of the correlation function. As the NMR intensity has been found to follow a Curie law for $T<T_g/10$, it can be concluded that the same Mn spins are probed in this temperature range. The observed transverse nuclear magnetization decays are only slightly non-exponential (fig.9), i.e.the distribution of T_2 within the sample does not exceed a factor 5 for a given shell of near neighbours of the Mn atoms. Therefore the local dynamic susceptibility is quite uniform throughout the sample. The T_2^{-1} relaxation rates have been found independent of σ, of the applied field and of the magnetic anisotropy, in the experimental range investigated. Therefore the local dynamic susceptibility is the same in the various energy wells in which the system can be trapped, at least at very low T, and the $q = 0$ modes of the magnetization which should be affected by a modification of the macroscopic anisotropy do not dominate the observed relaxation.Conversely the broadening of the ESR line which is linked with the damping of these $q = 0$ modes, do depend on the macroscopic anisotropy [47].

The data for T_2, taken as the time for which the transverse nuclear magnetization reaches 1/e of its initial value is found to scale linearly with T^{-1}, and as can be seen on fig (10)

$$T_2 T \propto T_g\ . \tag{11}$$

For an RKKY spin glass it can be easily shown, from scaling arguments that

$$(T_2\ \omega_n^2)^{-1} \propto J(\omega_n, T) \propto \frac{1}{T_g}\ J'\left(\frac{T}{T_g}, \frac{\omega_n}{T_g}\right)\ . \tag{12}$$

Combining this result with the data of Eq(11) implies

$$J'\left(\frac{T}{T_g}, \frac{\omega}{T_g}\right) \propto T/\omega\ .$$

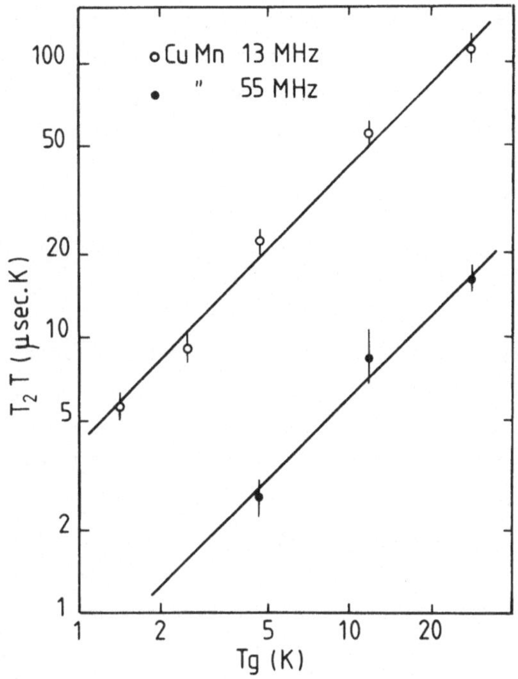

Fig.10: T_2T is plotted versus T_g for the first and fourth nearest neighbour shells of Mn in $Cu_{1-x}Mn_x$ alloys for x = 0.1%, 0.2%, 0.4%, 1.35% and 4.7%. The full lines are the best fits to a linear T_g dependence(46). The data have been taken for $T \lesssim 0.1\ T_g$.

Such a $1/\omega$ noise power spectrum would correspond to a log t decay of the local correlation function. If we directly consider the data taken on the two distinct resonance lines (fig.10) to probe the frequency dependence of $J(\omega,T)$ it can be immediately seen that $(T_2T)^{-1}$ does not scale with $\omega_n^{\ 2}$ and rather correspond to a $\omega^{-0.7}$ dependence of $J(\omega)$.

Similarly, the muon spin longitudinal relaxation data [48] taken for $T/T_g > 0.3$ as a function of an applied field H is an average relaxation rate of the muon spins randomly distributed in the lattice (equivalent to the nuclear relaxation of the bulk Cu or Ag nuclei). As $J(\omega)$ is found independent of the external field from the above NMR data, the muon relaxation directly yields the frequency dependence of $J(\omega_\mu)$ through the field variation of the muon Larmor frequency ω_μ . Mc Laughlin et al.[48] found $J(\omega) \propto \omega^{-0.4}$ for $T/T_g \sim 0.3$. The μSR and NMR data which are not taken in the same temperature range do not yield exactly the same exponents for $J(\omega)$, and further work is certainly necessary to perform more detailed comparisons. However from both experiments it can be concluded that the local correlation function does not decay exponentially for $T \ll T_g$ and rather has a long time tail as has been found around T_g from neutron scattering [45]. From the NMR data it is also clear that this behaviour is not associated for $T \ll T_g$ with a microscopic distribution of relaxation rates, as is the case near T_g (In that case the transverse relaxation T_2 is highly non-exponential and a severe loss of NMR signal intensity has been detected for $T \sim 0.7\ T_g$[49]). It would certainly be interesting to decide whether indeed $1/\omega$ noise takes place in spin glasses as suggested by some authors[50]. Long time tails for the correlation function are also predicted from various mean field solutions of the SK model[51][52].

V. CONCLUSION

It can be seen from this short review that real progress has been done in the understanding of the low T properties of *experimental* spin glasses. These progresses are certainly as important as that performed by our theoretician colleagues, who are now able to provide for instance a mean field solution for the SK model, and the interactions between theory and experiment have been by no way negligible. The experimental studies on the macroscopic anisotropy not only have allowed to discard the conjecture that a spin glass is a mere independent magnetic grain system, they have also shown *that collective rigid behaviour of the spin system* occurs and that the isotropy of the spin glass state yields a *triadic character* of the anisotropy. Although the macroscopic anisotropy plays no role in determining the magnitude of the remanence and its time decay in RKKY spin glasses, the anisotropy is indeed required to maintain the remanence in a fixed direction and is a real parameter of the experimental spin glass problem. From Walstedt and Walker's [35] Monte Carlo simulations, the very existence of anisotropy is even a necessary feature for a cusp in the susceptibility to occur. Although complete studies of the anisotropy cannot be performed if K is very large, a simple measurement of the transverse susceptibility gives an estimate of K. What happens to the experimental properties of spin glasses if anisotropic interactions become comparable to exchange? This is certainly a question to be answered in the future. The extensive work performed on the anisotropy also has taught us to pay attention to other observables than the magnetization. In that sense the recent attempt to probe the crossover from spin glass to paramagnetism through the disappearance of the anisotropy [53] is rather original.

The low T magnetization curves, and irreversibility of spin glasses can be described with a phenomenology derived from the Néel model, which essentially tells us that the decay of energy and remanence of spin glasses occur through activated jumps over energy barriers, widely distributed in energy heights. Although a given jump over an energy barrier might imply a rearrangement of about 1000 spins [24], such an ensemble of spins will not keep any coherence in further barrier jumps. Such a picture has been used by C.Henley [54] and also results from numerical simulations [35]. Furthermore such a phenomenology is bound to accept as experimental facts the existence of a *near equilibrium field cooled state* as well as the *disappearance of the energy barriers at the field dependent crossover from spin glass to paramagnetic* regime. Hereagain it has been shown that new kinds of irreversibilities, associated with the anisotropy occur in spin glasses, such as irreversible rotations of the anisotropy triad, rearrangements which reduce K without change of $|\vec{\sigma}|$, etc...[(14)(18)]. Further studies of these properties could give some insight on the low T properties of spin glasses.

It has usually been assumed that properties of spins glasses were very sensitive to the applied field, which seems indeed to be true near T_g. However at low T it has been seen that the quasi equilibrium state of spin glasses is rather insensitive not only to a moderate applied field, but even to the magnetothermal history. For instance the macroscopic anisotropy, the initial susceptibility $\chi(0)$, the local spin correlation function, the attempt frequency τ_0^{-1} are rather independent of the remanence, that is of the configuration in which the system has been trapped. Although the low T excitations are not really understood, it is interesting to find out that the *local spin correlation functions which decay slowly in time* ($1/\omega$ noise?), are *somewhat uniform throughout the spin glass*. This would be coherent with collective excitations (such as damped magnon modes?) being at the origin of the localspin fluctuations.

The experimental results do tell us that breaking of ergodicity in spin glasses is not of the trivial fine grain type, and that collective phenomena occur in spin glasses. This is certainly not sufficient to decide between a true equilibrium phase transition and a glass like freezing in spin glasses. Although we cannot imagine for spin glasses an equilibrium thermodynamic state with lower energy (as is the crystal for the glass) some recent experiments [(33)] show that something in spin glasses (as in glasses) depends on the rate of cooling of the spin glass through T_g in a constant applied field. The magnetization is for instance found to decrease slowly in time. This is certainly not the trend towards a Curie law expected by those sticking on the fine grain model, or for the case of the 2D Ising model [(34)]. However this challenging experiment still has to pass some tests to allow to yield a conclusion. As only a very small amount of energy is involved in such rearrangements, can we be sure that defects are not playing a role? Would the same experiment on a ferromagnet give a different answer (what about pinning centers for the walls etc...)? Although all the studies performed around T_g are challenging and somewhat spectacular I personally believe that more efforts should be done to clarify the low T properties of the spin glass state. Indeed there is probably no example of a phase transition for which the details of the critical behaviour could be understood before knowing the (order) parameters which characterize the low T phase, the nature of the broken symmetry etc...Would spin glasses be an exception to this general rule? And even if the paramagnetic spin glass crossover is some day found not to be a phase transition, it would not have been vain to study the low T properties.

Acknowledgements

My knowledge of spin glasses has strongly benefited from the intensive work and continuous interactions between the members of the magnetism group in Orsay. I therefore would like to acknowledge particularly I.A.Campbell, J.Ferré, A.Fert, F. Hippert, T.Garel, P.Monod, J.P.Renard, S.Senoussi as well as D.E.Mac Laughlin, J.J. Préjean, J.Souletie, W.Saslow and G.Toulouse for illuminating discussions.

Références

* Laboratoire associé au Centre National de la Recherche Scientifique.

(1) For recent review articles see A.Blandin,J.Phys.Colloq 39, 1499(1978), R.Rammal and J.Souletie, Magnetism of metal and alloys , M.Cyrot Editor (North Holland 1982) p.379.
(2) J.Mydosh, this conference.
(3) P.Monod and H.Bouchiat, J.Physique Lettres 43, 45 (1982).
B.Barbara, A.P.Malozemoff and Y.Imry, Phys.Rev.Letters 47, 1852 (1981).
A.Berton, J.Chaussy, J.Odin, R.Rammal and R.Tournier , J.Physique Lettres 43, 153 (1982).
(4) For a recent review on the macroscopic anisotropy of spin glasses see H.Alloul and F.Hippert, J.Mag.Mag.Mat.31-34,1321 (1983).
(5) See for instance R.A.Brand, V.Manns and W.Keune, this conference and H.Maletta this conference.
(6) L.Néel,Ann.Geophys.5, 99 (1949).
(7) J.S.Kouvel, J.Phys.Chem.Solids 21, 57 (1961).
(8) H.Alloul, Phys.Rev.Letter 42, 603 (1979); J.Appl.Phys.50, 7330 (1979).
(9) F.Hippert and H.Alloul, J.Physique 43, 691 (1982).
(10) P.Monod and Y.Berthier, J.Mag.Mag.Mat.15-18, 149 (1980)
(11) A.Fert and F.Hippert, Phys.Rev.Letters 49, 1508 (1982).
(12) S.Schultz, E.M.Gullikson, D.R.Fredkin and M.Tovar, J.Appl.Phys. 52, 1776 (1981)
(13) W.Saslow, Phys.Rev.Letters 48, 505 (1982).
(14) F.Hippert, thèse de doctorat es Sciences, Orsay 1983 (unpublished).
(15) C.L.Henley, H.Sompolinsky and B.I.Halperin, Phys.Rev.B25 5849 (1982).
(16) A.J.Leggett,private communication.
(17) E.M.Gullikson, D.R.Fredkin and S.Schultz , Phys.Rev.Letters 50, 537 (1983).
(18) F.Hippert, H.Alloul and A.Fert, J.Appl.Physics 53, 7702 (1983).
(19) J.J.Préjean, M.J.Joliclerc and P.Monod, J.Physique 41, 427 (1980)
(20) A.Fert and P.Levy, Phys.Rev.B23, 4667 (1981).
(21) E.Velu, J.P.Renard and J.P.Miranday, J.Physique Lettres 42, 237 (1981).
(22) P.Monod , J.J.Préjean and B.Tissier, J.Appl.Phys.50, 7324 (1979).
(23) F.Holtzberg, J.L.Tholence and R.Tournier, in Amorphous Magnetism II,Edit. R.A.Levy and R.Hasegawa (Plenum New York 1977) p 155
(24) J.J.Préjean and J.Souletie, J.Physique 41, 1335 (1980).

(25) J.Souletie, J.Physique 44, (1983).

(26) A.Berton, J.Chaussy, J.Odin, R.Ramual, J.Souletie and R.Tournier, J.Physique Lettres 40, 931 (1979).

(27) S.Senoussi, J.Phys.F 10, 2491 (1980).

(28) E.P.Wohlfarth, J.Phys.F 10, L241 (1980).

(29) C.A.M.Mulder, A.J.Van Duynevelt and J.A.Mydosh, Phys.Rev.B25, 515 (1982).

(30) J.L.Tholence, Solid State Commun.35, 113 (1980).

(31) G.J.Nieuwenhuys and J.A.Mydosh, Physica 86-88b, 880 (1977).

(32) J.Odin, Thèse de doctorat es Sciences, Grenoble (1982).

(33) L.Lundgren, P.Svedlindh and O.Beckman, Phys.Rev.B26 3990 (1982).

(34) K.Binder and W.Kinzel, Proceedings of this conference.

(35) R.E.Walstedt, Proceedings of this conference.

(36) L.E.Wenger and P.M.Keesom, Phys.Rev.B11, 3497 (1975).

(37) L.R.Walker and R.E.Walstedt, Phys.Rev.B22 4503 (1980).

(38) I.A.Campbell, Phys.Rev.Letters 47, 1473 (1981).

(39) P.J.Ford and J.A.Mydosh, Phys.Rev.B14, 2057 (1976).

(40) D.L.Martin, Phys.Rev.B21, 1906 (1980).

(41) R.Caudron, P.Costa, J.C.Lasjaunias and B.Levesque, J.Phys.F.11, 451 (1981).

(42) P.W.Anderson, B.I.Halperin and C.M.Varma, Phil.Mag.25, 1(1972).

(43) A.P.Murani, Phys.Rev.B 28, 432 (1983).

(44) B.Hennion, M.Hennion, F.Hippert and A.P.Murani, to be published in Phys.Rev.B (rapid communications).

(45) F.Mezei and A.P.Murani, J.Magn.Magn.Mat.14, 211 (1979).

(46) H.Alloul, S.Murayama and M.Chapellier, J.Magn.Magn.Mat.31-34, 1353 (1983).

(47) J.Cowen and P.Monod, private communication.

(48) D.E.Mac Laughlin, L.C.Gupta, D.W.Cooke, R.M.Heffner, M.Léon and M.E.Schillaci Phys.Rev.Letter 51, 927 (1983).

(49) D.E.Mac Laughlin and H.Alloul, Phys.Rev.Letters 36, 1158 (1976) and 38, 181(1977) D.A.Levitt and R.E.Walstedt, Phys.Rev.Letters 38, 177 (1977).

(50) E.Marinari, G.Paladin, G.Parisi and A.Vulpiani (to be published).

(51) H.Sompolinsky and A.Zippelius, Phys.Rev.B25, 6860 (1982).

(52) A.J.Bray and M.A.Moore, J.Phys.C.15 2417 (1982).

(53) I.A.Campbell, D.Arvanitis and A.Fert, Phys.Rev.Letters 51, 57 (1983).

(54) C.Henley, to be published.

SOME RECENT HIGH-TEMPERATURE EXPERIMENTS ON SPIN-GLASSES

J.A. Mydosh
Kamerlingh Onnes Laboratorium
der Rijks-Universiteit Leiden
Leiden, The Netherlands

Abstract

The experimental behavior of spin-glasses will be examined as the temperature is reduced towards the freezing temperature T_f. First we will consider the effects of atomic short-range order, which unfortunately always occur in real systems, on the local magnetic correlations and magnetic phase diagrams. Then a number of high-temperature ($T \gtrsim T_f$) experiments will be reviewed beginning with some static measurements such as resistivity, dc susceptibility and specific heat, and extending to a few dynamical ones such as neutron spin echo, muon spin relaxation and ac susceptibility. The static experiments give us some insight into the development of local, spatial spin-correlations, while the dynamic ones can be related to time-correlation functions which are temperature dependent. Based upon these results many experimentalists interpret their data in terms of the growth of magnetic clusters, followed by cluster relaxation leading to a broad distribution of relaxation times which significantly shifts to longer time scales as $T \to T_f^+$. Hence a spin-glass has similar properties to an ordinary glass.

I. INTRODUCTION

Intentional spin-glass research is now entering its second decade with seemingly no diminution in its intensity. For at least the past five years an enormous amount of effort has been devoted to this problem with over 500 spin-glass papers appearing each year in a ratio of roughly two theoretical works for every experimental one. Thus far, a few hundred different, real systems have been claimed to exhibit spin-glass-like behavior and one can always find an experimental effect to "satisfy" just about every theoretical prediction (or vice versa as has more often been the case). Yet despite this vast concentration of endeavor no generally accepted solution or complete description of spin-glass freezing and the low temperature properties has been obtained.

Most of the theoretical struggle began with the Edwards-Anderson model[1] which attempted to calculate phase transition behavior in the mean field approximation. Although recently, after much difficulty, a viable solution[2] has been found for the Sherrington-Kirkpatrick mean field model[3], it is not able to confront all the vastly different experimental effects which have been observed. Nevertheless, the new results from the mean field theory do seem applicable in describing a number of experimental properties[4], especially relating to the field-cooled behavior — the experimentalists' trick of replicating the equilibrium (time independent?) spin-glass state. However, there is a limited amount of experimentation that one can do in constant field, and once the field is varied by sweeping or oscillating it, a whole gamut of time-dependent, seemingly nonequilibrium effects start to appear. This has caused a second school of theorists[5] to treat the problem from a dynamic, metastable point of view. Led mainly by the results of Monte-Carlo simulations, this school claims the nonexistence of a static or equilibrium spin-glass phase transition and treats the freezing as dynamic phenomenon with time dependent order parameter and susceptibility[6].

At temperatures far below the freezing temperature T_f, a peculiar displaced hysteresis loop occurs in some spin-glasses after field-cooling. Outside the loop the field-cooled magnetization varies linear with field beginning from a remanent value M_r, i.e. $M=M_r+\chi H$. This simple dependence and the anisotropy (unidirectional-like) associated with the displacement of the hysteresis loops allow one to construct a free energy[7] which can then be used to calculate not only the magnetization behavior, but also torque, transverse susceptibility, NMR and ESR[8]. Two such hydrodynamics models — a vector[9] one and á traid[10] one — have been proposed and are presently being used to probe the experimental data. The first involved comparisons indicate that these models are inadequate to fully describe the different low temperature measurements, and the exact nature of the macroscopic anisotropy is still cloudy[11].

For the review now at hand I would start with a most unpleasant and annoying topic, namely, the occurrence and necessity to include atomic (or chemical) short-range order (ASRO) in the spin-glass systems. Spin-glasses are not completely random as so often idealized, and the influence of ASRO will profoundly affect the magnetic SRO which in turn creates the basic building blocks out of which the spin-glass state is formed. Then I should like to concentrate upon an often neglected part of the spin-glass problem, viz. the high temperature one, $T \gg T_f$. Here all is not simply paramagnetic, i.e. Curie-like, and both static (resistivity, dc susceptibility and specific heat) and dynamic (neutron spin echo, muon spin relaxation and ac susceptibility) experiments give a particular insight into the early stages of spin-glass evolution — spatially and temporally. Finally, the results of these measurements will be collected in terms of a model involving local ferromagnetic clusters with a broad distribution of relaxation times and a dramatic slowing down of the relaxation as $T \to T_f$ from above. Unfortunately, space and time will not permit me to go into any discussion of the unusual low temperature ($T \ll T_f$) spin-glass properties mentioned above. However, they will be considered by other authors[12] at this Conference.

II. ATOMIC SHORT-RANGE ORDER

All alloy systems possess some degree of atomic short-range order (ASRO). This may be seen by simply looking the tables compiled by Miedema et al.[13] for the heats of formation ΔH for many combinations of solid solutions. Except for a few fortuitous cases ΔH is always finite meaning a tendency towards ASRO. Indeed, for amorphous materials including the Metglasses, many of which are spin-glasses, there is certainly ASRO[14]. Also for the "real" inorganic glasses (borates, silicates and chalcogenides) one speaks of a molecular cluster network with cluster diameters reaching many tens of Å. Since the magnetic short-range interactions (MSRO) are an important ingredient in the spin-glass problem the ASRO will work hand in hand with the MSRO, even at low concentrations ($c \gtrsim 1$ at.%), to produce local magnetic correlations. These correlations will appear well above the freezing temperature and they will be an important part of the freezing process and low temperature spin-glass state.

But how to measure and analyze the ASRO? Perhaps the best experimental method to determine ASRO is via diffuse neutron scattering[15], although recently X-rays (EXAFS)[16] and Mössbauer effect[17] have also been employed. The usual theoretical treatment of ASRO is by correlation functions in terms of the Warren-Cowley parameters[18] α_n which are defined by

$$\alpha_n(c) = [4Nc(1-c)]^{-1} \langle \hat{\sigma}_i \hat{\sigma}_{i+n} \rangle \quad .$$

Here $\hat{\sigma}_i$ are the Flinn operators which describe whether a site is occupied by an A or B atom and the angular brackets denote a thermodynamic average. Positive values of α_n mean that the probability for a like atom as neighbor in the nth neighboring shell, with respect to an atom at origin i, is greater than for a purely random distribution (clustering) and negative α_n-values result in a greater probability for finding a dissimilar atom as neighbor (anticlustering).

Experimentally the diffuse neutron scattering cross-section is directly related to the α_n's:

$$\frac{d\sigma}{d\Omega} = c(1-c)(b_A - b_B)^2 \Sigma Z_n \alpha_n [\sin(kr_n)/(kr_n)] \quad ,$$

where b_A and b_B are the nuclear scattering lengths of the alloy components A and B, Z_n is the coordination number of the nth shell, k the neutron scattering wave vector and r_n the distance to the nth shell. As a typical example let us consider the <u>Pt</u>Mn system which has recently been studied by Morgownik and Mydosh[19]. In Fig. 1 the diffuse cross-section is shown for a <u>Pt</u>Mn (8 at.%) alloy as a function of k. The usual corrections have been taken into account for neutron background, paramagnetic scattering, multiple scattering and atomic size effect. The spectrum exhibits a pronounced minimum in the low k region which means a strong anticlustering

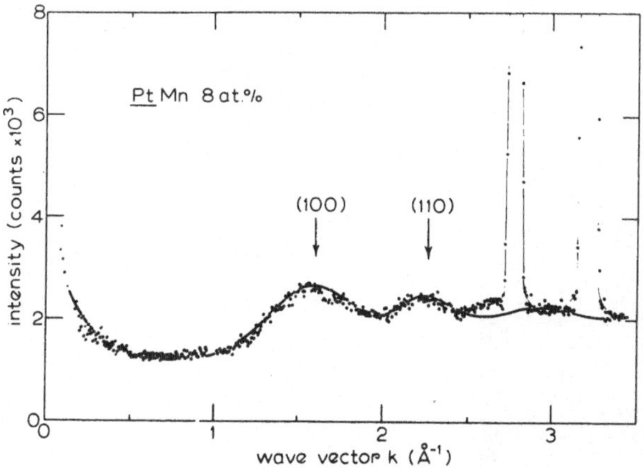

Fig. 1. Neutron scattering cross-section for PtMn (8at.%) as a function of scattering vector k. The heavy solid line is the fit to the data used to determine the atomic short-range order parameters.

ASRO. Moreover, the oscillating cross-section also contains small period contributions around the maxima at the (100) and (110) positions which indicate a relatively large correlation length ≈10Å or 12th neighbor for this concentration alloy. The solid line gives the fit of the $d\sigma/d\Omega$ equation with n=12. The nearest and next nearest neighbor α-values are $\alpha_1 = -0.056$ and $\alpha_2 = +0.047$ and there is an alternating sequence in the sign of the α_n's (odd-n α's are negative, even ones positive). For comparison X-ray measurements on the atomic long-range ordered (ALRO) intermetallic compound Pt_3Mn, which exhibits pronounced superlattice reflections of the Cu_3Au structure, are shown in Fig.1 by the arrows. These superlattice lines coincide exactly with the ASRO maxima of the 8 at.% alloy. This requires that between 8 and 25 at.% Mn the Mn-atoms take predominantly corner positions in the fcc lattice as in Cu_3Au. Thus low concentration PtMn alloys are definitely not random and display strong anticlustering, and further, precursors of the long-range (ALRO) Pt_3Mn structure.

By using previous measurements[20] of ASRO and introducting the concept of partial (or quasi) long-range order (APLRO) we can compare a number of canonical spin-glass alloys. This latter quantity, APLRO, marks the beginning of a correlated-site percolation on the superlattice and designates the transition regime between ASRO and ALRO. The results for four alloy systems are schematically displayed in Fig.2 in terms of atomic ordering diagrams. The exceptional system in this group is CuMn (and its Doppelgänger AgMn) with their large concentration region of ASRO. CuMn has a comparatively weak anticlustering ASRO and no ALRO intermetallic compounds[21]. Above ≈30 at.% the anticlustering in CuMn starts to change into clustering and segregation of α-Mn occurs.

In contrast PtMn and AuMn both have much stronger anticlustering ASRO and each system has intermetallic compounds, e.g. Pt_3Mn and Au_4Mn, which are essentially second neighbor ordered structures of Mn atoms. The correlated second neighbor percolation for these superlattices is approximately 8 at.% Mn. Thus the appearance of an infinite cluster of ordered Mn first occurs around this concentration, and this will have a profound effect on the magnetic properties which will be discussed below.

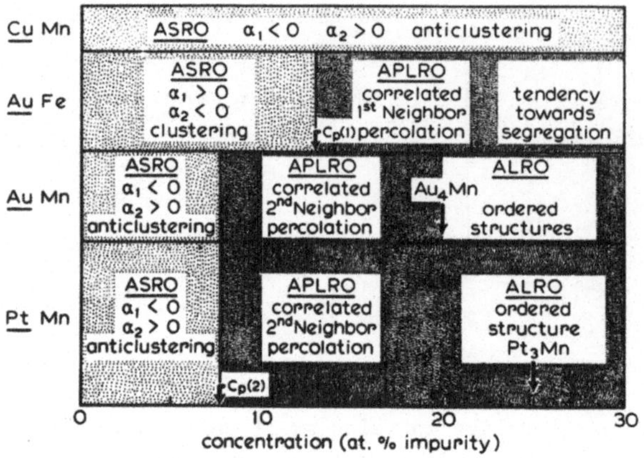

Fig. 2. Atomic ordering diagram for PtMn, AuMn, AuFe and CuMn.

AuFe is totally unlike the former systems, due to its strong nearest neighbor clustering. Here the APLRO (correlated-site percolation for first neighbor) begins at ≈13 at.% Fe and there is a powerful tendency towards Fe segregation. Macroscopic precipates of Fe would correspond to the ALRO. Indeed, after almost 20 years of Mössbauer effect measurements on AuFe alloys Violet and Borg conclude[(22)]: "In AuFe alloys with Fe concentration of about 10 at.% or more it appears to be virtually impossible to quench in a single-phase, random solid solution.... A rapidly forming second phase, more concentrated in Fe, is evident in the Mössbauer spectra at 10.5 at.% Fe and increases with Fe-concentration of the sample".

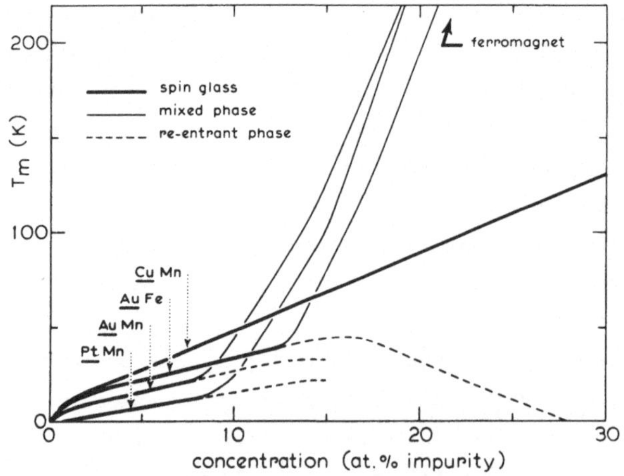

Fig. 3. Magnetic phase diagram for PtMn, AuMn, AuFe and CuMn.

The magnetic phase diagram for the above four systems is given in Fig.3. "Mixed phase" is used to denote a random or partially frustrated ferromagnet[23] where truely uniform ferromagnetic order is not yet present. Note the intimate relation between the atomic ordering diagram and the magnetic phase diagram (Figs.2 and 3). Furthermore, since the spin-glass phase only exists at low enough concentration we now find that the destruction mechanism of the spin-glass phase, which often was believed to be a random nearest neighbor percolation, is in fact strongly controlled by the ALRO.

We should also mention here with respect to the special behavior of CuMn (and most probably AgMn) the results of a recent neutron polarization analysis[24] on CuMn. This study has revealed in addition to the ASRO two distinct types of MSRO. The most pronounced of these is a long-period spin modulation with a correlation length of about 35Å. Also there are shorter-range ferromagnetic correlations or clusters which are directly related to the ASRO. The large modulated regions seem to be a "medium" in which the smaller ferromagnetic clusters are imbedded. The coupling between these two MSRO could provide the basic mechanism for the strange low temperature behavior of CuMn regarding the unusual unidirectional anisotropy. Indeed, back in 1957 Meiklejohn and Bean[25] used a ferromagnetic particle, Co, surrounded by antiferromagnetic medium, CoO, to create a field-cooled, displaced hysteresis loop, exactly as in CuMn. This required a new type of magnetic anisotropy which they introduced and called unidirectional exchange anisotropy. So now after 26 years we apparently have a similar picture emerging for CuMn with the focus of study shifting from the ferromagnetic clusters to the peculiar structure of the modulated region. The medium is the message for understanding CuMn.

III. STATIC EXPERIMENTS AT HIGH TEMPERATURES

a) Resistivity

Considerable resistivity data already exist for the noble-metal based spin-glasses covering a wide range of T and c[26]. The magnetic resistivity $\Delta\rho(T)$ clearly shows precursor effects far above T_f with a slowly varying behavior extending up to room temperature. Only a small fraction ($\approx 1/4$) of the total spin disorder resistivity $\Delta\rho_\infty(T\to\infty)$ usually remains at T_f. However, until recently a quantitative description of $\Delta\rho(T)$ has been hampered by the lack of a scattering model and complications with the experimental data such as the Kondo effect. Campbell et al.[27] have proposed a model for the conduction electron — local-moment excitation interaction using the Walker-Walstedt excitation spectrum $P(\Delta)$[28]. The $P(\Delta)$ is generated by computer simulation of a Heisenberg, RKKY coupled, dilute spin system. The resistivity can be simply calculated by integrating over the excitation spectrum with Bose statistics,

$$\Delta\rho(T) \propto \int \frac{\Delta P(\Delta)/kT}{\exp(\Delta/kT)-1}\, d\Delta \ .$$

Good agreement has been found with the above formular for many alloy systems at low T using a two parameter fit.

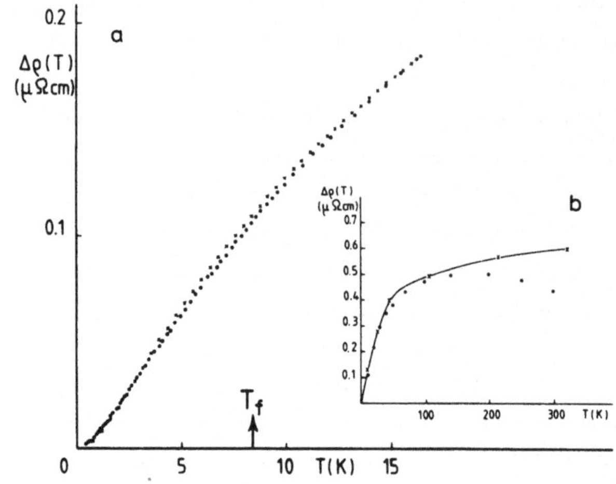

Fig. 4. The magnetic resistivity versus temperature for a AuFe (1 at.%) alloy. Circles represent experimental values and crosses the model calculations. After Campbell et al., Ref.27.

Fig.4 shows a typical comparison between experimental and calculated data. Notice how the calculated resistivity is practically constant above 100 K and the ever increasing fall off as T is reduced. For this AuFe (1 at.%) alloy about 80% of the total spin disorder resistivity has disappeared by the time T_f=8.5 K is reached. This means there is an enormous abatement in the paramagnetic spin flip scattering far above T_f, and this can only be caused by local interactions, i.e. MSRO. Also note the complete absence of critical effects spanning T_f.

The resistivity behavior is quite similar to the magnetic specific heat C_m and the above excitation spectrum model[27] gives $C_m(T) \propto d[T\Delta\rho(T)/\Delta\rho_\infty]/dT$. In general the two sets of data nicely superpose at low temperatures. However, the maxima in $d(\Delta\rho)/dT$ are usually well below T_f while those in C_m are above T_f. These deviations suggest that a different type of excitation or even static correlations are more important in $C_m(T)$ than in $\Delta\rho(T)$. We shall return to the analysis of the specific heat below.

According to Campbell[27] the resistivity with decreasing temperature strongly reflects a particular dynamical process, namely small-angle vibrations of spins within a cluster or complex. These scattering excitations are related to a relaxation time which is not the same as the intercluster spin flips usually measured in dynamical experiments, e.g. μSR — see Sec.IVb. Only in the limit of $T\to\infty$, all paramagnetic spins, will the two relaxation times be similar.

b) dc Susceptibility

There are numerous measurements of the high-T susceptibility on large concentration (10-25 at.%) spin-glass alloys[29]. However, most of these low temperature experiments stop at 300 K and the authors try to interpret their results in terms of Curie or Curie-Weiss analysis. This in many cases is incorrect and the "forced" Curie-Weiss fits have led to wrong conclusions concerning the local magnetic interactions. We now realize that gradual deviations from Curie-Weiss, beginning at temperatures 5 to 10 times T_f, occur and these prohibit calling a <u>limited</u> linear region of χ^{-1} vs T the Curie-Weiss regime. A proper determination requires measurements far above room temperature or at reduced concentrations so that T_f lies below about 30 K.

A systematic susceptibility study of such low concentration $\approx$0.5 to 6 at.% alloys has been performed by Morgownik et al.[30] on several noble-metal-based spin-glasses. First and foremost a Curie-Weiss regime is truely established by comparing the slopes of the χ^{-1} vs T curves for the various concentrations. Then the paramagnetic Curie temperature θ may be reliably extrapolated from high temperatures. In Fig.5 the concentration dependence of $\theta(c)$ is exhibited for four systems. Note that three distinct types of behavior are found. <u>Cu</u>Mn and <u>Au</u>Mn are more "ferromagnetic" than <u>Au</u>Fe. This clearly calls into question the traditionally accepted belief, based on the incorrect high-c, χ^{-1} vs T analysis, that in <u>Cu</u>Mn the first neighbor is antiferromagnetically coupled.

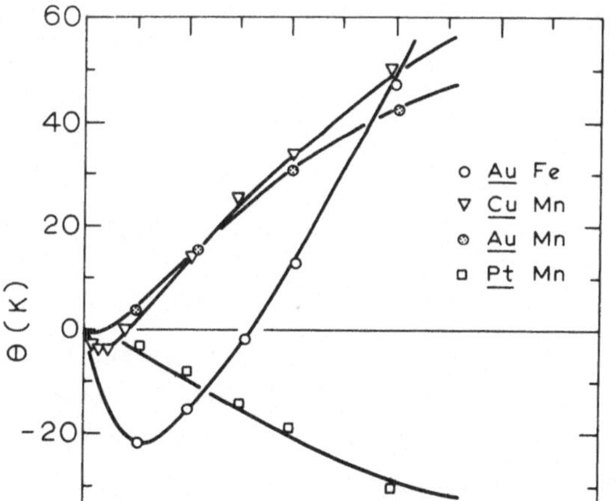

Fig. 5. Paramagnetic Curie temperature θ versus concentration for <u>Pt</u>Mn, <u>Au</u>Mn, <u>Cu</u>Mn and <u>Au</u>Fe. The solid lines represent the configuration cluster model's calculated values.

Secondly Morgownik et al.[30] have developed an ad hoc configurational cluster model which determines the various neighbor exchange interactions J_n between the magnetic impurities. The model consists in computer calculating the canonical ensemble for all possible states and energy levels of the first 14 configurations (1 singlet, 5 pairs, 8 triplets) in which magnetic impurities can be grouped in a fcc lattice. The distribution functions of these "clusters" are computed including the ASRO (see Sec.II). From the resulting partition function we can calculate $\chi(T,H,c)$ and compare with experiment. Fig.6 makes such a comparison for CuMn, where the data, plotted as χT vs T, are fitted to the model. This in turn gives values of J_n which are found to be concentration independent up to 6 at.% with $n \leq 5$ for the various systems.

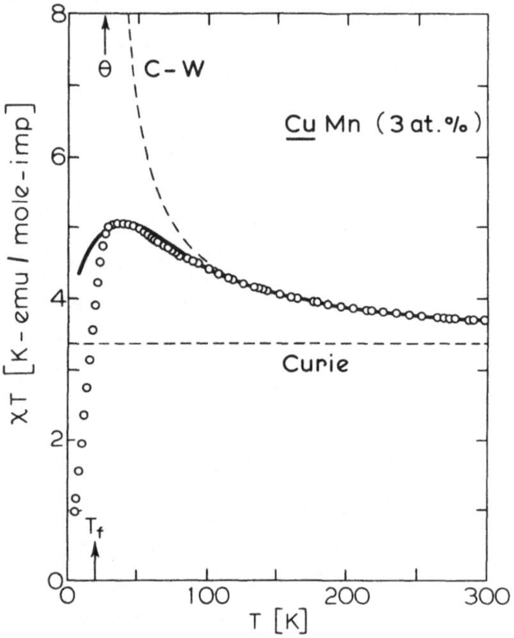

Fig. 6.
χT versus T for a CuMn (3 at.%) alloy. The solid line represents a fit to the configuration cluster model. After Morgownik et al., Ref.30.

The resulting values of J_n (+ for ferromagnetic and - for anti-ferromagnetic) are collected as a function of distance in Fig.7 for the four different alloy systems considered in Fig.5. Again the four systems are separated by their various behavior into three catagories. In Fig.7 the dashed line represents the RKKY-like conduction electron polarization around a single (Mn) impurity as determined from Cu^{63}-NMR measurements[31]. The interaction parameters J_n found from the susceptibility analysis for CuMn, AuMn and AuFe do not follow this isolated impurity polarization of conduction electrons. However, a similar, damped oscillatory behavior cannot be excluded from these results, but with the Fermi wave vector k_F much smaller than the free electron values for CuMn, AuMn and AuFe. This suggests that, for interacting

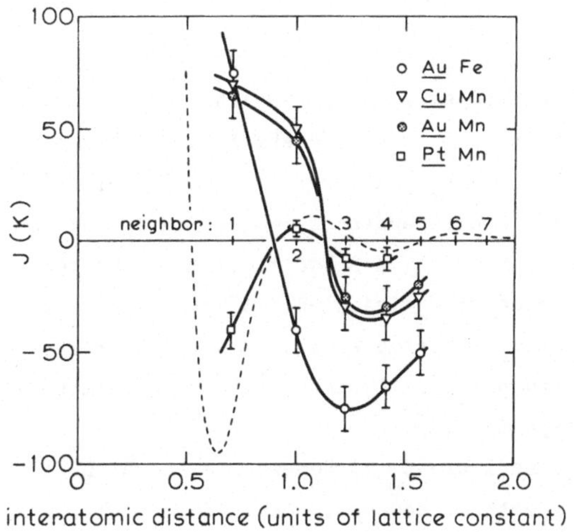

Fig. 7. Exchange parameters J_n as function of neighbor distance. The dashed line represents the RKKY-conduction electron polarization.

impurities, d-electrons, which are delocalized or itinerant over a few lattice lengths, are mediating the interaction between the impurities[32].

In addition by comparing the sign of the Cowley-Warren α_n-ASRO parameters (Fig.2) and the J_n in Fig.7 there is a clear relation between α_n and J_n to favor ferromagnetic clusters. These local ferromagnetic regions appear well above T_f and seem to be a general property of metallic spin-glasses. They should certainly be taken into account in any complete theory of spin-glasses.

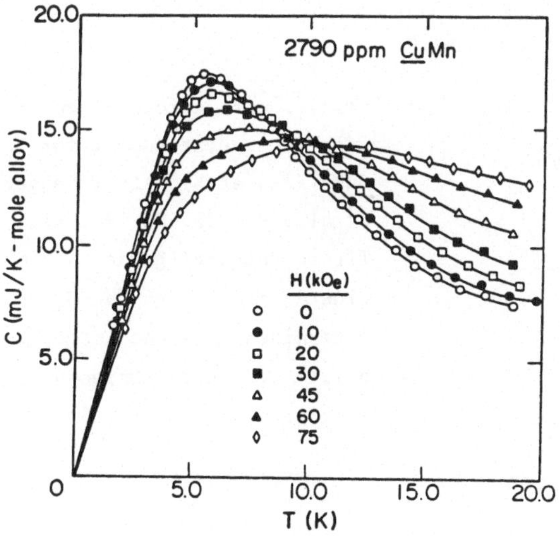

Fig. 8. Temperature and magnetic field dependences of the magnetic specific heat for a CuMn (0.3 at.%) alloy. After Brodale et al., Ref.33.

c) Specific heat

The magnetic specific heat C_m is difficult to extract at temperatures above T_f due to the large electron and phonon contributions. Nevertheless for a number of systems data for C_m are available at a few times T_f. The overall behavior of $C_m(T,H)$ is remarkable with respect to its smooth T and H dependences — see Fig.8. The most precise measurements[33] exhibited in Fig.8 show only a broad minimum in the temperature coefficient B(T) of the field dependence of $C_m/T=A(T)+B(T)H^2$ around T_f and there is a slight "anomaly" in C_m/T vs T at T_f relative to a somewhat arbitrary, smooth background curve. Consequently, it is practically impossible to determine T_f and any critical phenomena from $C_m(T,H)$ experiments.

As discussed in the Sec.IIIa the low temperature $C_m(T)$ dependence can be nicely fit using the Walker-Walstead excitation spectrum. However, near and above T_f, while the resistivity can be further described by these "simulated" excitations, the specific heat can not. Here there is a maximum in C_m just above T_f and a long tail extending to much higher temperatures (as in Fig.8). In order to calculate the behavior of $C_m(T,H)$ in this $T>T_f$ regime, Morgownik[34] has employed the same model (configurational cluster) as in the dc susceptibility (Sec.IIIb). Once the partition function is known, taking temperature derivatives instead of field derivatives will determine $C_m(T,H)$. By using the same values of $J_1,\dots,J_5$ found from the susceptibility to fit the C_m data an addition check on the validity of the method is gained. In Fig.9 we show the measured specific heat[35] for <u>Cu</u>Mn (2.4 at.%) and the model predictions with the susceptibility values of J_1 to J_5 given in Fig.7.

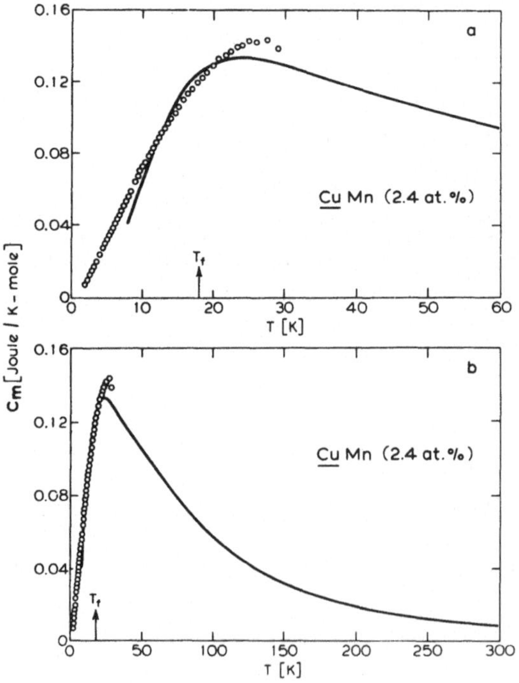

Fig. 9. Open circles: measured magnetic specific heat for <u>Cu</u>Mn (2.4 at.%). After Wenger et al., Ref.35. Solid line: fit to the configuration cluster model with the J_n's determined from the susceptibility fit. After Morgownik, Ref.34.

In contrast to the susceptibility fit which breaks down at $\approx 2T_f$ (see Fig.6), the specific heat model results are in excellent agreement without any adjustable parameters down through T_f. This means that the local exchange interactions J_n (static, spatial correlations) are able to describe the complete specific heat behavior, yet something more is required for the susceptibility as T is lowered towards T_f. This something extra is the time correlations that strongly affect $\chi(T)$ and which do not play a role in $C_m(T)$, since most of the entropy $\approx 80\%$ has already gone into the local spatial correlations for $T>T_f$. We now proceed to consider the dynamical experiments in order to understand more fully these time correlations.

IV. DYNAMIC EXPERIMENTS AT HIGH TEMPERATURES

a) Neutron spin echo

A special type of neutron scattering, called neutron spin echo (NSE) spectroscopy is especially suited to studying the spin-glasses. This technique[36] of high resolution inelastic scattering is able to detect time correlations from 10^{-12} sec up to 10^{-8} sec, and simultaneously to measure spatial correlations as a function of q (the scattering wave vector). Thus one can present the experimental NSE information[37] in terms of a static correlation function or structure factor $S(q)=S(q,t=0)$ and a dynamical or time-dependent correlation function $s(q,t)$ where $s(q,t=0)=1$ such that $S(q,t)=S(q)s(q,t)$.

For <u>CuMn</u> measurements of S(q) show little q dependence at small q values as the temperature is lowered through T_f. Oppositely, for <u>AuFe</u> there is a significant change in S(q) over a similar T region. These results have been interpreted[37] in terms of ferromagnetic short-range order (MSRO). In the case of <u>CuMn</u> the MSRO is already present at a few times T_f and does not grow excessively as T_f is reached. On the other hand in <u>AuFe</u> there is a large development of MSRO just above T_f. This interpretation is complementary to that given in Sec.II and is certainly related to the very different ASRO and $J_n(r)$ found in <u>CuMn</u> and <u>AuFe</u>.

The time-dependent part of the correlation function is exhibited in Fig.10, where $s(q\to 0,t)$ is plotted versus log(t) for <u>CuMn</u> (5 at.%). At high temperatures $T \gg T_f$ an extrapolated exponential decay $e^{-t/\tau}$ (τ is the spin relaxation time) is expected. However, as the temperature is lowered much slower decays are observed. Here for $T>T_f$ the observed spectra can be attributed to a broad distribution of relaxation times $P(\tau)$. Thus

$$s(q,t) = \int P(\tau)e^{-t/\tau}d\tau \ .$$

The spin relaxation above T_f can also be described by an Arrhenius law with a large distribution of activation energies. But when $T \leq T_f$ strong deviations from the Arrhenius law appear and the correlation function changes to a power law relation,

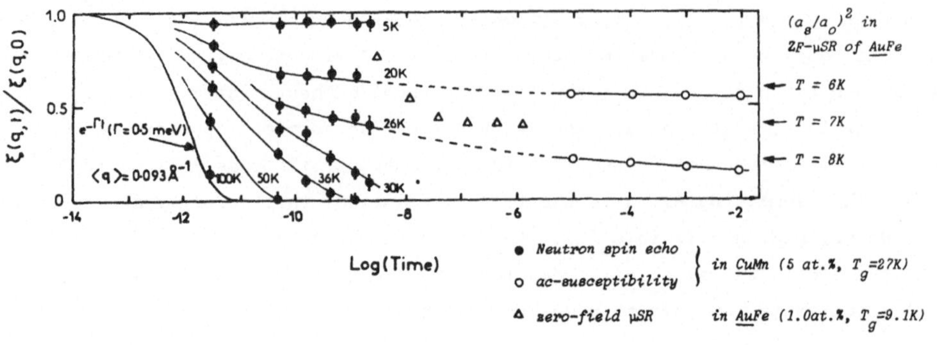

Fig. 10. Spin correlation function versus log of the time. Solid circles: NSE results. After Mezzei, Ref.(37). Open triangles: zero field μSR.After Uemura et al., Ref.38. Open circles: ac susceptibility. After Murani et al., Ref. 46.

i.e. $s(q,t) \propto t^{-\nu}$ which becomes nearly constant or very slow as $T \to 0$.

The main conclusions (37) drawn from the NSE investigations are: a) The existence of ferromagnetic clusters (MSRO) which are either already present at $T \gg T_f$ or which grow in size as $T \to T_f^+$. b) Continuous slowing down of spin relaxation which is dramatic at T_f. c) Cooperative character of the spin freezing. d) Finally, the gradual appearance of larger and larger energy barriers as the spin relaxation spectrum becomes "hard" in the frozen state.

b) Muon spin relaxation

Another technique which can determine the static and dynamical properties, S(q,t), of spin-glasses is the measurement of muon spin relaxation (μSR). Here the q dependences are not tunable as in the NSE, but are broadly averaged over all q and the effective range of correlation or relaxation times span the 10^{-11} to 10^{-4} sec regime — complementary to the NSE. The μSR method has been extensively employed to investigate a number of spin-glass alloys by three groups: TRIUMF-Tokyo (38), LAMPF et al. (39) and SIN-Braunschweig (40).

Let us begin with the newly developed longitudinal, zero field μSR. This technique effectively measures a spin relaxation function $G_z(t)$ which in different limits (slow and fast modulations) determines a mean spin correlation time τ_c of the impurity moments (38). A plot of τ_c against T_f/T is shown in Fig.11 for AuFe, CuMn and AgMn. Note the rapid increase (two orders of magnitude between 0.8 T_f and 1.2 T_f) in τ_c for all systems around T_f. The use of a single relaxation time to describe the spin-glass freezing is, of course, a very crude approximation, but nevertheless it illustrates a number of basic characteristics such as precursor effects above T_f, the cooperative but incomplete freezing at T_f, and the existence of a spectrum of slower relaxation times (10^{-6}–10^{-5}sec) for $T<T_f$.

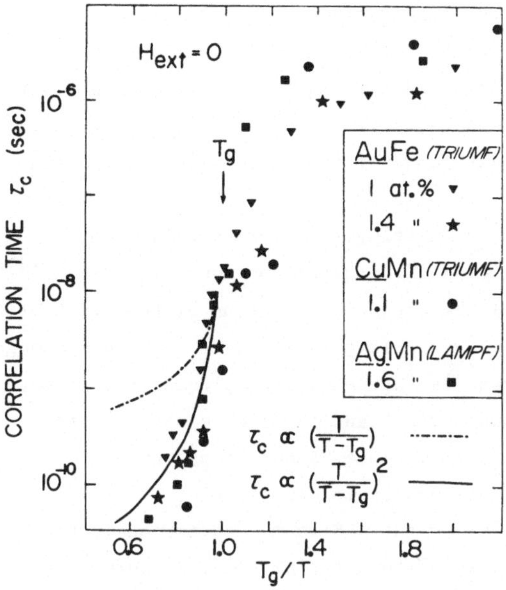

Fig. 11. Spin correlation time as a function of temperature determined from zero field μSR experiments. After Uemura et al., Ref.38.

In order to further study the dynamical behavior of the spin-glass freezing it is possible to apply a small longitudinal field H and measure $G_z(t)$ for various H. This decouples static random fields from rapidly fluctuating ones. Here Uemura et al.(38) have attempted to fit the field dependences by various models as shown in Fig.12. Now the single relaxation time and exponential auto-correlation function approximation clearly cannot explain the large decoupling effect of the field. Even a distribution

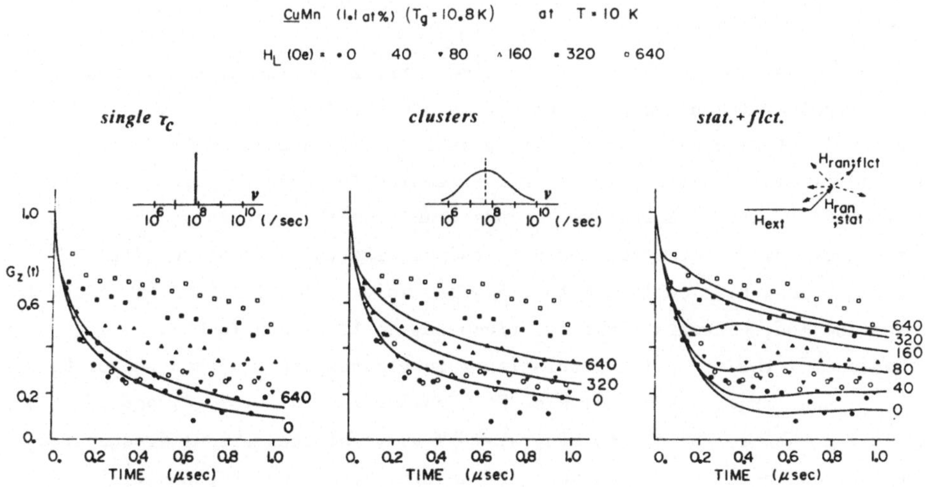

Fig. 12. Comparison of some model attempts to fit the field dependences of longitudinal field μSR for a CuMn spin-glass. After Uemura et al., Ref. 38.

of relaxation times corresponding to various cluster sizes does not significantly improve the agreements. It is only when a static (with respect to the μSR time scales) random field related to the spin-glass order parameter q_{EA} is combined with a dynamical fluctuating component that a reasonable fit is obtained (see Fig.12). This indicates that below T_f there is a coexistence of frozen (static) fields and still rapidly fluctuating dynamical ones.

A similar analysis can be performed on the existing transverse-field muon depolarization data which also exhibit a strong field dependence near T_f. Uemura[38] has suggested two possible explanations: a) The frequency spectrum of the fluctuating spins is shifted towards lower frequency by the applied field, i.e. τ_c becomes longer for higher H, and b) A small static component is induced in the fluctuating moment spectrum by H. A comparison between the longitudinal and transverse μSR results suggests b). Moreover, even in zero field there is some "spontaneous" onset of static components and this is enhanced by the application of a field.

If we relate the time correlation function in the NSE to τ_c determined by the μSR[38], we can liken the results of these two different time scale measurements. This is illustrated in Fig.10 by triangles for the μSR determination of the auto correlation function. Notice how reasonable the μSR data are and how they nicely span the time gap between the NSE and ac susceptibility. A more meaningful test is to compare τ_c from μSR with τ_e (decay time where the correlation function is 1/e is original value) from NSE. This has been carried out[38,39] by plotting log $(\tau_c T_f)$ and log $(\tau_e T_f)$ versus T/T_f for a number of spin-glasses. Excellent agreements exist between the two measurements in the temperature range where the time scales overlap.

c) ac Susceptibility

At the slower end (10^{-1}-10^{-4}sec) of the time correlation range, the ac susceptibility offers another technique with which to study the dynamical behavior of spin-glasses. For all metallic systems there is a small (≈1%), but distinct frequency dependence of $\chi'(T)$(dispersion) near T_f and extending to lower temperatures[41]. Thus the freezing or cusp temperature depends weekly on the frequency, $T_f(\omega)$, in this audio range. More information could be gained from the absorption $\chi''(\omega)$, however, it is a most difficult task to try and resolve $\chi''(\omega,T)$ since the effect is so small and the eddy currents limit the frequency to less than 10 kHz.

Nevertheless for the so-called insulating spin-glasses the ω dependence of both χ' and χ'' are much larger and measurable[42]. Knowing the full $\chi'(\omega,T)$ and $\chi''(\omega,T)$ behavior one can then employ Argand diagrams (χ'' vs χ' at the various frequencies) and the well-known equations for magnetic relaxation. As a recent example[42] in Fig.13 we show the temperature dependences of χ' and χ'' at three different frequencies for a $Eu_{0.2}Sr_{0.8}S$ sample. The data from Fig.13 can be transformed into the Argand diagram at various temperatures thereby ascertaining an average relaxation time τ. Deviations from a single τ, i.e. χ''-χ' semicircle, mean a distribution of

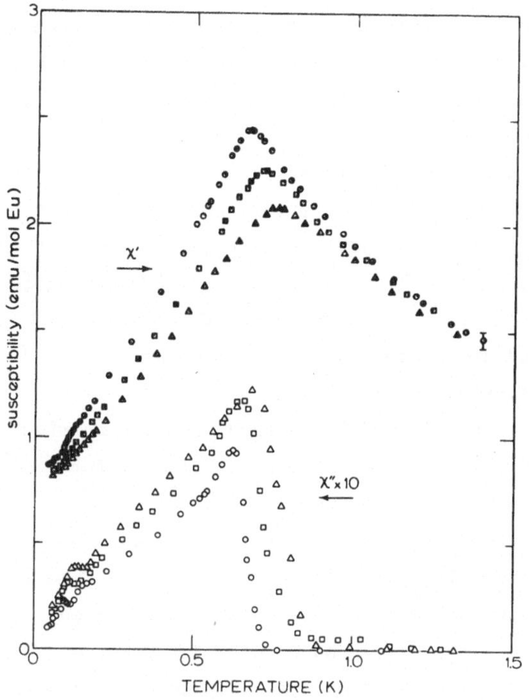

Fig. 13. Temperature dependence of the dispersion χ' and absorption χ'' for $Eu_{0.2}Sr_{0.8}S$. Circles: 10.9 Hz, squares: 261 Hz and triangles 1969 Hz. After Hüser et al., Ref. 42.

relaxation times and the strength of the deviations estimates the width of the distribution function. For the above example of (EuSr)S near T_f there is only a gradual shift in τ_{AV} which remains in the 10^{-4} sec range, but the width dramatically broadens into a long time tail for T just above T_f to T just below. A more detailed treatment and comparison of relaxation times and their distributions for different systems as determined by ac susceptibility measurements will be given by Wenger[43].

The $\chi(\omega)$ behavior requires much faster relaxation times and is more strongly correlated in the case of metallic spin-glasses. Recently Lundgren et al.[44] have developed a model bases on a wide spectrum of relaxation times to relate $\chi'(\omega)$ to $\chi''(\omega)$. This model considers the time variation of cluster moments $m_o(\tau)$ which respond to an applied ac field h and which become blocked at an individual blocking temperature $T_B(\tau)$. A broad spectrum of relation times is assumed to exist between τ_{min} and τ_{max}. Thus

$$\chi'(\omega) = \frac{1}{h}\int_{\tau_{min}}^{\tau_{max}} \frac{1}{1+(\omega\tau)^2}\, m_o(\tau)g(\tau)d(\ln\tau)$$

$$\chi''(\omega) = \frac{1}{h}\int_{\tau_{min}}^{\tau_{max}} \frac{\omega\tau}{1+(\omega\tau)^2}\, m_o(\tau)g(\tau)d(\ln\tau)$$

Depending upon the distribution function $g(\tau)$ there is a proportionality between χ'' and $\frac{\partial\chi'}{\partial \ln\omega}$ with a slightly different constant for each distribution function. Although this proportionality has been verified for many alloy spin-glasses the form of $g(\tau)$ can only be estimated in the large τ limit. According to Lundgren et al.(44) as the temperature is reduced $g(\tau)$ shifts to longer time scales and enters the time-window of the ac susceptibility measurement. This appearance of relaxation times within the time of measurement causes the characteristic temperature and frequency behavior of χ' and χ'' around T_f (see Fig.13). At a yet lower T a spontaneous local ordering is said to occur originating from strong intercluster interactions. This creates relaxation times which go to infinite and are therefore responsible for the dc susceptibility and magnetization properties of spin-glasses.

Further consideration and verification of this model approach has been carried out by Van Duyneveldt and Mulder(45) and the complete calculation of $g(\tau)$ for different types of "spin-glasses" will be described by Wenger(43).

For not too low temperature the ac susceptibility may be related to the NSE $S(q=0)$ and $s(q=0,t)$ by(37,46) $\chi'(\omega)\propto\frac{1}{kT}\,S(q=0)\,[1-s(q=0,t)]$ and $t=0.7/\omega$. The term in square brackets gives the fraction of spins which is mobile on time scale t and therefore responding to the h-field. With $S(q=0)$ independent of T for CuMn, $s(q=0,t)$ may be determined from measuring $\chi'(\omega)$ at different temperatures. The data from this analysis are shown in Fig.10 for $10\text{Hz}<\omega<10$ kHz. Notice the smooth extrapolation to shorter times scales and the nice connection with the faster μSR and NSE results.

Another use of the ac susceptibility has been in measuring the higher order harmonics $\chi(3\omega)$, $\chi(5\omega)$ etc. by Miyako et al.(47), and thereby determining the non-linear susceptibility $\chi_{ac}^{(2)}(T)$. This susceptibility exhibits a very sharp peak and it shows a logarithmic $(T-T_f)$ behavior rather than a simple power law near T_f. The surprising sharpness in $\chi_{ac}^{(2)}(T)$ has been interpreted as lending support to a type of phase transition and critical phenomenon in spin-glasses.

V. THE FREEZING PROCESS

Let us first recall some of the salient features from the three static experiments discussed in Sec.III. The magnetic resistivity $\Delta\rho$ can be quantitatively described over a fairly large temperature range by the computer simulated excitation spectrum. This causes a local scattering between the interacting local moments and the conduction electrons. At $T>T_g$ $\Delta\rho$ is more complicated to analyze due to other (Kondo) effects, but strong correlations among the local moments remain well above T_f and produce similar types of excitations. Here a qualitative description of the data is obtained, if the high temperature ($\approx 10\ T_f$) interactions are taken into account. The estimation of relaxation times from resistivity measurements give a faster, more local (intracluster), small-angle excitation than the intercluster relaxations found in the dynamic measurements (Sec.IV). Consequently, the high T resistivity seems more related to the growth and development of local cluster which can support these special excitations at the expense of the paramagnetic spin flip or spin disorder scattering.

The dc susceptibility in an external field can be fully treated down to $\approx 2T_f$ in terms of local exchange interactions J_n which oscillate in sign, i.e. parallel and antiparallel couplings. The combination of the MSRO with the ASRO produces mainly ferromagnetic clusters and these freeze out (become dynamic) at lower temperatures. This model of a limited number of J_n breaks down for the susceptibility as $T \to T_f$. The predicted susceptibility continues to rise, whereas the experimental one has a peak or a shoulder. This descrepancy could mean that the exchange interactions are becoming longer-ranged and cooperative. However, it could also signify the dominance of dynamical freezing or relaxation processes, and the onset of the T-dependent anisotropies. The dc susceptibility in an external field which strongly affects the relaxation is not a good measurement of such dynamical effects.

In contrast, the zero field specific heat can be fit with a limited number of J_n both above and below T_f, see Fig.9. This indicates that the relatively smooth $C_m(T)$ behavior with no indication of T_f is not sensitive to time-dependent effects and simply traces the loss of entropy as reflected in the development of local, spatial spin correlations. At T_f most ($\approx$70%) of the degrees of freedom for the spin system have been removed by the formation of clusters and the dynamical freezing processes are not clearly perceived from C_m measurements.

In summary, these three static experiments consistently demonstrate the formation of local correlations far above T_f. These, dynamically evolving, mainly ferromagnetic clusters are the basic building blocks out of which the spin-glass state is formed. In mean field theory they are simply omitted, something equivalent to throwing the baby out with the bath water.

Recollecting the three dynamical experiments, all give information concerning the time correlation function $s(q \to 0,t)$ or average relaxation time τ. In addition the NSE can also determine the static spatial spin correlations through the q dependence. These dynamic measurements have tracted the time correlations in metallic spin-glasses from $\approx 10^{-12}$sec to ≈ 1 sec — twelve orders of magnitude! Fig.10 displays the time correlation function over this vast region of time. A single relaxation time approximation was found too naive and better agreement with the data is obtained with a distribution of relaxation times $P(\tau)$ which evolves with temperature.

In Fig.14 we schematically sketch the qualitative character of such a $P(\tau,T)$ for a metallic spin-glass. The true paramagnetic (noninteracting spins) regime ($T > 10\ T_f$) is distinguished by a nearly single Korringa relaxation time at $\approx 10^{-12}$sec. As T is lower, the spatial correlations begin to form and this perturbes the distribution function. Then as $T \to T_f^+$ there is a sudden shift of $P(\tau)$ to slower times due to longer-ranged, cooperation interactions. Finally at T_f a very long-time tail appears. Here it should be mentioned that the spatial correlation usually become fixed, i.e. do not grow any further around T_f, and the dominant effects are caused by the time dependences or relaxations of these well-defined clusters. The extension of $P(\tau)$ to include static times ($\tau \to \infty$) signalizes the onset of the frozen spin-glass state.

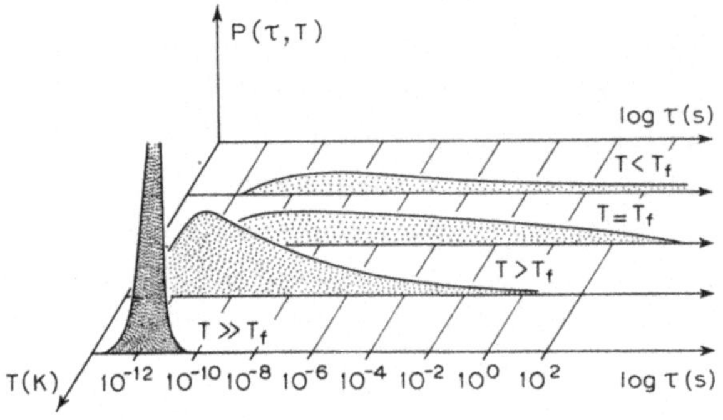

Fig. 14. Schematic representation of the probability distribution for spin relaxation times with its evolution as a function of temperature.

The actual dependence of the shift and distortion of $P(\tau)$ on T is a function of the longer-range exchange couplings between the local clusters, viz. the inter-cluster interactions. For metallic spin-glasses with 3d impurities these are quite strong and we have the strong coupling limit, where a sudden shift in $P(\tau)$ occurs in a small range of temperature near T_f. Table I indicates how this coupling is reduced for other types of spin-glasses, until finally, for a random superparamagnet with no interactions between the clusters $P(\tau)$ is governed solely by the Arrhenius law $\tau=\tau_o \exp(E/k_B T)$, and a more gradual, completely "thermally activated" freezing occurs.

Table I. Examples of differently coupled spin-glasses .

Strong-coupling	Intermediate-coupling	Weak-coupling	No coupling
metallic 3d-impurities	metallic 4f-impurities	insulating systems	superparamagnetics rock magnets
<u>Cu</u>Mn <u>Au</u>Fe <u>Pt</u>Mn <u>Ag</u>Mn	$(\underline{Y}Gd)Al_2$ $(\underline{La}Gd)B_6$ $(\underline{La}Gd)Al_2$ $(\underline{La}Er)Al_2$	(EuSr)S Mn/Co-aluminosilicate glasses	CoO Holmium borate glasses

VI. CONCLUSIONS

The atomic short-range order works together with the magnetic short-range order to produce mainly ferromagnetic clusters far above T_f. The three static measurements discussed above give incontrovertible evidence for the existence of MSRO which may be quantitatively described by the local exchange interactions J_n. These clusters are similar to the networks encounted in ordinary glasses.

The necessity for a distribution of relaxation times which depends upon the temperature $P(\tau,T)$ has been amply demonstrated from the three dynamical experiments considered above. A significant transformation in $P(\tau)$ occurs around T_f, the degree of which is related to the strength of the intercluster coupling. For strongly coupled spin-glasses the dramatic changes of $P(\tau)$ in a narrow T-interval surrounding T_f give the appearance of a phase transition. An analogous $P(\tau,T)$ is also encountered in ordinary glasses. Cooperative effects between free volumes give rise to similar time-temperature dependent effects[48] in the viscosity of ordinary glasses as those found in the ac-susceptibility of spin-glasses.

Monte Carlo simulations of 2-dimensional, Gaussian, Ising spin-glasses[49] have shown the existences of the high temperature clusters and naturally result in a distribution function $P(\tau,T)$ with the same behavior as in the real spin-glasses. Also this model simulation[50] exhibits the same characteristic field-cooled properties, without a phase transition, as the mean field theory does with a phase transition.

In conclusion a spin-glass is a glass, is a glass, is a glass[51].

I wish to acknowledge L.E. Wenger for almost a year of stimulating discussion in Leiden and to thank the Nederlandse Stichting voor Fundamenteel Onderzoek der Materie (FOM) for their financial support of spin-glass research at the Kamerlingh Onnes.

REFERENCES

1. S.F. Edwards and P.W. Anderson, J. Phys. F5, 965 (1975) and F6, 1927 (1976).
2. G. Parisi, Phys. Rev. Lett. 43, 1754 (1979) and 50, 1946 (1983), and H. Sompolinsky, Phys. Rev. Lett. 47, 935 (1981).
3. D. Sherrington and S. Kirkpatrick, Phys. Rev. Lett. 32, 1792 (1975).
4. See for example P. Monod and H. Bouchiat, J. Phys. (Paris) Lett. 43, 45 (1982); B. Barbara, A.P. Malozemoff and Y. Imry, Phys. Rev. Lett. 47, 1852 (1981), and R.V. Chamberlin, M. Hardiman, L.A. Turkevich and R. Orbach, Phys. Rev. B25, 6720 (1982).
5. Members include K. Binder, W. Kinzel, I. Morgenstern, A.M. Moore, A.J. Bray, J.A. Hertz et al.
6. Such time dependences also occur in the mean field theory's dynamical solutions, see Sompolinsky Ref.2.
7. See for example S. Schultz, E.M. Gullikson, D.R. Fredkin and M. Tovar, Phys. Rev. Lett. 45, 1508 (1980).

8. For a review see H. Alloul and F. Hippert, J. Magn. Magn. Mater. 31-34, 1321 (1983).
9. E.M. Gullikson, D.R. Fredkin and S. Schultz, Phys. Rev. Lett. 50, 537 (1983).
10. B.I. Halperin and W.M. Saslow, Phys. Rev. B16, 2154 (1977); C.L. Henley, H. Sompolinsky and B.I. Halperin, Phys. Rev. B25, 5849 (1982), and W.M. Saslow, Phys. Rev. Lett. 48, 505 (1982).
11. See for example A. Fert, S. Senoussi and D. Arvanitis, J. Phys. (Paris) Lett. 44, L-345 (1983).
12. H. Alloul, in these Conference Proceedings.
13. A.R. Miedema, P.F. de Chatel and F.R. de Boer, Physica 100B, 1 (1980).
14. See for example T. Egami, J. Magn. Magn. Mater. 31-34, 1571 (1983).
15. For some early neutron work on CuMn see H. Sato, S.A. Werner and R. Kikuchi, J. Phys. (Paris) 35, C-4, 23 (1974).
16. T.M. Hayes, J.W. Allen, J.B. Boyce and J.J. Hauser, Phys. Rev. B22, 4503 (1980), and E. Dartyge, H. Bouchiat and P. Monod, Phys. Rev. B25, 6995 (1982).
17. G.L. Whittle and S.J. Campbell, J. Magn. Magn. Mater. 31-34, 1337 (1983), and C.E. Violet and R.J. Borg, Bull. Amer. Phys. Soc. 28, 543 (1983).
18. P.A. Flinn, Phys. Rev. 104, 350 (1956), and P.C. Clapp and S.C. Moss, Phys. Rev. 142, 418 (1966).
19. A.F.J. Morgownik and J.A. Mydosh, to be published in Solid State Commun.
20. For CuMn: J.R. Davis, S.K. Burke and B.D. Rainford, J. Magn. Magn. Mater. 15-18, 151 (1980); for AuMn: A.F.J. Morgownik, J.A. Mydosh and C. van Dijk, J. Magn. Magn. Mater. 31-34, 1334 (1983); and for AuFe: E. Dartyge et al. Ref. 16.
21. N. Cowlam and A.M. Shamah, J. Phys. F10, 1357 (1980).
22. C.E. Violet and R.J. Borg, Bull. Amer. Phys. Soc. 28, 720 (1983).
23. "Spatially disordered ferromagnet" was also used recently by W.M. Saslow, Phys. Rev. Lett. 50, 1320 (1983).
24. J.W. Cable, S.A. Werner, G.P. Felcher and N. Wakabayashi, Phys. Rev. Lett. 49, 829 (1982).
25. W.H. Meiklejohn and C.P. Bean, Phys. Rev. 105, 904 (1957).
26. J.A. Mydosh, P.J. Ford, M.P. Kawatra and T.E. Whall, Phys. Rev. B10, 2845 (1974) and P.J. Ford and J.A. Mydosh, Phys. Rev. B14, 2057 (1976).
27. I.A. Campbell, Phys. Rev. Lett. 47, 1473 (1981), and I.A. Campbell, P.J. Ford and A. Hamzic, Phys. Rev. B26, 5195 (1982).
28. L.R. Walker and R.E. Walstedt, Phys. Rev. Lett. 38, 514 (1977); Phys. Rev. B22, 3816 (1980).
29. See for example J. Kouvel, J. Phys. Chem. Solids 21, 57 (1961).
30. A.F.J. Morgownik and J.A. Mydosh, to be published in Solid State Commun. and A.F.J. Morgownik, Ph.D. Thesis, University of Leiden, 1983.
31. J.D. Cohen and C.P. Slichter, Phys. Rev. Lett. 40, 129 (1978).
32. For the opposite case of a very concentrated 3d alloy, see M.B. Stearns, Physica 91B, 37 (1977).
33. G.E. Brodale, R.A. Fisher, W.E. Fogle, N.E. Philips and J. van Curen, J. Magn. Magn. Mater. 31-34, 1331 (1983).
34. A.F.J. Morgownik, to be published and Ph.D. Thesis, University of Leiden, 1983.
35. L.E. Wenger and P.H. Keesom, Phys. Rev. B13, 4053 (1976).
36. F. Mezei in Neutron Spin Echo-Lecture Notes in Physics Vol.128, edited by F. Mezei (Springer Verlag, Heidelberg, 1980).
37. F. Mezei, J. Appl. Phys. 53, 7654 (1982).
38. See for example Y.J. Uemura and T. Yamazaki, Physica 109-110B, 1915 (1982), and Y.J. Uemura, Ph.D. Thesis, University of Tokyo, 1981.
39. See for example R.H. Heffner, M. Leon and D.E. MacLaughlin in Proceedings of the Yamada Conference on Muon Spin Rotation, Shimoda, Japan, 1983.
40. See for example K. Emmerich, F.N. Gygax, A. Hintermann, H. Pinkvos, A. Schenck, Ch. Schwink and W. Studer, J. Magn. Magn. Mater. 31-34, 1363 (1983).
41. C.A.M. Mulder, A.J. van Duyneveldt and J.A. Mydosh, Phys. Rev. B25, 515 (1982).
42. D. Hüser, L.E. Wenger, A.J. van Duyneveldt and J.A. Mydosh, Phys. Rev. B27, 3100 (1983).
43. L.E. Wenger in these Conference Proceedings.
44. L. Lundgren, P. Svedlindh and O. Beckman, J. Magn. Magn. Mater 25, 33 (1981), J. Phys. F12, 2663 (1982) and Phys. Rev. B26, 3990 (1982).

45. A.J. van Duyneveldt and C.A.M. Mulder, Physica 114B+C, 82 (1982).
46. A.P. Murani, F. Mezei and J.L. Tholence, Physica 108B, 1283 (1981).
47. Y. Miyako, S. Chikozawa, T. Saito and Y.G. Yuochunas, J. Appl. Phys. 52, 1779 (1981), J. Phys. Soc. Jpn. 51, 1394 (1981) and S. Chikazawa, S. Taniguchi, H. Matsuyama and Y. Miyako, J. Magn. Magn. Mater. 31-34, 1355 (1983).
48. M.H. Cohen and G.S. Grest, Phys. Rev. B20, 1077 (1979).
49. H. Takayama, K. Nemoto and H. Matsukawa, Solid State Commun. 44, 1281 (1982), J. Magn. Magn. Mater. 31-34, 1303 (1983), J. Phys. Soc. Jpn. 52 Suppl. 109 (1983) and K. Nemoto and H. Takayama, to be published.
50. W. Kinzel and K. Binder, Phys. Rev. Lett. 50, 1509 (1983).
51. With acknowledgements to J.A. Hertz (Copenhagen) and Gertrude Stein (Paris).

RELAXATION EFFECTS IN SPIN GLASSES AROUND THE FREEZING TEMPERATURE

L.E. Wenger

Department of Physics, Wayne State University,
Detroit, MI 48202, U.S.A. *
Kamerlingh Onnes Laboratorium, Rijks-Universiteit Leiden,
Leiden, The Netherlands

Abstract

Measurements of the ac magnetic susceptibility in zero field for several insulating spin-glasses are reported. By resolving the frequency dependence of both the in-phase component χ' and the out-of-phase component χ'', the dynamical behavior of these spin systems around their freezing temperatures T_f can be clearly characterized. For two insulating systems: (EuSr)S and cobalt aluminosilicate, the susceptibility behavior can be described in terms of a wide distribution of relaxation times for temperatures well above T_f and whose width increases dramatically as the temperature nears T_f. Furthermore by employing a simple phenomenological relaxation model, a quantitative picture of the relaxation time distribution can be ascertained as it evolves. From these pictures for the insulating as well as metallic spin-glass system, the dynamic behavior near T_f is qualitatively different than that of a long-range magnetically-ordered solid.

A. Introduction

Although ac magnetic susceptibility measurements have been routinely performed on spin-glass systems for over a decade with the cusplike peak still remaining as the characteristic signature of the spin-glass freezing temperature T_f, only recently have frequency dependent studies reached their full potential in providing a clear picture of the dynamical behavior of the spin freezing process. The earliest studies primarily focused on the frequency dependence of the susceptibility peaks in an attempt to answer the question of whether the spin-glass transition represented a cooperative phase transition of the Edwards-Anderson type [1] (no frequency dependence) or just a manifestation of a strong relaxation process analogous to the Néel superparamagnetic model [2]. Needless to say, the experimental results for different

* Permanent address

spin-glass (SG) systems showed varying degrees of frequency dependence such that neither representation appeared to be totally correct. In fact, the main contribution to the understanding of any dynamical behavior was the degree of failure or "unphysicalness" of an Arrhenius law to describe these frequency effects. With present susceptibility measurements [3-5] of the frequency dependence of the out-of-phase component χ'' (absorption) as well as the in-phase component χ' (dispersion) an opportunity exists for clearly characterizing the dynamical behavior near T_f and for even deducing a quantitative picture of the actual distribution of spin relaxation times. This paper will focus on χ' and χ'' results for three insulating SG systems: (EuSr)S, cobalt aluminosolicate and holmium borate glasses. From the strong dynamical behavior observed near T_f in these insulators, comparisons between each as well as to the behavior of metallic spin-glasses and more cooperative-ordered solids can be made.

B. Holmium borate glass

Fig. 1 shows the temperature dependence of both χ' and χ'' for a holmium borate glass in which the predominate interactions are short-range and antiferromagnetic. The characteristic susceptibility peaks in χ' are observed at low temperatures with a very strong frequency dependence. Also χ'' is nonnegligible for $T>T_f(\nu)$ and an

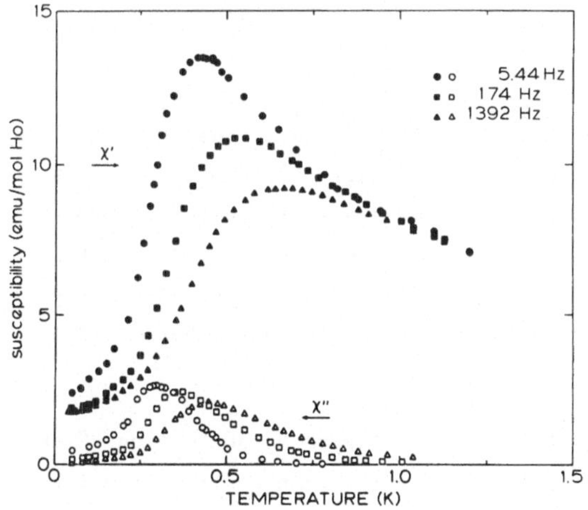

Figure 1. Temperature dependence of the dispersion χ' (solid symbols) and absorption χ'' (open symbols) for $(Ho_2O_3)_{0.08}(B_2O_3)_{0.92}$ glass.

inflection point seems to occur at a temperature very close to that corresponding to the χ' maximum. In order to interpret the observed frequency dependence in this glass as well as the other insulators, the analysis will follow the generally accepted approach based on a phenomenological model analogous to the work of Néel [2] on superparamagnetic particles. The model assumes that a random distribution of magnetic impurities in the host material with substantial short-range interactions leads to the formation of clusters of highly-correlated spins at temperatures well above any freezing temperature. Each cluster has a local anisotropy energy associated with it such that the transition rate for the clusters to surmount this energy barrier is equal to the inverse of the relaxation time

$$\tau^{-1} = \tau_o^{-1} \exp(-E/kT) \tag{1}$$

where τ_o^{-1} is the characteristic transition rate and E is the energy barrier height. If these clusters are identical, a simple expression (an Arrhenius law) is derivable for the frequency (ν) dependence of T_f:

$$kT_f = -E/\ln 2\pi\nu\tau_o \quad . \tag{2}$$

For this Ho glass, values of 6.6 K and $4.5\text{x}10^{-9}$s for E/k and τ_o are determined, which are quite reasonable. This suggests the observed spin freezing is teneable within the Néel framework; however, the full possibilities of these χ measurements are yet to be examined. The frequency dependence of the dispersion χ' and the absorption χ'' can be given in the case of magnetic relaxation by the Casimir and du Pré equations [6]:

$$\chi' = \chi_s + \frac{\chi_T-\chi_s}{1+\omega^2\tau^2} \tag{3a}$$

$$\chi'' = \omega\tau \frac{\chi_T-\chi_s}{1+\omega^2\tau^2} \tag{3b}$$

where χ_T is the isothermal susceptibility in the limit $\omega\to 0$ and χ_s the adiabatic one in the limit $\omega\to\infty$. At $\omega=1/\tau$ the dispersion will have an inflection point, whereas the absorption will show a maximum. Thus this maximum provides a method for determining an average relaxation-time constant τ_{av} for each temperature. Also according to Eq.(3b), the absorption should follow a sech (ln $\omega\tau$) functional dependence for a single relaxation time and will be considerably broadened if a distribution of

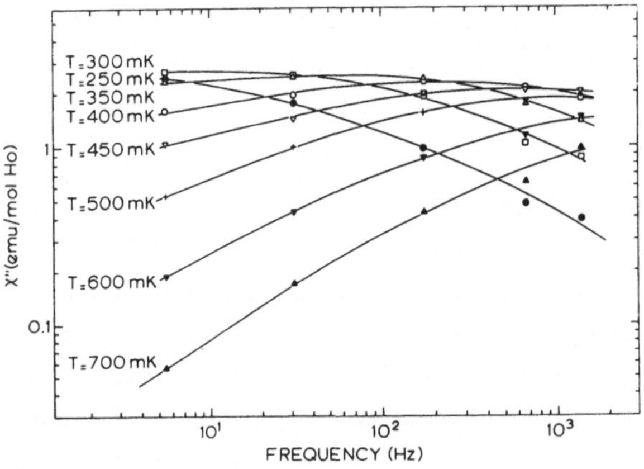

Figure 2. Absorption χ'' as a function of frequency for different temperatures. The solid lines are a visual guide.

relaxation times $g(\tau)$ is present. Therefore, the absorption usually provides more information about the dynamics of the spin freezing around an ordering temperature than the dispersion. In Fig.2, the absorption is shown for the same Ho glass. Clear indications of absorptive maxima for several temperatures are observed with the maxima shifting to lower frequency with decreasing temperature. The width of the χ'' curves, however, exceeds the sech ($\ln \omega\tau$) dependence for a single relaxation time. To further illustrate this relaxation time distribution, the susceptibility data can be plotted in the complex plane as χ'' vs χ'. These so-called Argand diagrams are shown in Fig.3 for several temperatures around T_f. Clearly the curves cannot be

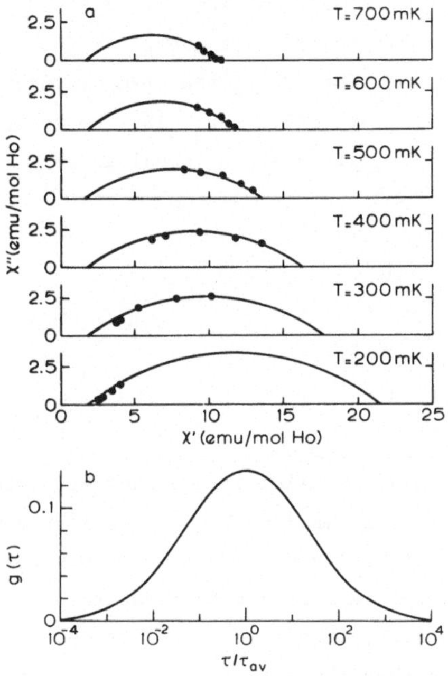

Figure 3. (a) Argand diagrams for several temperatures. The lines are arcs of semicircles. (b) The distribution of relaxation times $g(\tau)$. See text for details.

described as semicircles which is indicative of a single relaxation time but as arcs of semicircles. Following analyzes of Coles plots for the dielectric susceptibility [7], essentially a single Gaussian distribution is determined for all temperatures with the following form:

$$g(\tau) = \frac{b}{\sqrt{\pi}} \exp[-(b \ln \tau/\tau_{av})^2] \tag{4}$$

with b=0.23 and $\tau_{av}=1.5\times10^{-8}\exp(4.4/T)$. Thus the spin freezing in this Ho glass can easily be interpreted within the Néel picture of superparamagnetic relaxation, and the potential of these complex χ measurements is readily seen [8].

C. (EuSr)S

For comparison, Fig.4 shows the absorption χ" of a $(Eu_{0.2}Sr_{0.8})S$ spin-glass, where competing ferromagnetic and antiferromagnetic interactions exist. No clear maximum in χ" is observable over the frequency range investigated. Furthermore, the absorption at the lowest frequencies dramatically increases in the temperature interval from 700 to 600 mK [T_f(10 Hz)~640 mK]. For T<600 mK, the absorption is

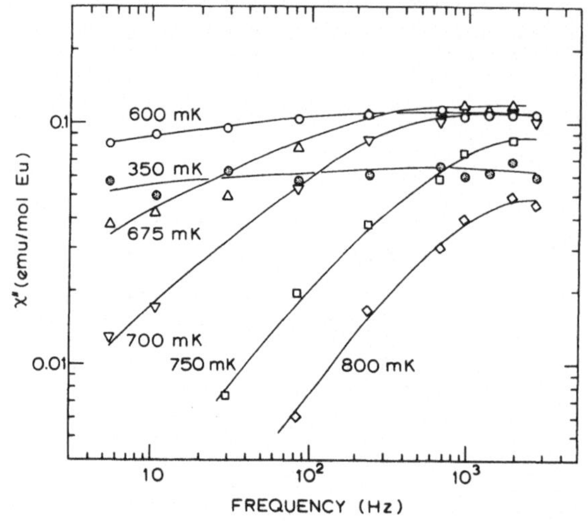

Figure 4. Absorption χ" as a function of frequency for three different temperatures for $(Eu_{0.2}Sr_{0.8})S$. The solid lines are a visual guide.

essentially flat with all curves remaining essentially parallel to one another. These absorption features in combination with the Argand diagrams [see Fig.3 of Ref.5] can be qualitatively described in terms of a broad distribution of relaxation times for temperatures above T_f which broadens even at a faster rate as the temperature approaches T_f and remains very broad for $T<T_f$. In fact, this feature is present in all spin-glasses and may be a more definitive test than the peak in χ' for characterizing a material as a spin-glass. The absence of clear maxima in the absorption does not allow a direct determination of the average relaxation time, although an upper limit of 10^{-4}s may be deduced for all temperatures. Comparing to frequency studies on insulating ferromagnetic or antiferromagnetic materials, the distribution of relaxation times for these ordered solids is much more narrow for all temperatures although some broadening may occur for temperatures below the ordering temperature [9]. Furthermore the average relaxation time τ_{av} will typically increase 4 to 8 orders of magnitude very rapidly over a small temperature interval around the ordering temperature such that a clear absorption maximum can be observed. Thus the (EuSr)S spin-glass shows more of a dynamical freezing behavior near T_f with no clear indication of a rapidly increasing τ_{av} nor close resemblance to the narrow relaxation time distribution observed in a cooperative-type freezing of magnetic spins.

D. Cobalt aluminosilicate glass

To demonstrate a more quantitative picture of the relaxation times for (EuSr)S or any other spin-glass, measurements over a much wider frequency range are required. Studies of the more traditional metallic spin-glasses which show weaker frequency dependences have been limited for two reasons: (i) the absorptive signal is only ~1% of the dispersive signal thus making analysis difficult, and (ii) the inherent eddy current effect in metals restricts the usable frequency range. Thus one is forced to study insulating spin-glasses where χ''/χ'~5-10% and the frequency range is only limited by the available experimental apparatus. Thus the remainder of this paper focuses on a study of a cobalt aluminosilicate glass $[(CoO)_{0.4}(Al_2O_3)_{0.1}(SiO_2)_{0.5}]$ in the frequency range 0.64 Hz to 30 MHz [10-11].

As with all insulating spin-glasses, the temperature dependence of χ' [see Fig.1 of Ref.10] shows a strong frequency dependence with the cusplike maxima [T_f(0.64 Hz)=6.25 K] becoming progressively more rounded and shifting to higher temperatures as the frequency is increased. Fitting this frequency effect to Eq.(2) yields unphysical values of $\tau_o=4\times10^{-20}$ s and E/k=286 K. Thus the spin freezing of this glass is vastly different from the Ho glass although the T_f's are only an order of magnitude different. For a better comparison, the absorption for the Co glass is shown in Fig.5. Even though the results are seemingly more complex than that of the

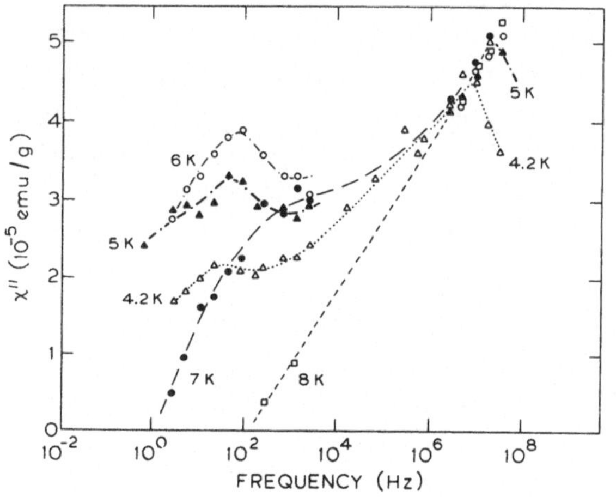

Figure 5. Absorption χ'' of a cobalt aluminosilicate glass as a function of frequency for several temperatures. T_f (ν=0.64 Hz)≈6.25 K. The lines are a visual guide.

preceding two insulators or of a simple sech (ln $\omega\tau$) behavior, the dramatic increase of the χ'' signal at the lower frequencies in the temperature range of 8 to 6 K is qualitatively similar to the (EuSr)S. The most pronounced increase in the absorption over the 8 K results is for T=6 K and 5 K in the frequency range below 1 kHz, with a sizeable absorptive signal still remaining for T=4.2 K. (The small maxima are within the experimental error and consequently may not be of great significance.) Furthermore the absorption for frequencies above 1 kHz is increasing and significantly larger, especially in the 1 MHz to 30 MHz range at all temperatures. Thus the "average" or most probable relaxation time corresponds to a frequency in the 5 MHz to 100 MHz range. Only for the 4.2 K results is a clear maximum in the megahertz frequency range observable. This frequency of 6 MHz corresponds to an average relaxation time $\tau_{av}=1/2\pi\nu \sim 3\times10^{-8}$s. An estimate for the 3 K data gives a lower limit of 2×10^{-7}s while the 5 K data yields 1×10^{-8}s for an upper limit. Thus τ_{av} is still very small, even for $T<T_f$, and apparently does not rapidly change several orders of magnitude near T_f. However as T approaches T_f, the sudden increase of the χ'' signal, corresponding to the appearance of some long spin relaxation times ($>10^{-3}$s), is characteristic of spin-glass freezing, not of a cooperative-type freezing. Finally, the time scale characteristic of this spin freezing is $10^{0}-10^{-3}$s which is ideal for χ measurements and probably too long for shorter-time resolution experiments, such as μSR [12].

Further information can be deduced from the absorption by determining the frequency at which χ'' decreases to zero. These frequencies would give an estimate of the minimum and maximum relaxation times, τ_{min} and τ_{max}. For 7 K and 8 K, values for τ_{max} of 160 ms and 1.6 ms are deduced. τ_{min} values could only be estimated for 2.4, 3.0, and 4.2 K data (2.4 and 3 K data are not shown in Fig.5) ranging between 10^{-9} and 10^{-10}s. In Fig.6, an Arrhenius-type of plot for several relaxation times is

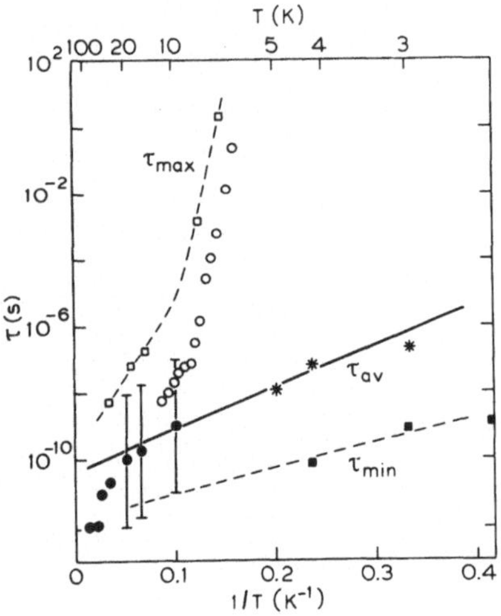

Figure 6. The relaxation time spectrum for the Co aluminosilicate glass as a function of inverse temperature. τ_{max}(□), τ_{av}(*), and τ_{min}(■) determined from χ'' measurements. Open circles (○) determined from frequency dependence of T_f from χ' peaks and solid circles (●) from μSR measurements of Ref.12.

shown. First, the open circles represent the temperatures of the χ' peaks for a particular measuring frequency ($\tau=1/2\pi\nu$) and are equivalent of the usual $\ln\nu$ vs $1/T_f$ plots. The stars represent τ_{av} determined from the absorption maxima in Fig.5. The solid squares indicate τ_{min} and the open squares τ_{max}. Note that τ_{max} increases at about the same rate as the open circles for $T_f(\nu)$, but are a factor of $\approx10^3$ different. Lastly, the solid circles represent the correlation times from zero-field μSR measurements [12]. The "error" bars on those solid circles below 20 K correspond to the halfwidth of the Gaussian distribution required to fit the μSR data. Combining τ_{av} values from μSR (●) and $\chi''(*)$ a linear fit of Eq.(1) can be produced with very physical values of τ_o=3x10^{-11}s and E=30 K. Similarly from fitting Eq.(1) to τ_{min} values, τ_o^{min}=4.2x10^{-12}s and E^{min}=14.4 K. Thus from this 2-dimensional plot, the relaxation time distribution has a τ_{av} that nicely follows a simple energy barrier relaxation picture and a width that increases rather slowly for decreasing temperatures (T>10 K). The lower limit, τ_{min}, of this distribution continues to follow an activation type relaxation for T<10 K; however τ_{max} dramatically increases below 10 K.

Thus, the remaining step of the analysis is to construct a quantitative 3-dimensional picture of the evolution of this spin relaxation time distribution for the Co glass. However to do so, a couple of simplifying assumptions about the phenomenological model in conjunction with Eq.(3) must be realized. Since a distribution of relaxation times $g(\tau)$ must be included, Eq.(3) will be modified as follows:

$$\chi' = \chi_s + \int_{\tau_{min}}^{\tau_{max}} \frac{\Delta\chi(\tau)}{1+\omega^2\tau^2}\, g(\tau)\, d\ln\tau \tag{5a}$$

$$\chi'' = \int_{\tau_{min}}^{\tau_{max}} \frac{\omega\tau\, \Delta\chi(\tau)}{1+\omega^2\tau^2}\, g(\tau)\, d\ln\tau \tag{5b}$$

where $\Delta\chi(\tau)=\chi_T(\tau)-\chi_s(\tau)$ and $g(\tau)$ is a normalized function. The frequency dependence of χ' may be more easily seen by taking the partial derivative of Eq.(5a) with respect to ln ω:

$$\frac{\partial(\chi'-\chi_s)}{\partial \ln\omega} = -\int_{\tau_{min}}^{\tau_{max}} \frac{2\omega^2\tau^2\Delta\chi(\tau)}{(1+\omega^2\tau^2)^2}\, g(\tau)\, d\ln\tau \quad . \tag{6}$$

These preceding equations are identical to those used by Lundgren,et al.[3] in their derivation of χ' and χ'' if $\chi_s(\tau)=0$ and $\chi_T(\tau)=m_o(\tau)/h_o$. By assuming $g(\tau)$ to vary very slowly in $\ln\tau$, such that in the region of $\omega\approx 1/\tau_m$ $g(\tau)$ is a constant; the integration of Eqs.(5) and (6) can be greatly simplified. Lundgren et al.[3] then were able to derive the following expression:

$$\chi'' \simeq -\frac{\pi}{2}\frac{\partial\chi'}{\partial\ln\omega} \approx -\frac{\pi}{2} \cdot \frac{m_o(\tau_m)}{h_o} \cdot g(\tau_m) \tag{7}$$

for the spin-glass regime ($\tau_{min} \ll 1/\omega_m \ll \tau_{max}$). This relationship was experimentally confirmed by those authors and more quantitatively for several metallic spin-glasses by van Duyneveldt and Mulder [4]. However, our χ'' results tend to suggest that $g(\tau)$ is not that slowly varying function. Instead, the following approach is envisioned. Each cluster of spins will be assumed to be noninteracting and the <u>difference</u> between $\chi_T(\tau)$ and $\chi_s(\tau)$ to be independent of its relaxation time. Then Eqs.(5b) and (6) become

$$\chi''/(\chi_T-\chi_s) = \int_{\tau_{min}}^{\tau_{max}} \frac{\omega\tau}{1+\omega^2\tau^2}\, g(\tau)\, d\ln\tau \tag{8a}$$

$$\frac{\partial}{\partial \ln\omega}\left(\frac{\chi'-\chi_s}{\chi_T-\chi_s}\right) = -\int_{\tau_{min}}^{\tau_{max}} \frac{2(\omega\tau)^2}{(1+\omega^2\tau^2)^2}\, g(\tau) d\ln\tau \ . \tag{8b}$$

In general the distribution $g(\tau)$ cannot be directly deduced from these preceding equations unless a further assumption is made about $g(\tau)$ or the hyperbolic function. Since the $\mathrm{sech}^2(\ln\omega\tau)$ function in Eq.(8b) has a smaller half-width in $\omega\tau$ space than the $\mathrm{sech}(\ln\omega\tau)$ function in Eq.(8a), the range $g(\tau)$ must be constant is reduced. Thus the dispersion relation of Eq.(8b) with $g(\tau) \sim$ constant will be used to deduce an initial relaxation time distribution. Subsequently this $g(\tau)$ and the absorptive relation of Eq.(8a) are utilized in an iterative fashion until a more accurate $g(\tau)$ can be determined. The result of this determination is shown in Fig.7 as well as

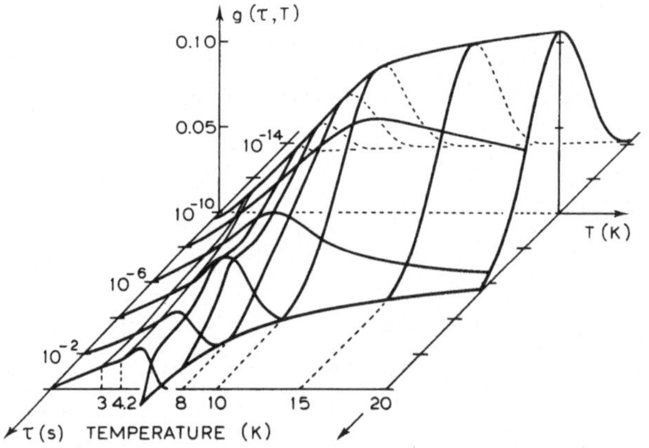

Figure 7. The distribution of relaxation times $g(\tau,T)$ as a function of time (10^0–10^{-14}s) and of temperature (0–20 K). The solid lines show several isotemporal and isothermal lines.

extensions of τ towards τ_{min} by recalling $g(\tau,T)$ must be normalized and using τ_{av} and τ_{min} from Fig.6. This allows an evaluation of $g(\tau,T)$ from 3 K to 20 K and 10^0s–10^{-14}s. Clearly the qualitative feature of a "small" dramatic increase of long relaxation times for $T \lesssim 6$ K is visible with the average relaxation time always less than 10^{-8}s. Furthermore, $g(\tau,T)$ is essentially constant over $\ln\tau$ at $T \lesssim 4.2$ K so that logarithmic time dependences can be expected for long-time relaxation measurements, such as for the time decay observed in the magnetic remanence. A similar approach to χ' results of CuMn give qualitative similar results but are limited to $10^{-1}s>\tau>10^{-4}s$ range, and thus do not allow a full comparison.

A final concluding remark about the relaxation effects observed in these insulating spin-glasses and its correlation with field-cooled dc-magnetization (FCM) measurements. With the rapid increase in relaxation times near T_f, another parameter — the cooling rate — should be an important factor in determining the magnetic properties and magnetic state of the spin-glass below T_f. Thus various cooling rates could result in different long-time (and possibly reversible) metastable states. Within the present relaxation model, a cooling rate less than the inverse of the temperature rate of change of the longest relaxation times $(\partial\tau_{max}/\partial T)^{-1}$ would be required in order for the FCM to trully be a thermodynamic equilibrium magnetization. Recently Kinzel and Binder [13] have shown this effect by computer simulation and this author has observed cooling rates effects experimentally for the Co alumino-silicate glass [14]. Thus a magnetic spin-glass may be more analogous to a real glass than previously thought with the spin-glass state resulting solely from dynamical processes.

Acknowledgements

The author wishes to thank Professor J.A. Mydosh for his critical reading of this manuscript and stimulating discussions throughout this past year as well as helpful discussions from Dr. A.J. van Duyneveldt, Dr. J.C. Verstelle, and D. Hüser about relaxation phenomena in various magnetic systems. Also part of this research was a result of the financial support by the National Science Foundation (DMR-7921298) and the Nederlands-American Fulbright Commission. Finally the author wishes to thank his wife and family for the sacrifice of several holidays and weekends during his sabbatical year in Leiden for completion of this work.

References

1. S.F. Edwards and P.W. Anderson, J. Phys. F5, 965 (1975).
2. L. Néel, Adv. Phys. 4, 191 (1955).
3. L. Lundgren, P. Svedlindh and O. Beckman, J. Magn. Magn. Mater.22, 271 (1981).
4. A.J. van Duyneveldt and C.A.M. Mulder, Physica 114B, 82 (1982).
5. D. Hüser, L.E. Wenger, A.J. van Duyneveldt and J.A. Mydosh, Phys. Rev. B27, 3100 (1983).
6. H.B.G. Casimir and F.K. du Pré, Physica 5, 507 (1938).
7. V.V. Daniel, Dielectric Relaxation, (Academic, London, 1967), pp. 95-109.
8. D. Hüser, T.A. Meert, L.E. Wenger, and J.A. Mydosh, in preparation.
9. H.A. Groenendijk, Ph.D. thesis (Leiden, 1981) unpublished.
10. L.E. Wenger, C.A.M. Mulder, A.J. van Duyneveldt and M. Hardiman, Phys. Lett. 87A, 439 (1982).
11. These results are combined measurements with A.J. van Duyneveldt, J.C. Verstelle, and N.J. Verhaar.
12. Y.J. Uemura, C.Y. Huang, C.W. Clawson, J.H. Brewer, R.F. Kiefl, D.P. Spencer, and A.M. de Graaf, Hyper. Inter.8, 757 (1981).
13. W. Kinzel and K. Binder, Phys. Rev. Lett. 50, 1509 (1983).
14. L.E. Wenger and J.A. Mydosh, in preparation.

NON-LINEAR SUSCEPTIBILITIES AND SPIN-GLASS TRANSITION IN CuMn

R. Omari*, J.J. Préjean, and J. Souletie
Centre de Recherches sur les Très Basses Températures,**
C.N.R.S., BP 166 X, 38042 Grenoble-Cédex/FRANCE

1. Theoretical background

The conventional scaling hypothesis [1] states that the Gibbs potential is a generalised homogeneous function. By differentiating with respect to the field and changing sign we obtain the scaling equation

$$\mathcal{R} = t^{\beta} f\left(\frac{\eta}{t^{\gamma+\beta}}\right) \tag{1}$$

where, for a conventional ferromagnetic transition $\mathcal{R}$ is the magnetization and

$$\eta = H/T_c \text{ and } t = (T-T_c)/T_c \tag{2}$$

are the normalised fields coupled to the order parameter which, in this case, is a moment. In the close vicinity of T_c (say $T_c < T < 1.1\ T_c$) where we are told to check the scaling hypothesis, the determination of the exponents is not affected if we use the non linear variables

$$\eta' = \frac{H}{T} = \frac{\eta}{1+t} \quad \text{and} \quad t' = \frac{T-T_c}{T} = \frac{t}{1+t} \tag{3}$$

which tend to η and t in the domain of interest where η and t tend to zero. We now can write equation (1) as

$$\mathcal{R} = M = n\mu f(\mu H/kT) \tag{4}$$

where $\mu = p_{eff} t'^{-(\gamma+\beta)}$ and $n^{-1} \cong N^{-1} t'^{-(\gamma+2\beta)} \cong (a\xi)^{d\nu}$ describe respectively the divergence of the moment and of the size of the Kadanoff cells in dimension d, when the coherence length ξ diverges on approaching T_c from above. An advantage of the variables η' and t' is that they insure a natural atomic high temperature limit for the moment when the correlated volume shrinks to one site ($\mu \to p_{eff}$ when $n \to N$ and $T \to \infty$). Equation (4), at the same time, transforms naturally into a Curie law

$$M = \alpha H/k(T-\gamma T_c) \text{ when } T \to \infty \tag{5}$$

In this spirit we were able to fit the susceptibility data in nickel between T_c and $3T_c$ [2] with the same values of the exponent γ, determined for $T_c < T < 1.1\,T_c$. Figure 1 shows the experimental variations of M(H/T) i.e. of $\mathcal{R}(\eta')$ in nickel near T_c. M, here, is the anomalous part of the magnetization deduced by subtracting a temperature independent paramagnetic contributions $\chi_o H$ ($\chi_o = 2.43\times10^{-6}$ u.e.m ; see reference [2]) from the classical Weiss and Forrer data [3].

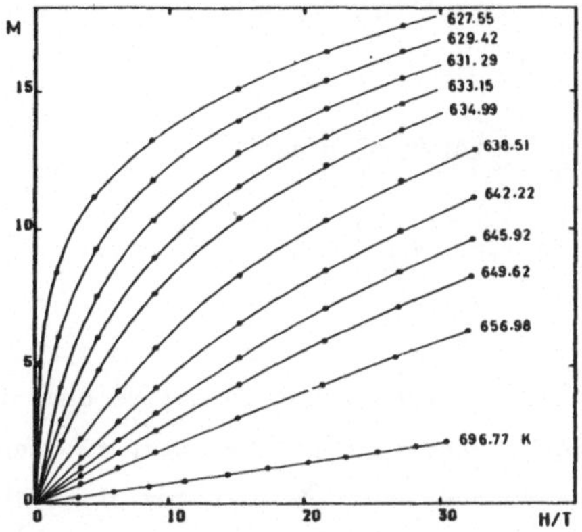

Figure 1 : Plot of $\mathcal{R}(\eta')$ in the case of the ferromagnetic transition of nickel. In this case $\mathcal{R}$ is the magnetization and $\eta' = H/T$ ($M/\text{u.e.m.} = M_{exp} - 2.43\times10^{-6}\ H/\text{Oe}$, see ref. [2]).

In spin glasses, the order parameter is the square of a moment: $q = \overline{\langle S^2\rangle}$ [4]. The associated quantity η which plays the role of the field is then

$$\eta^* = \left(\frac{H}{T_c}\right)^2 \tag{6}$$

Differentiating the free energy with respect to η^* we obtain the quantity $\mathcal{R}$ which plays the same role as the magnetization in the ferromagnetic transition and is equivalent to a susceptibility. Instead of Eq. (1) we have the relation of Suzuki [5] and Chalupa [6]

$$\mathcal{R} = \frac{M}{CH/T_c} - 1 = \left(\frac{T-T_c}{T_c}\right)^{\beta} F\left[\frac{(H/T_c)^2}{((T-T_c)/T_c)^{\gamma+\beta}}\right] \tag{7}$$

On substituting, as above, the variables H/T and $(T-T_c)/T$ and with $\eta' = (H/T)^2$ we obtain

$$\mathcal{R} = \frac{M}{CH/T} - 1 = \left(\frac{T-T_c}{T}\right)^{\beta} F\left[\left(\frac{T}{T-T_c}\right)^{\gamma+\beta}\left(\frac{H}{T}\right)^2\right] \tag{8}$$

which is a possible generalisation of Equation (7) at temperatures larger than T_c.

Notice that, as in the case of the ferromagnetic transition, this formulation is consistent with the mean field result (here the S.K. result [4])

$$\frac{M(H)}{CH/T} \cong 1 - \frac{H^2}{T(T-\gamma T_c)} \qquad \text{when } T \to \infty \tag{9}$$

2. Principle of the experiment and results

Equation (1) can be expanded in terms of η as

$$= A_1\eta + A_2\eta^2 + \dots A_n\eta^n + \dots \tag{10}$$

with $A_n \sim t^{\beta-n(\gamma+\beta)}$.

The two exponents γ and β which determine the anomalous behaviour of $\mathcal{R}$ near T_c can in principle be obtained from a knowledge of the first two non zero susceptibilities A_1 and A_i which enter in the expansion of $\mathcal{R}$ in terms of η. They can be derived from the intercept and the slope of the initial variation on an $\mathcal{R}/\eta$ vs. η^{i-1} plot. Since $\mathcal{R}$ is already a susceptibility in spin glasses and $\eta^* \sim H^2$, the determination of A_1 requires the extension of the development of M in terms of H to higher order than would have been necessary for a ferromagnet. This accounts for the motivation to study the so-called non-linear susceptibilities [7,8,9,10].

In the spirit of the arguments which led us to Equation (5), we have chosen to expand the magnetization in terms of H/T rather than in terms of H/T_c. We write

$$\frac{M}{M_{sat}} = \sum_{n=0}^{\infty} a_{2n+1} L_{2n+1} \left(\frac{\mu H}{kT}\right)^{2n+1} \tag{11}$$

where L_{2n+1} is the coefficient of $\left(\frac{\mu H}{kT}\right)^{2n+1}$ in the Taylor expansion of the Langevin function. The identification of Equations (11) and (8) imposes

$$\begin{cases} a_1 = 1 \\ a_m = \left(\dfrac{T-T_c}{T}\right)^{-m(\beta+\gamma)+\beta} \qquad \text{for} \quad m = 2n+1 > 1 \end{cases} \tag{12}$$

and the Langevin function is the natural high temperature limit of the magnetization ($a_{2n+1} \to 1$ when $T \to \infty$).

We have measured the magnetization of a 1 at% CuMn sample in fields up to 7 Teslas and at temperatures between T_{c_0} and 4 T_{c_0} where T_{c_0} = 10.05±0.05 K is the low field, low frequency limit of the temperature of the susceptibility cusp. More details about the experiment and the analysis of the data will be found in reference [11].

None of the pathological aspects generally associated with a phase transition are visible on the traditional plots of M(H) shown in Figure 2 vs. H and H/T (We will not in this paper justify the small correction T_o = .8 K which appears in the figures : See reference [11] for more details).

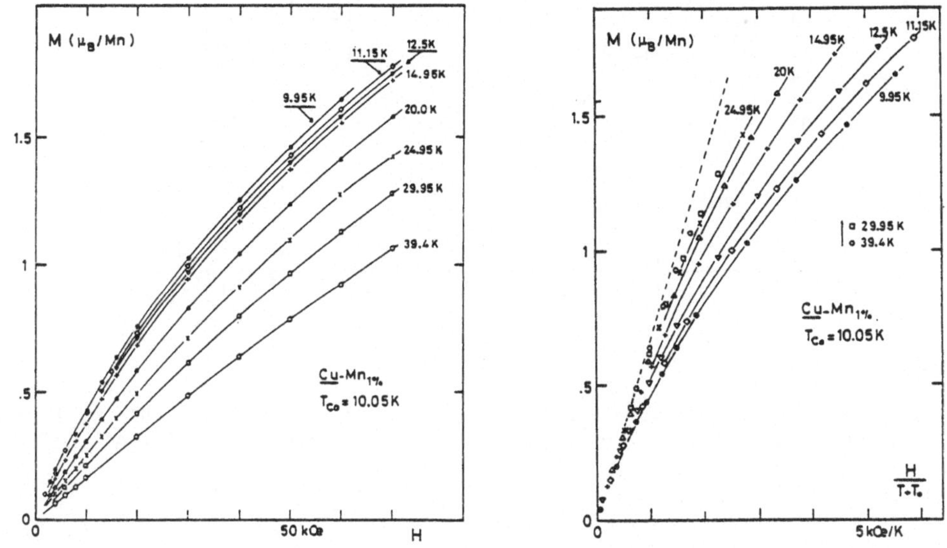

Figure 2 : Plot of the magnetization versus field and versus H/T (r.h.s.) in CuMn 1 at% at different temperatures between $\sim T_{c_o}$ and 4 T_c. T_{c_o} is the temperature of the susceptibility cusp. T_o will be neglected in this and the following figures (see ref. [11]).

When we plot, as in Figure 3, $\mathcal{R}(\eta)$ vs. η [i.e. $\frac{M}{CH/T}$ - 1 vs. $(\frac{\mu H}{kT})^2$] we recognise the same pathology which is observed on the M(H/T) plot of Figure 1 for nickel : essentially the initial susceptibility diverges on approaching T_c ; here it is the non-linear susceptibility a_3. A log-log representation of the same data stresses that the domain where $\mathcal{R}(\eta)$ is linear in η (i.e. M(H)- $\frac{CH}{T}$ is cubic in $\frac{H}{T}$) is sufficiently well defined so that we can get reliable estimates of a_3(T). Making a further step, we can determine a_5(T) using the equivalent of an $\frac{\mathcal{R}(\eta)}{\eta}$ vs η plot at the different measurement temperatures (see reference [11]).

Our values for a_3 and a_5 are represented in terms of temperature on the right-hand side of figure 4. On the left-hand side Log a_5 is represented in terms of Log a_3. We have, within the experimental accuracy,

$$a_5 \sim a_3^{2.25\pm0.05} \quad \text{for } T \geqq 1.1\ T_c \tag{13}$$

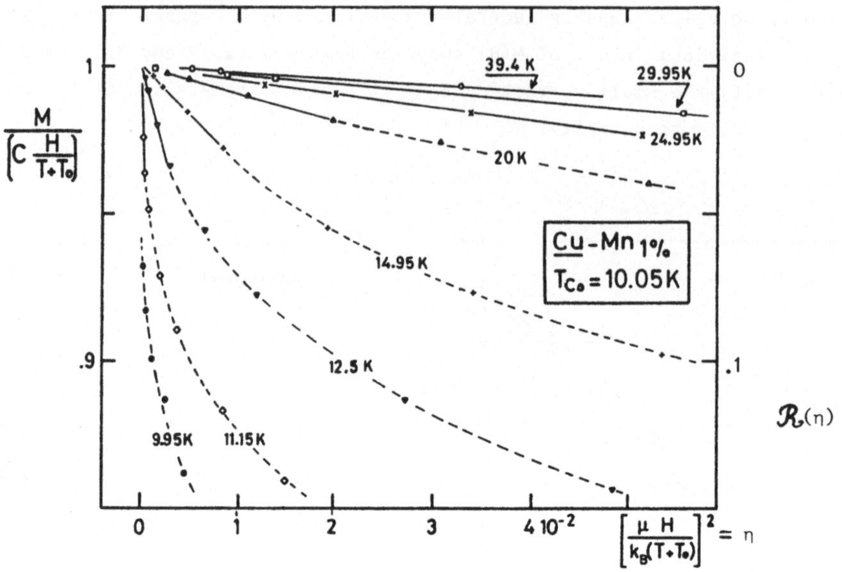

Figure 3 : Same data as in Figure 2 represented in a [M/(CH/T)] vs. [(μH)/(kT)]^2 plot. This is directly 1-$\mathcal{R}(\eta)$ vs. η. The slope of the initial part is proportional to a_3.

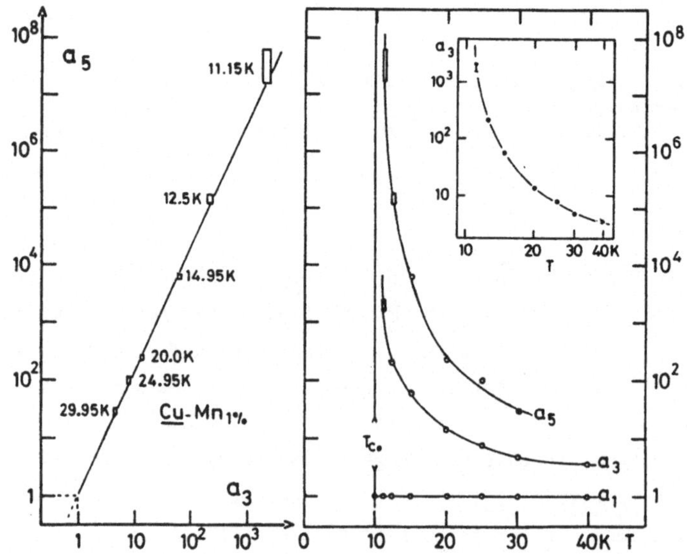

Figure 4 : The coefficients a_1,a_3,a_5 involved in the series expansion of the magnetization in terms of μH/kT are represented vs. temperature in a semi-log plot on the right-hand side. The insert shows a log-log representation of the a_3(T) dependence. On the left-hand side we have represented Log a_5 vs. Log a_3.

over a range of measurement where a_3 and a_5 vary, respectively, by 3 and 7 orders of magnitude. This figure is a plain representation of experimental evidence quite independent of any model or interpretation. The fact that $a_5 \neq (a_3)^2$ implies that two exponents, γ and β, are necessary to describe this transition with Eq. (9) or (10). This eliminates the possibility of a transition occuring at $T_c = 0$.

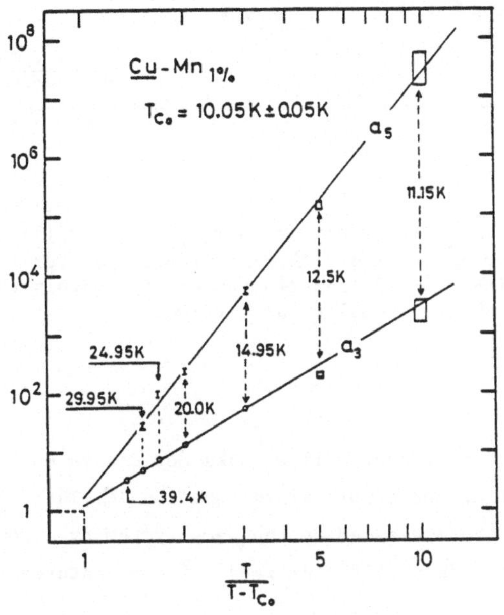

Figure 5 : Log-Log plot of a_3 and a_5 vs. $T/(T-T_c)$. We have $a_3 = [T/(T-T_c)]^{3.25\pm0.05}$ and $a_5 = [T/(T-T_c)]^{7.25\pm0.05}$.

We have represented Log a_3 and Log a_5 vs. $\mathrm{Log}[T/(T-T_{c_o})]$ on the figure 5. The divergence of both a_3 and a_5 can be described over the whole range of temperatures by a power law in $T/(T-T_{c_o})$, consistent with the result expected from Equation 12, provided we take

$$\begin{cases} \gamma = 3.25 \pm 0.10 \\ \beta = 0.75 \pm 0.25 \end{cases} \tag{14}$$

The success of Equation 12 in accounting also for the terms of higher order in H^{2n+1} is illustrated by the good superposition of all our experimental data ($T \geq T_c$, $H \leq 7$ Teslas) onto a universal curve in the reduced diagramme of Figure 6.

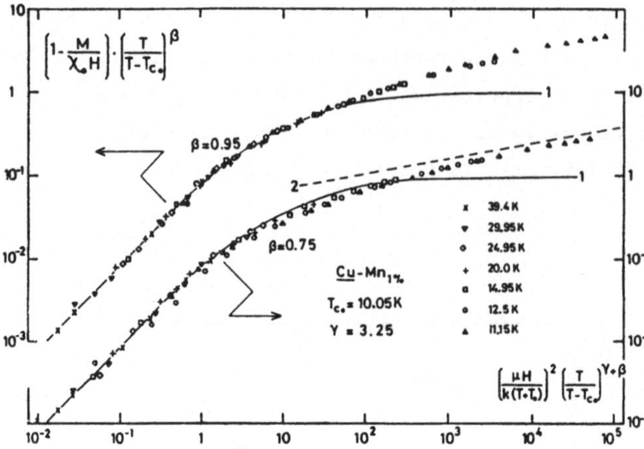

Figure 6 : Universal plot of $[1-(M/CH/T)]\ [T/(T-T_c)]^{\beta}$ vs. $(\mu H/kT)^2\ [T/(T-T_c)]^{\gamma+\beta}$ showing the success of the scaling argument $[\mathfrak{M}/t^{\beta} = f(\eta^2/t^{\gamma+\beta})]$ for all our data points ($T < 4T_c$, $H < 7$ Teslas) with $\gamma = 3.25$ and two values of β within the error bars of Equation 14. For $\beta \sim 1$ we obtain a smaller dispersion.

The dispersion in this plot is improved if we take for β a value close to one i.e. closer to the higher limit of our error bar in Equation (14). These values of the exponents are consistent with those deduced on the same system by a completely different technique [9] and in a completely different range of temperatures [9,10].

3. Discussion

The theory tells us that the S.K. model presents a phase transition. For short range models the present opinion seems to be that there might be no phase transition at d=3 and that there is not at d = 2 [6,12] . Simulation data at d = 2 show a lot of similarities with real experimental data, including ours . Such similarities are the behaviour of the magnetization and of the specific heat, the occurrence of time-dependent effects and of a cusp in the susceptibility at $T_c(\omega)$. $T_c(\omega)$ though is claimed to collapse to zero in the static limit and everything seems to support the claim that $T_c=0$ in this dimensionality. A scaling plot can be made with the data, consistent with this hypothesis [13].

The similarities between the d = 3 and d = 2 cases should by no means prompt us to reason (rapidly) that $T_c = 0$ at d = 3 also. We conclude that a transition exists at finite temperature in a R.K.K.Y. spin glasses on the basis of the following evidence

i) The finite value of the low frequency, low field limit of the temperature of the susceptibility cusp has been very convincingly and accurately established in those systems [14]. ii) The divergence of $a_3(T)$ is observed and described by a unique exponent over 3 decades in the present data (7 if we include also the data of reference [8]). This range (3 decades) makes our data in favour of a phase transition in R.K.K.Y. spin glasses, just slightly more convincing than the evidence classically accepted for the occurence of a ferromagnetic transition in nickel considering magnetic data (see Figure 1 and references [1] and [15]). In two-dimensional spin glasses the magnitudes are by no means comparable. iii) The Log a_3 vs. Log a_5 plot of Figure 5 eliminates the possibility of describing our data with a single exponent and hence to construct a scaling fit like that of Figure 6, consistent with $T_c = 0$. Of course much will be gained from further comparison of the experiment with the simulation results, and in particular with the results of the simulation at d = 3 [16].

Finally it should be noted that we have obtained values of the exponents which are very different from those familiar, in other phase transitions. In the spirit of the derivation proposed here for the Suzuki-Chalupa expression (Equation (7)), it would have been logical, from the point of view of dimensional analysis, to introduce in Equation (6) together with $\eta^* = (\frac{H}{T_c})^2$ the quantity $t^* = (\frac{T-T_c}{T_c})^2$ which would play the role of the temperature associated with our order parameter $\overline{\langle S^2 \rangle}$. Equation (7) would then be exactly equal to Equation (1) except that the new exponents would be

$$\gamma^* = \gamma/2 \sim 1.62$$
$$\beta^* = \beta/2 \sim 0.5$$

The latter values are much closer to theoretical values in classical ferromagnetism, and very close to experimental determinations, not near T_c, in amorphous ferromagnets [17] . We take these as the correct values since Eq. (1) should be valid, if the Gibbs potential is a generalised homogeneous function, provided the relevant parameters (η and t) are correctly identified. To finish we believe that the use of the variables η' and t' of Equation (3) does extend the range of temperatures where one can determine significant effective exponents in spin glasses and in ferromagnetics as well [2].

* Present address of R. OMARI : Université Hassan II, Faculté des Sciences, BP 5366 Maarif, Casablanca, Morocco

** Associated to Université Scientifique et Médicale de Grenoble

References

[1] See for example : H.E. STANLEY in "Introduction to phase transitions and critical phenomena", Clarendon Press, Oxford 1971

[2] J. SOULETIE and J.L. THOLENCE, to appear in Solid State Commun. 1983

[3] P. WEISS et R. FORRER, Ann. Phys. (Paris) 5, 153 (1926)

[4] For a review of theory and experiment in spin glasses see R. RAMMAL and J. SOULETIE in "Magnetism of Metals and Alloys (Les Houches, Winter School proceedings (1980)), M. Cyrot Editor (North-Holland) 1981, p. 342

[5] M. SUZUKI, Progr. Theor. Phys. 58, 1151 (1977)

[6] J. CHALUPA, Solid State Commun. 22, 315 (1977)

[7] S. CHIKAZAWA, C.J. SANDBERG and Y. MIYAKO, J. Phys. Soc. Japan 50 (1981) and references therein

[8] P. MONOD and H. BOUCHIAT in J. Physique-Lettres 43, L-45 (1982)

[9] A. BERTON, J. CHAUSSY, J. ODIN, R. RAMMAL and R. TOURNIER, J. Physique-Lettres 43, L-153 (1982)

[10] B. BARBARA, A.P. MALOZEMOFF and Y. IMRY, Phys. Rev. Lett. 47, 1852 (1981) and M.M.M. Conf. (Montréal 1982), J. Appl. Phys. 53, 7672 (1982)

[11] R. OMARI, J.J. PREJEAN and J. SOULETIE, J. Physique 44, 25 (1983)

[12] H. SOMPOLINSKY and A. ZIPPELIUS, Phys. Rev. Lett. 50, 1294 (1983)

[13] K. BINDER and W. KINZEL, this conference and references therein

[14] L. LUNDGREN, P. SVEDLINH and O. BECKMAN, J. Phys. F 12, 2663 (1982) and Phys. Rev. B 26, 3990 (1982)

[15] J.S. KOUVEL and M.E. FISHER, Phys. Rev. A 136, 1626 (1964)
J.S. KOUVEL and J.B. COMLY, Phys. Rev. Letters 20, 1237 (1968)

[16] A.P. YOUNG, this conference

[17] R. MEYER and H. KRONMÜLLER, Phys. Stat. Sol. (b) 109, 693 (1982) and M. FAHNLE, G. HERZER, H. KRONMÜLLER, R. MAYER, M. SAILE and T. EGAMI (preprint)

THE FERRO- AND FERRIMAGNETIC - SPIN GLASS TRANSITION AS STUDIED BY MÖSSBAUER SPECTROSCOPY

R.A. Brand, V. Manns and W. Keune,
Laboratorium für Angewandte Physik, Universität Duisburg,
D-4100 Duisburg, Fed. Rep. Germany

Abstract

We show in the examples of Au-16.8 at% Fe, the metallic glass $Fe_xNi_{78-x}Si_9B_{13}$ and the spinel $Mg_{1+t}Fe_{2-2t}Ti_tO_4$ that the transition from ferro-(first two), or ferrimagnetic (last) to a spin-glass-like state can be determined by Mössbauer spectroscopy. Spectra taken in external magnetic field show that strong spin canting starts at this transition, while in zero external field the transition is accompanied by an anomalous increase in the saturation average hyperfine field in the low temperature state, proportional to the average magnetic moment. These results indicate that the transition is not simply a spin rotation, but is the condensation of transverse spin degrees of freedom in the spin-glass-like state, as in recent models for Heisenberg infinite ranged and short ranged spin glasses.

Introduction

Recently there has been much interest in the magnetic phase diagram in systems which show a crossover from a spin glass state to conventional (periodic) magnetic order. We consider here ferro and ferrimagnetic systems which show reentrant behavior in temperature: a transition well below the critical temperature T_c (or T_N) to a spin-glass-like state at T_f (for an experimental review of metallic systems, see [1]). This question has received much theoretical attention within the infinite ranged model of Sherrington and Kirkpatrick (SK)[2] in a recent series of mean field calculations[3]. For insulating nearest neighbor systems, a similar model has been proposed by Villain [4]. We will confine ourselves here to giving a physical picture of how (^{57}Fe-)Mössbauer hyperfine measurements can be used to study this question. For other aspects of this question, see [5]. The resonant emission (in a source), and absorption (in an absorber, normally the sample) of nuclear gamma rays is only practical in solids due to the possibility of recoil-free processes. The question of the correct hyperfine Hamiltonian is complicated by the nuclear transition between the nuclear ground and an excited state, but generally one takes for magnetic hyperfine interactions [6]: $H_M = g_N \mu_N H_{eff} I$ where g_N is the nuclear g-factor, and μ_N the nuclear magneton. I is the nuclear spin. H_{eff} is the magnetic (hyperfine) field created at the Mössbauer nucleus by the local and neighboring magnetic moments, and any external magnetic field: $H_{eff} = H_{ext} + H_{HF}$ where H_{ext} is the external field, and the internal field H_{HF} is

$H_{HF} = H_L + H_d + H_c$. In most cases, the orbital term H_L is negligible(if the orbital moment is quenched). H_d is the dipolar term for spins outside the nucleus, and H_c is the Fermi contact term coming from spin density at the nucleus. In Fe, this is separated into a 4s (conduction) electron polarisation, and a 3d-s core polarisation (generally the largest term). It is found that for ^{57}Fe in most metals and in not very dilute insulators for spin-only moments the average H_{HF} closely follows the empirical expression $H_{HF}/\mu = -125$ kOe/μ_B [7] where μ is the local magnetic moment (in μ_B). The magnetic hyperfine field splits the nuclear ground and excited states (nuclear Zeeman effect) resulting in 6 possible transitions (^{57}Fe). The line separation (Fig. 1a) is proportional to $/H_{HF}/ \propto \mu$. Thus only the (scalar) thermal spin $S \propto \mu$ can be measured. In almost random magnetic systems one observes a pronounced distribution of hyperfine fields. Because of the local spectroscopic character of the Mössbauer effect, this is seen as the superposition of (6 line) spectra for different values of H_{HF}, of relative intensity proportional to the probability $P(H_{HF})$, which then can be calculated from the spectrum. If $P(H_{HF})$ is caused solely by the distribution of spin magnitude $P(S)$, then this can be obtained, but generally $P(H_{HF})$ is stongly influenced from hyperfine terms from neighboring atoms, making this difficult. Even in this case, the average $\overline{H_{HF}}$ is taken as proportional to the average magnetic moment.

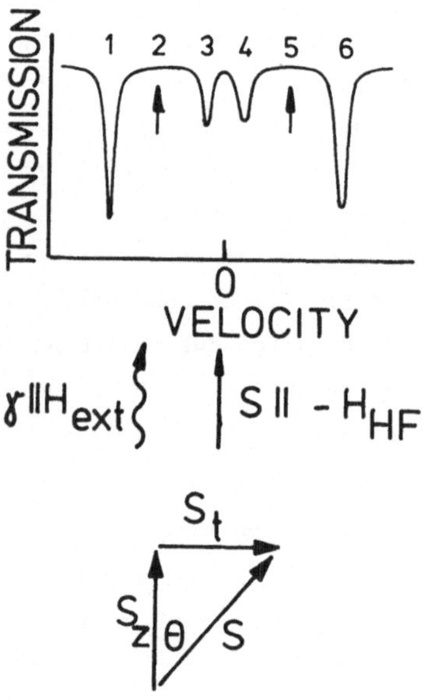

Figure 1.

a) spectrum for $\gamma // H_{HF}$ (lines 2 and 5 absent).

b) Canting transition in spin glasses. $S_z // H_{ext}$ exists in ferro, S_t only in the spin-glass-like phase.

The spectrum is sensitive to the orientation between total hyperfine field H_{eff} and the γ ray direction (here, // to H_{ext}) through the relative intensities of lines 2 and 5, as is shown in Fig. 1a. Denoting line intensities as 3:x:1:1:x:3, it is known[6] that $\sin^2\Theta = 2x/(4+x)$ for one given spin (// to γ, x=0; $\perp$, x=4; and random, x=2). In the presence of spin texture (distribution of Θ), only the variance of $\sin\Theta$, $\langle\sin^2\Theta\rangle$, averaged over the sample can be measured (but see Gonser [6], Ch. 3). As seen in Fig. 1b, for a thermal spin S at angle Θ from H_{ext}, and stationary transverse component S_t, $\sin^2\Theta$ is equal to $(S_t/S)^2$ so that $\langle(S_t/S)^2\rangle$ (averaged over the sample) is similar to the transverse order parameter $q_t = \langle S_t^2\rangle_J$ introduced in mean field calculations [3] and Monte Carlo computations [8]. This shows the interest in Mössbauer measurements in external magnetic field in the reentrant and in the pure spin glass regions. q_t seems to be a better order parameter than the original Edwards-Anderson $q_{EA} = \langle S^2\rangle_J$. The spectra discussed here have been analyzed by a histogram [9] and by a Fourier series expansion method [10], yielding $P(H_{HF})$. A complication is the strong electric quadrupole hyperfine splitting ΔE_Q which is included in these programs only as a first order perturbation to the hyperfine field Hamiltonian. This means that $\Delta E_Q/H_{HF} < 5.5\ 10^{-3}$ mms^{-1}/kOe should be true, which over the low field part of $P(H_{HF})$ is not the case. A "confidence region" for $P(H_{HF})$ is relatively easy to define. Writing for a typical (low field) hyperfine element, $H_{HF} = \overline{H_{HF}} - n\sigma_H$, where σ_H is the standard deviation of the distribution of H_{HF}, then n=2 gives a reasonable region of confidence over most of the calculated $P(H_{HF})$: $\Delta E_Q/\overline{H_{HF}} = 5.5(1 - n\sigma_H/\overline{H_{HF}})\ 10^{-3}$ is shown in Fig. 2 for n=1 and 2. A further complication occurs from excessive line overlap (at $\sigma_H/\overline{H_{HF}}$ = 0.13, lines 1 and 2 overlap; = 0.29, lines 2 and 3; = 0.36, all lines). The prefered region is shown as cross-hatched. This diagram is due to Le Caër [11]. The worst case for the computation is for the nickel metallic glasses, and in any case, at increasing temperatures.

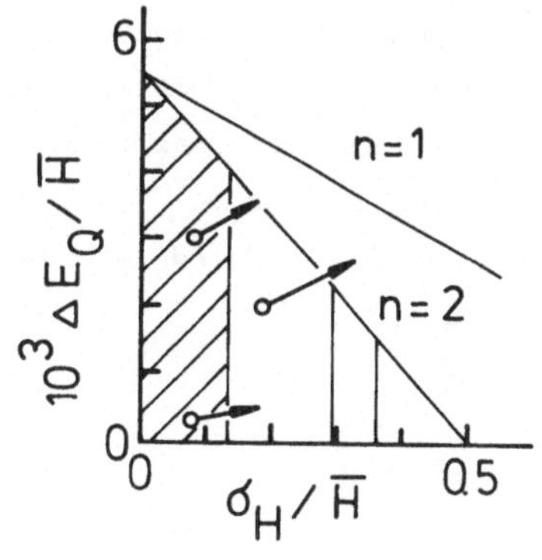

Figure 2.
Confidence diagram from LeCaër[11].
Arrows from top: Au-16.8% Fe, $Fe_5Ni_{73}Si_9B_{13}$, $Mg_{1-t}Fe_{2-2t}Ti_tO_4$, for increasing temperatures.

Discussion

Due to the fact that several similar systems are to be presented here, for simplicity the analysis in light of the present spin glass models is postponed for the last section.

The system $Au_{1-x}Fe_x$ shows crossover to ferromagnetism at about the percolation threshold [1]. For a sample with concentration x = 16.8 at% Fe, spectra in the presence of an external field of 2T, published previously in [12], show clearly the ferromagnetic region above about 5oK: lines 2 and 5 are absent. Below this temperature, canting grows progressively, as indicated by the intensity of these lines. This is shown in the lower part of Fig. 3, where $\langle \sin^2\Theta \rangle$ has been calculated. The value of $\langle \sin^2\Theta \rangle$ is also shown for a random distribution of spins. Associated with this canting is an increase in the average $\overline{H_{HF}}(T)$ shown in the upper part of Fig. 3 as solid points. The extrapolation to T = 0 gives a large difference in $\overline{H_{HF}}(0)$ in the two states, which we take as indicating a larger average moment μ in the spin-glass-like state at low temperature. The average $\overline{H_{eff}}(T)$ measured in H_{ext}=2T is shown as open circles in the figure (uncorrected for H_{ext}; usually H_{HF} and H_{ext} have opposite directions in a ferromagnet). No break is observable at T_f because of the very strong increase of H_{HF} ($=H_{eff}+H_{ext}$) in the ferro region. H_{HF} is insensitive to domain rotation (independent of Θ), so that this increase means an increase of the local magnetic moment in external field. Such an effect is not expected or usual in a ferromagnet well below T_c. A similar strong local susceptibility is seen well above T_c(Fig. 3). The simplest Curie-Weiss model for a ferromagnet above T_c would give $H_{HF}(H_{ext})/H_{HF}(T{=}0) = M(H)/M_o = 4\mu_B H_{ext}/3k_B(T-T_c)$, or about 5% at 2ooK, while we measure more than 3o%.

Figure 3.
Au-16.8 at % Fe:
Upper: average field $\overline{H_{eff}}$ measured in zero field (solid) and in an external field of 2 T (open circles).
Lower: calculated $\langle \sin^2\Theta \rangle$ at 2 T.

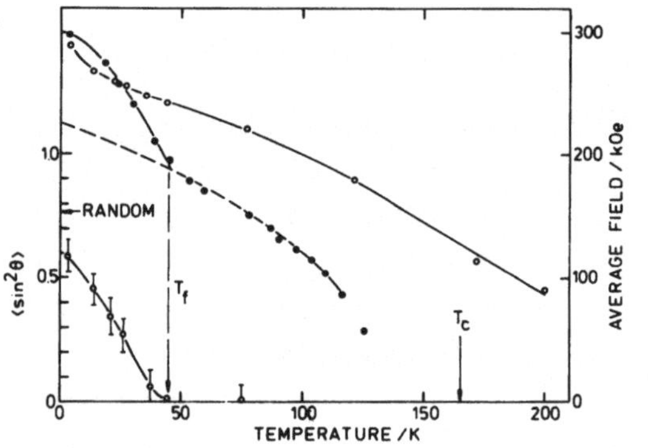

The spectra analyzed in more detail yield the distribution functions $P(H_{HF})$ shown in Fig. 4 for several temperatures in zero and in an external field of 2T. These results are not to be interpreted in detail because of the problems from ΔE_Q, but it is clear that the external field has little effect on $P(H_{HF})$ below T_f, but strongly suppresses the low field part which appears above. At the present time we are working on methods for making the evaluation of such spectra more reliable, so that more detailed information can be obtained.

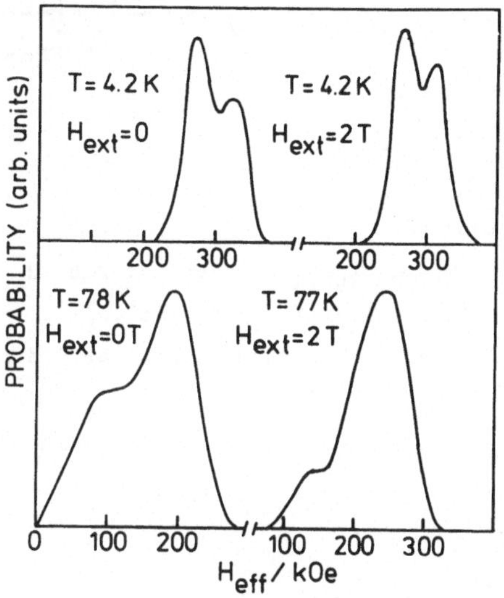

Figure 4.
Au-16.8 at % Fe:
Distribution functions in
upper: spin-glass-like and
lower: ferromagnetic states in
zero field and at 2 T.

We have studied the metallic glass $Fe_xNi_{78-x}Si_9B_{13}$ obtained from melt spinning in the range of x = o.3 to 9 % Fe. This system is similar to the $(M_xNi_{1-x})_{75}R_{25}$ system (M = transition metal, R = glass former) reviewed in [13]. The results will be published shortly elsewhere in more detail and will only be summarized here. By Mössbauer, specific heat, and χ_{DC} measurements we have shown that for $x < 4\%$, a spin glass transition is observed, and above this, crossover to reentrant ferromagnetism. The canting is shown in Fig. 5, with again a break in the average $\overline{H}_{HF}$ at T_f in zero field, but not in a field of o.8T. The phase diagram as determined from these measurements is shown in Fig. 6, and is similar to that of Durand [14] for a similar system.

In a similar metallic glass, $Pd_{77.5-x}Cu_6Si_{16.5}Fe_x$, similar measurements have been made (J. Lauer, private communication, and[9]), and have shown essentially equivalent results. An advantage of the Pd-based metallic glasses is a smaller quadrupole splitting for the iron.

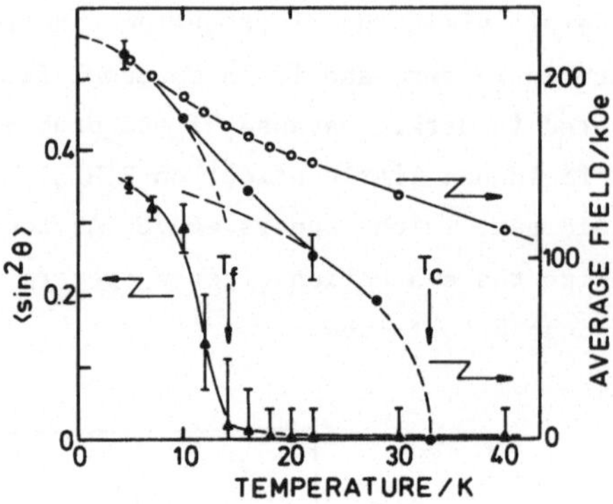

Figure 5.

$Fe_5Ni_{73}Si_9B_{13}$. Upper: average $\overline{H_{eff}}$ in zero field (solid) and in an external field of 0.8T (open circles).

Lower: calculated $<\sin^2\theta>$ at 0.8T.

Figure 6.

Magnetic phase diagram for $Fe_xNi_{78-x}Si_9B_{13}$.

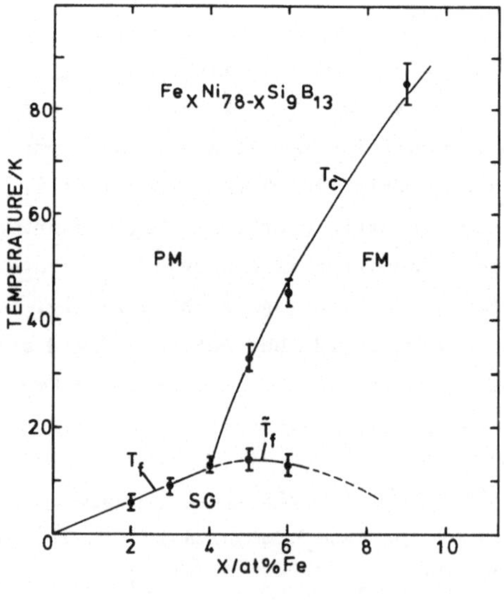

The third example is the insulating spinel $Mg_{1+t}Fe_{2-2t}Ti_tO_4$. Because of the presence of two sublattices (A and B), selective dilution is possible, and in this system the magnetic sublattice concentrations are given by[15]: $X_A = (2-2t)/(3-t)$, and $X_B = (2-3t+t^2)/(3-t)$ shown in Fig. 7 (X_i=1 for 1 magnetic ion i= A, or 2 ions i= B; for a general reference, see [16]). In this system, all exchange integrals $/J_{AB}/ >> /J_{BB}/ > /J_{AA}/$ are found to be antiferromagnetic [17]. Thus for small dilution t, there is ferrimagnetic ordering with all A-site moments antiparallel to all B-site moments. Point I in Fig. 7 is the J_{AB}-only percolation threshold[15], which should be the limit of ferrimagnetism. The presence of the unsatisfied J_{AA} and J_{BB} bonds should however lead to breakdown of this state before this point. Above I, a magnetic ordered state is possible only including these bonds, and point II is the percolation threshold for all bonds [18]. In the region I-II, spin glass behavior has been reported [18]. Reentrant behavior has been observed near point I by susceptibility measurements. Fig. 8 shows M(T) at 5o Oe measured after zero field cooling (lower curve) and cooling in field (upper), for the sample with t = o.6o. The Néel temperature T_N is obtained by extrapolating $\chi^{-1}(T)$. The onset of irreversibility is taken as T_f, and this will be compared to the Mössbauer measurements. The resulting phase diagram is shown in Fig. 9, with ferrimagnetic, spin glass, and canted spin-glass-like regions. This latter will be discussed using t = o.6o as an example.

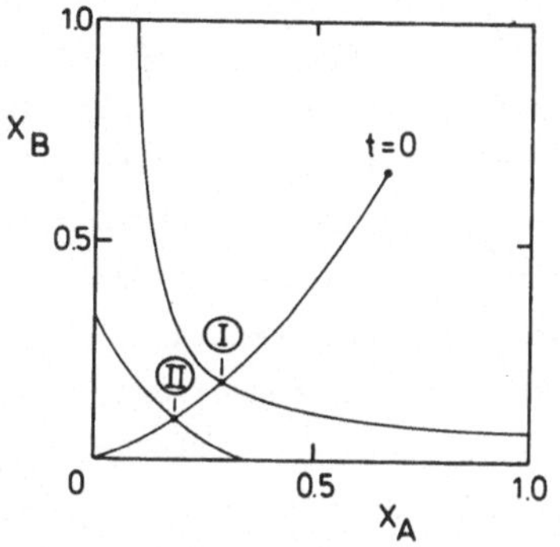

Figure 7.
Concentration plane for the spinel lattice. Line at t = o, dilution line for $Mg_{1+t}Fe_{2-2t}Ti_tO_4$. Point I = intersection with J_{AB}-only percolation limit. Point II = intersection with J_{AB}-J_{BB}-J_{AA} percolation limit.

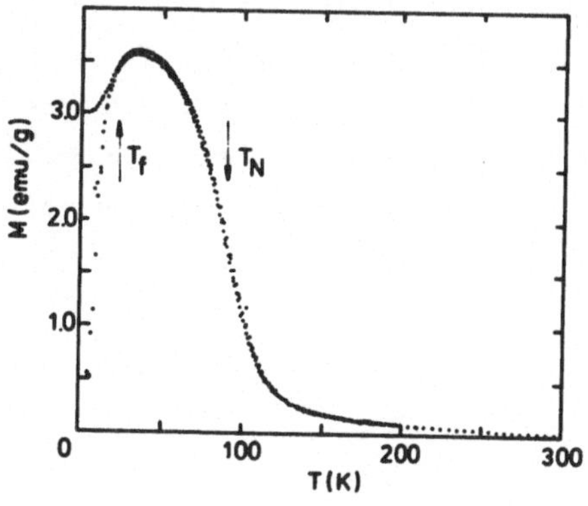

Figure 8.
Spinel for t = 0.60.
Magnetization at H_{ext} = 50 Oe.
T_N from $\chi(T)$.
Lower curve, cooled in zero field.
Upper curve, cooled in field.

Figure 9.
Magnetic phase diagram for $Mg_{1+t}Fe_{2-2t}Ti_tO_4$, as function of dilution t.

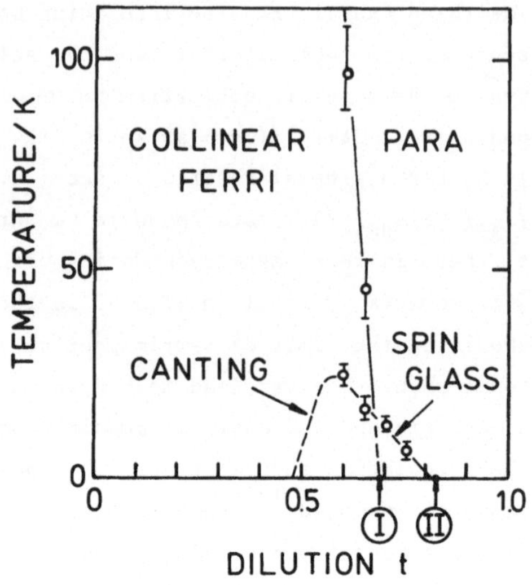

Spectra in zero and in external field up to 2T have been taken: the results are presented in Fig. 10. At temperatures below T_f, strong canting is observed, with $<\sin^2\Theta>$ extrapolating to zero in good agreement with the magnetization results. At higher temperatures reliable values of $<\sin^2\Theta>$ are difficult to obtain because in this region, σ_H becomes quite large (an effect seemingly associated with T_f). The average $\overline{H_{HF}}$ also shows a break at this temperature, and this break is strongly influenced by external fields (see Fig. 10). The distribution function $P(H_{HF})$ for two temperatures is shown in Fig. 11. Below T_f, in the canted state, $P(H_{HF})$ is composed of only one continuous distribution, but above, a second peak appears at very low fields. At T_N, the spectrum becomes a quadrupole doublet, but the value of ΔE_Q is small as compared to the metallic systems. The phase diagram from susceptibility and Mössbauer measurements, Fig. 9, reproduces the percolation limits as presented in Fig. 7. The spin glass and canted spin-glass-like regions are separated because preliminary neutron diffraction measurements indicate that in the latter the ferrimagnetic Bragg intensity remains unchanged, indicating a spontaneous ferrimagnetic magnetization[19].

Figure 10.

For spinel with t = 0.60

Upper: Average field $\overline{H}_{HF}(T)$ in zero field (solid points) and in H_{ext} = 1 T (open triangles), and standard deviation of $P(H_{HF})$ (open circles).

Lower: variation of $\langle \sin^2\theta \rangle$ in H_{ext} = 1 T.

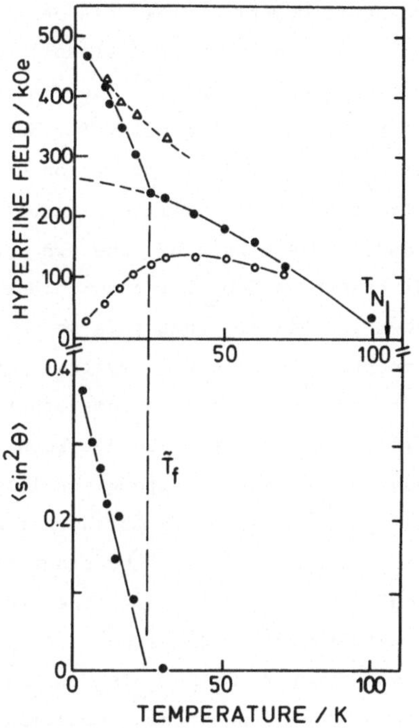

Figure 11.

Typical $P(H_{HF})$ distributions for spinel with t = 0.60, in spin-glass-like (left) and ferrimagnetic (right) states.

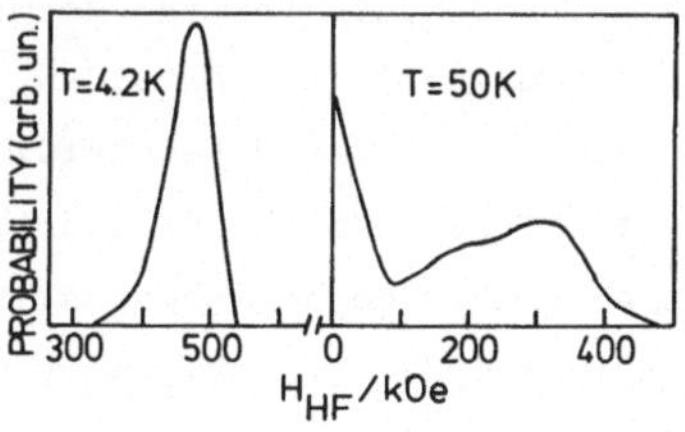

Conclusions

We have seen in two metallic and one insulating system that the reentrant behavior near the concentration limit for ferro- or ferrimagnetism can be studied by Mössbauer spectroscopy. The great advantage offered by this technique is the separation of the average magnetic moment, and the orientation of moments. Since the measurement is local, the distribution $P(H_{HF})$ can be obtained, as well as $\langle \sin^2\theta \rangle$, but global properties such as the spontaneous magnetization M_s cannot. We have shown that the low temperature transition well below T_c (or T_N) denoted as T_f, is a canting transition which is associated with an increase in the local magnetic moment μ. This can be interpreted as the freezing of transverse spin components into a new spin-glass-like state where each spin is canted with respect to the original direction. The (vector) sum of S_Z existing above T_f and S_t results in a larger total S below T_f. Note that in a normal canting transition, the value of S_z would decrease (rotation of S only). In the infinite ranges model[3], this effect occurs for all spins equally, but in real spin glasses this need not be the case, as each spin has its own specific environment. Indeed the results for the distribution $P(H_{HF})$ for Au-16.8at% Fe (Fig. 4), and $Mg_{1+t}Fe_{2-2t}Ti_tO_4$ (Fig. 11) seem to indicate that this may be the case. Local environments must then play an important role in some systems, but to be absolutely clear, there is no evidence in either of these two systems, that a separation takes place between ferro and spin glass regions. As we do not know from our measurements whether $M_S \to 0$ below T_f, we have termed this a spin-glass-line state. There is some evidence that below T_f in $Au_{1-x}Fe_x$[20] there is a further transition to a state with much higher coercivity. Consistent with this, the transition at T_f can be interpreted as the Gabay-Toulouse (GT line), while the transition at still lower temperatures as the de Almeida-Thouless (AT line)[3]. As yet, only Crane and Claus[21] have reported on measurements of M_S in $Au_{1-x}Fe_x$ in the reentrant domain; they report that in certain cases (depending on heat treatments), a transition with $M_S \to 0$ is seen. This would be then the transition to a pure spin glass state, as predicted by Sherrington and Kirkpatrick[2] (SK line). Since the system $Au_{1-x}Fe_x$ is so influenced by heat treatments (short range order, SRO), it is difficult to know where this transition is with respect to the canting and coercivity transitions. This SRO is probably the basis for the seeming separation into "weak" and "strong" moments in the ferromagnetic region of $Au_{1-x}Fe_x$, an effect which is absent in $Fe_xNi_{78-x}Si_9B_{13}$, for example. It would be useful to know how this difference reflects itself in the difference between the GT and AT lines (which in the infinite ranged model have been predicted to be identical) and the SK line ($M_S \to 0$).

The spinel system is seemingly completely different from metallic systems; only near neighbor interactions exist, and all are antiferromagnetic.

Spin glass states are here the result of dilution and frustration alone.

This is seen on the ferrimagnetic side of the J_{AB} percolation limit as reentrant ferrimagnetism. Above this limit, there is a transition to a pure spin glass state. Both transitions seem very similar to models proposed by Villain[4]. For the first, local canting aroung magnetic imperfections leads to an effective XY (transverse to M_S) spin. These can then interact, leading to a low temperature (below T_N) transition to what Villain calls a semi-spin-glass state. It is possible that the low field parts of $P(H_{HF})$ in the ferrimagnetic state reflect these freely rotating XY spins, which then freeze at T_f. Neutron diffraction studies indicate that for the t=0.60 sample at 4.2K (in the reentrant region), the ferrimagnetic M_S as seen by the magnetic part of the Bragg scattering, does not go to zero[19]. For this reason, we separate the canted reentrant region from the spin glass (also canted) part.

In conclusion, the canting transition gives a physically reasonable picture of the low temperature state in reentrant systems, and explaines the higher order associated with this state. The high temperature phase is disordered with respect to q_t and thus has a higher spin disorder entropy.

The authors would like to thank J. Lauer who performed some of these measurements, and I.A. Campbell and K. Usadel for helpful discussions.

REFERENCES

1. G. J. Nieuwenhuys, B.H. Verbeek and J.A. Mydosh, J. Appl.Phys. 50 (1979) 1685.
2. D. Sherrington and S. Kirkpatrick, Phys.Rev.Lett. 35 (1975) 1792, and S. Kirkpatrick and D. Sherrington, Phys.Rev. B17 (1978) 4384.
3. J.R.L. de Almeida and D.J. Thouless, J.Phys. A: Math.Gen 11 (1978) 983, and M. Gabay and G. Toulouse, Phys.Rev. Lett., 47 (1981) 201, and G. Toulouse and M. Gabay, J.Physique Lett. 42 (1981) L 103, and G. Toulouse and M. Gabay, T.C. Lubensky and J. Vannimenus, J.Physique Lett. 43 (1982) L109.
4. J. Villain, Z.Phys. B33 (1979) 31.
5. R.A. Brand, H. Georges-Gibert and C. Kovacic, J.Appl.Phys. 51 (1980) 2647 and R.A. Brand and H. Georges-Gibert, J.Phys.F: Met.Phys. 10 (1980) 2501.
6. P. Gütlich, R. Link and A. Trautwein, "Mössbauer Spectroscopy and Transition Metal Chemistry" (Springer Verlag 1978), and R.W. Grant in "Mössbauer Spectroscopy", U. Gonser, ed. (Springer Verlag 1975).
7. P. Panissod, J. Durand and J.I. Budnick, Nucl.Instr.Meth. 199 (1982) 99.
8. R.E. Walstedt and L.R. Walker, J.Appl.Phys. 53 (1982) 7985.
9. R.A. Brand, J. Lauer and D.M. Herlach, J.Phys. F: Met.Phys. 13 (1983) 675.
10. B. Window, J.Phys. E.: Sci.Instr. 4 (1971) 401.
11. J.M. Dubois and G. LeCaër, Second Internat. Conf: Struct. of Non-Crystalline Materials. (1982) (in press), and G. LeCaër, private communication.
12. J. Lauer and W. Keune, Phys.Rev.Lett. 48 (1982) 1850.
13. J.A. Geohegan and S.M. Bhagat, J.Magn.Magn.Mater. 25 (1981) 17.
14. J.Durand, Rev.Phys.Appl. (Paris) 15 (1980) 1036.
15. F. Scholl and K. Binder, Z. Phys. B39 (1980) 239.
16. A. Herpin, "Théorie du Magnétism", (Presses Universitaires de France. 1968), Ch. 19.
17. E. DeGrave, R. Vanleerberghe, C. Dauwe, J. de Sitter and A. Govaert, J.Physique C6-37 (1976) C6-97.
18. J. Hubsch, G. Gavoille and J. Bolfa, J.Appl.Phys. 49 (1978) 1363.
19. J. Hubsch, private communication.
20. F. Varret, A. Hamzić and I.A. Campbell, Phys.Rev. B26 (1982) 5195, and I.A. Campbell, S. Senoussi, F. Varret, J. Teillet, and A. Hamzić, Phys. Rev.Lett 50 (1983) 1615.
21. S. Crane and H. Claus, Phys.Rev.Lett 46 (1981) 1693.

THE FERROMAGNETIC TO SPIN GLASS CROSSOVER IN $Eu_xSr_{1-x}S$

H. Maletta
Institut für Festkörperforschung, Kernforschungsanlage Jülich,
D-5170 Jülich, W.-Germany

Abstract: This paper is a brief review of recent measurements performed on $Eu_xSr_{1-x}S$ just below and above x_c=0.51, i.e. within the ferromagnetic to spin glass crossover regime where quite interesting and novel effects are observed. Spin-glass freezing below x_c is studied in the high-frequency region and in dependence on a magnetic field. Neutron-diffraction experiments measure the spin correlations directly and provide evidence for a re-entrant ferromagnetic phase boundary above x_c, which can be explained in terms of random-field effects.

I. Introduction

Disordered magnetic systems with competing interactions have attracted a great deal of attention in recent years. Non-conventional magnetic order phenomena found in such random alloys have been studied systematically in $Eu_xSr_{1-x}S$ /1/ where the ferromagnetic insulator EuS is diluted with SrS. There is the advantage that the unique properties of $Eu_xSr_{1-x}S$ can be described quantitatively by a realistic site-disorder model of a Heisenberg system with wellknown short-range competing interactions between well-localized spin-only moments of Eu^{2+} ions /2/, namely a ferromagnetic nearest-neighbor exchange, J_1, and an antiferromagnetic next-nearest neighbor exchange interaction, J_2, with the ratio J_2/J_1 = -0.5.

Figure 1a shows the magnetic phase diagram of $Eu_xSr_{1-x}S$: long-range ferromagnetic order (FM) is unstable against dilution already at concentration x_c = 0.51, far above the percolation threshold x_p=0.13, and spin-glass (SG) behavior is observed for $0.13 \leq x \leq 0.65$. Insulating spin glasses $Eu_xSr_{1-x}S$, similar to metallic alloys like AuFe, exhibit a freezing transition to some new type of local magnetic order, the nature of which is still heavily debated. It is well established, however, that both disorder (due to the dilution effect) and frustration (due to the competition of exchange interactions of opposite sign) are considered as necessary and essential ingredients for a spin glass. Here, we are interested in studying the evolution of spin-glass behavior from the ferromagnetic phase by increasing the magnetic dilution in $Eu_xSr_{1-x}S$. Our results in the crossover regime can be explained by the interplay between ferromagnetic and spin-glass type of order phenomena.

First, we present experimental data of $Eu_xSr_{1-x}S$ as function of composition which provide evidence of a significant change of magnetic behavior in the diluted ferromagnets when approaching the ferromagnetic to spin glass boundary. As shown

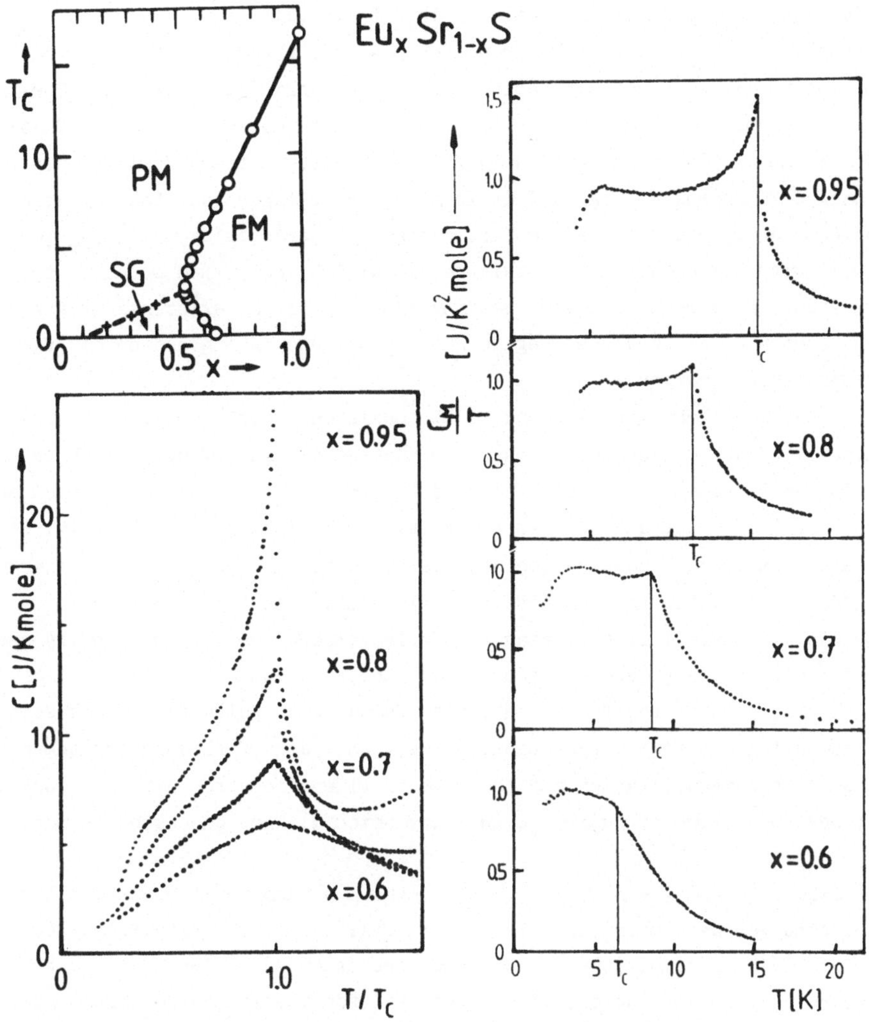

Fig. 1: (a) Magnetic phase diagram of $Eu_xSr_{1-x}S$.
(b) Specific heat of four dilute ferromagnets $Eu_xSr_{1-x}S$ /3/.
(c) Plot of C_M/T versus temperature, C_M = magnetic specific heat.

in specific heat measurements /3/ of Fig. 1b, 5% dilution (x=0.95) does not alter the λ-anomaly near T_c as expected for a ferromagnetic transition, whereas further dilution (up to x=0.6) reveals a rapid loss of sharp critical character. The broad maximum of the specific heat of $Eu_{0.6}Sr_{0.4}S$ around the Curie temperature T_c already strongly resembles the behavior of spin glasses (as e.g. observed with x=0.4 /4/). By plotting C_M/T versus T (Fig. 1c) it is made even clearer that the magnetic specific heat C_M near T_c undergoes a distinct change in character by comparing the curves for $x \geq 0.70$ with x = 0.60. Indeed, this result will be confirmed by other types of measurements as described below, lending support for the existence of a ferromagnetic to spin glass crossover regime with quite interesting and novel effects in magnetic order behavior.

This transient behavior is found to be consistent with corresponding results from numerical Monte Carlo simulations by Binder et al. /5/. They also observed significant deviations from a fully aligned ferromagnetic ground state below about x = 0.65, until at about x = 0.50 ferromagnetic order is destroyed totally. Figure 2 shows their normalized magnetization data versus concentration x for the exchange ratio J_2/J_1 = -0.5 which is a realistic model calculation for $Eu_xSr_{1-x}S$.

Obviously the specific heat near the Curie temperature is dominated by short-range order effects long before the critical concentration x_c = 0.51 is reached. Spatial fluctuations of the magnetization produced by the disorder of magnetic atoms and the competing exchange interactions between them also influence the temperature dependence of the magnetization and the line profile of neutron diffraction spectra in the crossover regime of $Eu_xSr_{1-x}S$ in a way which will be discussed in more detail here.

For dilute ferromagnets $Eu_xSr_{1-x}S$ the inverse paramagnetic susceptibility, $1/\chi_o$, plotted versus temperature, T/T_c-1, in Fig. 3a shows a significant curvature over a wide range of temperature which gets more extended by increasing magnetic dilution. An estimate of the critical exponent γ for the initial χ_o from these data reveals anomalously increasing values as the ferromagnetic to spin glass crossover regime is approached. One obtains an effective γ of 2.2 for x = 0.60, with a crossover very near T_c to the ordinary value of 1.38 known for Heisenberg ferromagnets (Fig. 3b).

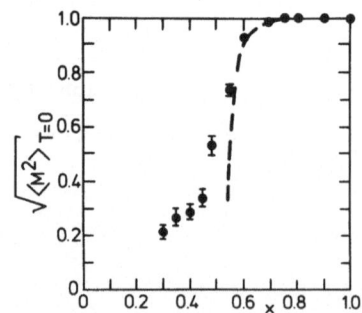

Fig. 2: Spontaneous magnetization versus magnetic concentration x as obtained by numerical simulations with J_2/J_1 = -0.5 /5/.

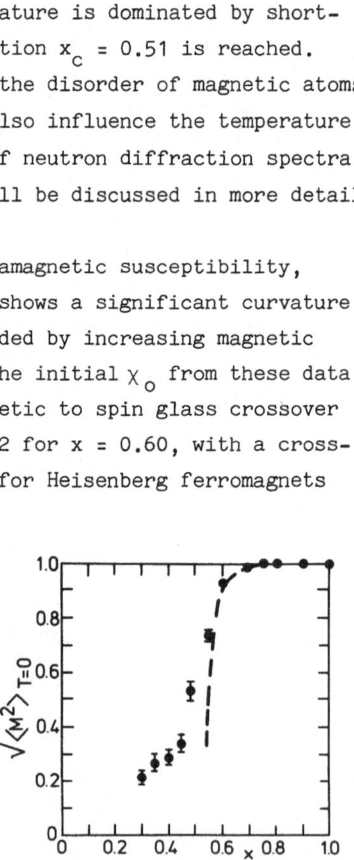

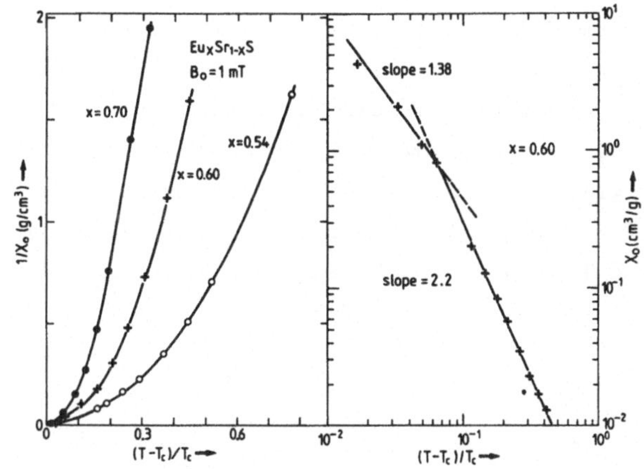

Fig. 3: (a) Inverse initial susceptibility χ_o of $Eu_xSr_{1-x}S$.
(b) Critical behavior of $\chi_o(T)$ for x=0.60.

Rather unusual properties are observed for concentrations $0.51 \leqslant x \leqslant 0.65$ at low temperatures, as shown with the ac-χ results in Fig. 4. The susceptibility reaches a plateau below T_c as expected for a ferromagnet but an anomalous drop-off in χ follows at lower temperature (Note that the different plateau values in Fig. 4 are only due to different sample geometries in the various measurements, which was checked experimentally /6/).

These results can be understood by assuming a re-entrant phase boundary between the ordered and disordered phase in $Eu_xSr_{1-x}S$, as indicated in the magnetic phase diagram of Fig. 1a. That means, the susceptibility curve in Fig. 4 is interpreted as showing a para-to-ferromagnetic and, subsequently, a ferromagnetic to spin glass transition. Even the observed high effective γ-values with the crossover behavior can be understood. Binder argues /7/ that at the re-entrancy point, $x = x_c$, one approaches the ferromagnetic phase boundary tangentially, and hence thermal exponents have twice their ordinary values. But somewhat further away from this point

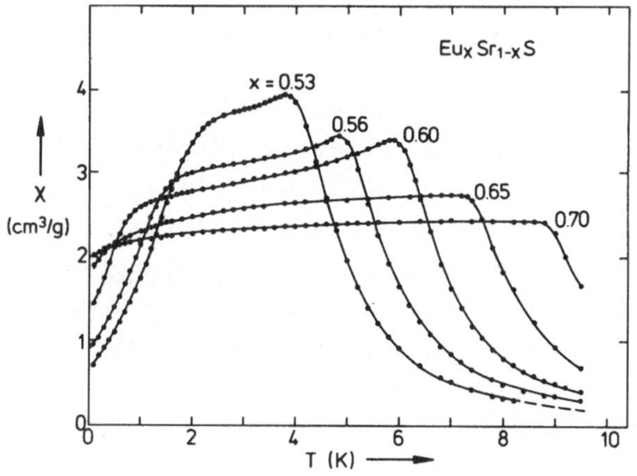

Fig. 4: Ac susceptibility of $Eu_xSr_{1-x}S$ in the cross-over regime /6/.

a crossover should occur similar to multicritical points, as indeed is observed with γ_{eff} (T) in $Eu_xSr_{1-x}S$.

In the next sections we present a brief review on recent measurements performed on $Eu_xSr_{1-x}S$ just below and above $x_c = 0.51$ in order to study the anomalous magnetic behavior in the ferromagnetic to spin glass crossover regime. The experiments have been performed in collaboration with G.P. Singh and M. von Schickfus /8/, J.A. Hamida, C. Paulsen and S.J. Williamson /9/, and G. Aeppli and S.M. Shapiro /10/.

II. Spin Glass

Spin-glass properties of insulating $Eu_xSr_{1-x}S$ for $x < 0.51$ have already been described elsewhere /11/. Here, very recent data of χ at high concentrations in the high-frequency region and in dependence on a magnetic field will be discussed.

The nature of the transition into the spin-glass state is still an unsolved problem. Although the temperature dependence of the ac-χ in spin glasses exhibits sharp maxima, their position T_f is often frequency dependent. $T_f(\nu)$ does not follow a simple Arrhenius law as in superparamagnets, but spin freezing is a collective process /11/ where spin correlations are strongly temperature dependent. The dynamics of this process can well be studied by χ-measurements over a broad range of frequencies, and we went up to 12 GHz in $Eu_xSr_{1-x}S$ (which is not possible in the usually studied metallic spin glasses because of large losses due to eddy current heating).

Measurements on the spin glass $Eu_{0.43}Sr_{0.57}S$ were carried out using cavity perturbation technique /8/; the data are displayed in Fig. 5. The temperature of the susceptibility maximum increases with frequency, as observed in many other spin glasses, but only up to a measuring frequency of about 500 MHz, and subsequently decreases. At still higher frequencies of 8 GHz and 12 GHz χ even exhibits a minimum (instead of a maximum) as function of temperature. Experiments at 12 GHz on several Eu concentrations in the $Eu_xSr_{1-x}S$ series provide evidence that the dip in χ is more pronounced for spin glasses near the ferromagnetic phase boundary (see curve for x = 0.43 in Fig. 6).

This anomalous high-frequency dependence at high Eu concentrations is simply not interpretable in terms of a transition temperature at present. It can be qualitatively understood by considering the coexistence of both fast and slow spin relaxations in the freezing process associated with ferromagnetically coupled spins in finite clusters and spin-glass type of order, respectively. The low-frequency experiment probes the response of the whole spin system. One expects a broad distribution of relaxation times due to cooperative freezing of the infinite cluster and blocking of small clusters, and hence the χ-maximum will be broadened and shifted to higher temperature with increasing frequency /11/. In concentrated spin glasses near $x_c = 0.51$, there is a strong tendency to ferromagnetic short-range order. By means of neutron scattering experiments we find a ferromagnetic correlation length,

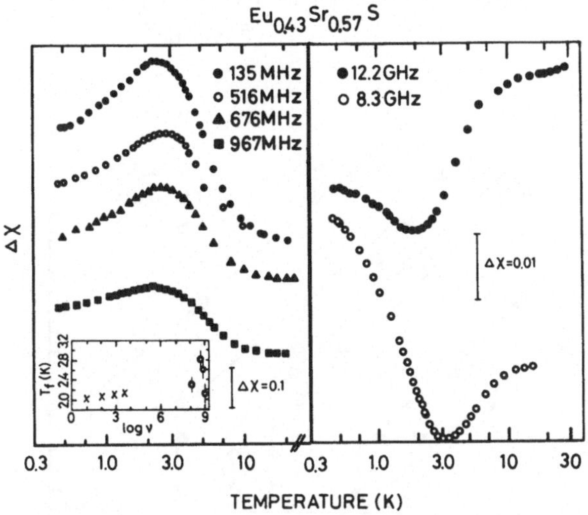

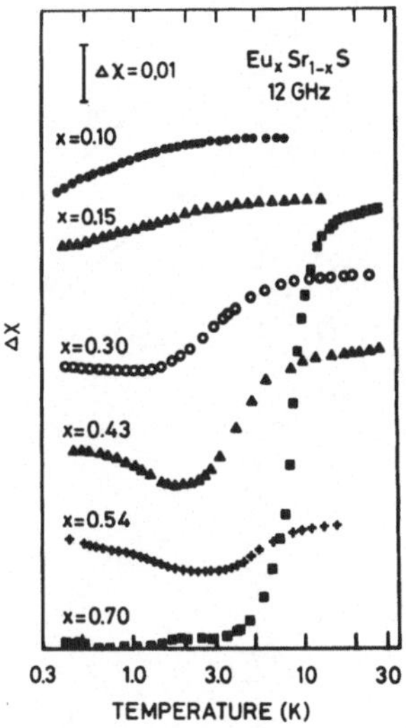

Fig. 5: $\Delta\chi$ versus temperature of $Eu_{0.43}Sr_{0.57}S$ in the high-frequency range /8/.

Fig. 6: $\Delta\chi(T)$ at 12 GHz for various concentrations x /8/.

ξ_{FM}, passing through a finite maximum versus temperature at T_f /6,10/, due to the interplay between spin-glass type and ferromagnetic order. Consequently, one also may expect the relaxation time τ of the strongly coupled spins to pass through a maximum according to $\tau(T) \sim d^{-y}$, since the normal distance d(T) from the re-entrant ferromagnetic phase boundary passes through a minimum. The high-frequency susceptibility, however, is dominated by the response of these fast spins, which then directly leads to a minimum in $\chi(T)$ near the re-entrant phase line. Thus, it appears important to consider the change of correlation length and relaxation time with temperature in order to understand the phenomenon of spin-glass freezing.

Another problem on which much attention has been focused recently concerns the nature of spin-glass freezing in a finite magnetic field. De Almeida and Thouless (AT) /12/ derived in the mean-field model of Sherrington and Kirkpatrick a transition line $H_c(T)$, with T_f being suppressed by $H^{2/3}$. Several static and dynamic experiments /13/ seem to lend support to such a transition line. Very recently, Monte Carlo simulations performed on a two-dimensional Ising spin-glass model with nearest-neighbor interactions (which has no phase transition at a nonzero T_f) obtained a dynamical critical field H_c (T,t) for fixed time t which is surprizingly similar to the AT-line of the infinite-range model (which has a phase transition at a nonzero T_f) /14/.

In order to elucidate this issue we have studied $Eu_xSr_{1-x}S$ spin glasses with x = 0.40 and 0.46 by measuring the real (χ') and imaginary (χ'') component of the magnetic susceptibility for frequencies 7 Hz $\leqslant \nu \leqslant$ 5 kHz in weak dc fields H up to

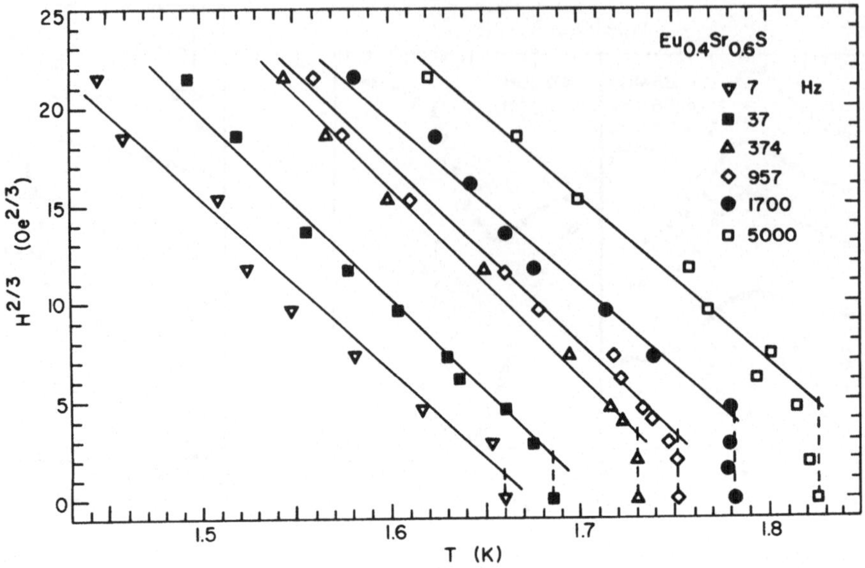

Fig. 7: Transition lines $H_c(T,\nu)$ for various frequencies ν as determined by the field dependence of the inflection point of $\chi''(T, \nu)$ /9/.

100 Oe /9/. On cooling through T_f where χ' displays a maximum, χ'' increases from the baseline indicating the appearance of a time-lag (hysteresis) in the response. For fields H above $\simeq$15 Oe, the temperature at the inflection point of χ'' (T) curves is suppressed in a way similar to the AT-line, as shown in Fig. 7. Two important properties are immediately evident: Firstly, the AT-line is frequency dependent which confirms the result of Bontemps et al. /15/, and secondly, for sufficiently weak fields H we find essentially no field dependence in the χ' and χ'' curves unless the field exceeds a threshold strength $H_o(\nu)$. The threshold marks the onset for nonlinear ac response in χ' and opposite temperature-shifts of $\chi'(T)$ and $\chi''(T)$. Its increase with frequency, which can be described by $H_o \sim \ln \nu$, is even more dramatic for the compound with x = 0.46 than for x = 0.40. This suggests that the behavior is related to the proximity of these samples to the crossover to ferromagnetic order. In Ref. 9 we propose a specific mechanism as possible explanation of the threshold based upon the notion that spin glasses have a ramified structure of domains and walls, in the sense as described by Gabay and Garel /16/, where the magnetic response is dominated by the movement of the walls.

III. "Frustrated Ferromagnet"

In this section we discuss magnetic properties of $Eu_xSr_{1-x}S$ just above x_c=0.51. Returning to the susceptibility behavior shown in Fig. 4, the question arises whether these curves really reflect a "double transition". Several quite different systems seem to exhibit such a re-entrant behavior /17/, mostly deduced by magnetization measurements. They reveal a strong sensitivity to even weak fields which accounts for the uncertainty in characterizing the various phases and their transition temperatures. Sherrington and Kirkpatrick /18/ first predicted a re-entrant behavior within the infinite-range model for an Ising spin system, while for a Heisenberg system within the same model Gabay and Toulouse /18/ found that unsaturated long-range ferromagnetic order coexists with spin-glass ordering among the transverse components of the spins at the lowest temperature.

Recently, we have performed high-resolution neutron-diffraction measurements on single crystals of $Eu_xSr_{1-x}S$ with x = 0.52 and 0.54 /10/. This technique allows zero-field measurements and provides microscopic information. Figure 8 displays line profiles in (200) direction for x = 0.52 at various temperatures which give evidence for an unusual high amount of diffuse scattering intensity from magnetic disorder below the Curie temperature T_c = 3 K. The solid lines in Fig. 8 represent fits to the data using the scattering function

$$S(\vec{Q}) = \frac{A}{q^2 + \kappa^2} + B \cdot \delta(\vec{q}) \qquad (1)$$

with momentum transfer $\vec{q} = \vec{Q} - \vec{\tau}$ and $\vec{\tau}$ as reciprocal-lattice vector, where the cross section in the Ornstein-Zernike approximation is the sum of Lorentzian fluctuation scattering together with Bragg scattering. The curve through the T = 6 K data is a resolution-limited Gaussian due to nuclear Bragg reflection. Analysis of the spectra gives very peculiar temperature dependencies of the fitting parameters, A and κ , as plotted in Fig. 9, where B stays constant at the nuclear level. Thus, to within experimental error (± 0.3 μ_B/Eu^{2+} ion) true long-range ferromagnetic order is never established in $Eu_{0.52}Sr_{0.48}S$. The Lorentzian amplitude A, which increases with decreasing temperature, indicates that an increasing number of spins are ferromagnetically correlated. The Lorentzian linewidth κ , which is equal to the inverse ferromagnetic correlation length ξ , passes through a minimum which is indistinguishable from zero (ξ>400 Å) around 2.5 K, and subsequently increases again at lower temperature.

In the light of these findings it is very interesting to study other samples with somewhat higher Eu concentration; data of $Eu_{0.54}Sr_{0.46}S$ are given in Fig. 10. As expected from the phase diagram in Fig. 1a and from data in the insert of Fig.10 there is a broader intermediate (ferromagnetic-like) state in x = 0.54, compared to x = 0.52, within the temperature range 1.5 K < T < 4 K. In order to separate quantitatively the contribution of magnetic diffuse scattering from Bragg scattering intensity, we again analyzed the line profiles. Typical parts of (200) scans in

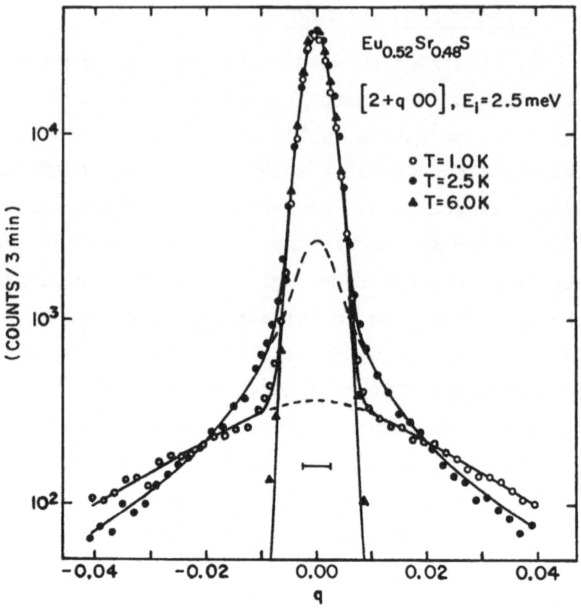

Fig. 8: (2+q,0,0) scans in momentum transfer q in $Eu_{0.52}Sr_{0.48}S$ at three temperatures. Solid lines represent fits using eq.1, dashed lines indicate Lorentzian scattering /10/.

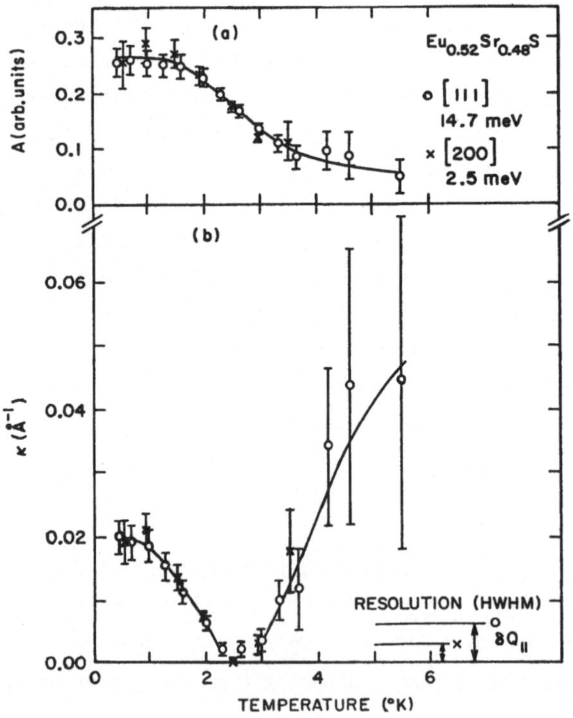

Fig. 9: Temperature dependence of the Lorentzian amplitude A and half-width κ obtained from fits to data as shown in Fig. 8 /10/.

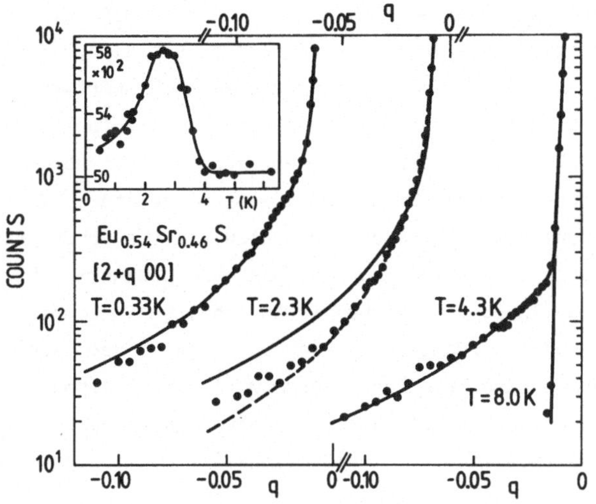

Fig. 10: Line profile of (2+q, 0,0) scans in $Eu_{0.54}Sr_{0.46}S$ at four temperatures. Solid and dashed lines represent fits to data using eqs. 1 and 2, respectively. Insert shows (200) peak intensity versus T in the same sample /10/.

momentum transfer q for x = 0.54 at four temperatures are shown in Fig. 10. Least-squares fits to eq. 1, represented by solid lines in Fig. 10, are reasonable only for temperatures $T \geqslant 4.0$ K and $T \lesssim 1.5$ K. They yield Lorentzian fitting parameters, A(T) and $\kappa(T)$, which behave as in the x=0.52 sample: κ decreases to zero at T=4.0 K and increases again from zero below 1.5 K to a value which is approximately six times smaller than κ for x = 0.52 at the lowest temperature. The fitting parameter B again has the same constant value in both T-regimes where eq. 1 gives an adequate account of the data.

For intermediate temperatures, $1.5\ K \lesssim T \leqslant 4.0$ K, the observed scattering profile shows clear deviations from the standard Lorentzian form eq. 1. A substantial improvement in the quality of the fit is achieved by using the form

$$S(\vec{Q}) = \frac{A}{(q^2 + \kappa^2)^{y/2}} + B \cdot \delta(\vec{q}) \qquad (2)$$

where the Lorentzian is raised to an arbitrary power y/2. If we fixed B at its nuclear value, the parameter κ is indistinguishable from zero and the exponent y(T) goes through a maximum, with a maximum value of 2.6 at T = 2.3 K. The dashed line in Fig. 10 represents such a fit of eq. 2 to the data. Thus, the ferromagnetic-like state in the crossover regime is strongly influenced by the proximity to the spin-glass phase (see also the interpretation of the data below), and we prefer to call it a "frustrated ferromagnetic" state.

Finally, it should be emphasized that all our neutron scattering results with x = 0.52 and 0.54 are found to be independent of the crystallographic direction.

One of our principal results is that in $Eu_xSr_{1-x}S$ with x = 0.52 and 0.54, there exists, within experimental error, no ferromagnetic moment in the crossover regime at the lowest temperature. The spin-glass regimes above and below $x_c = 0.51$ differ only in so far that the ferromagnetic correlations have different, but always finite extents. Consequently, our data of $Eu_xSr_{1-x}S$ (a Heisenberg system with short-range interactions) are not in agreement with the Gabay-Toulouse proposal /18/ for an infinite-range Heisenberg model to show coexistence of a net ferromagnetic moment with randomly frozen transverse degrees of freedom at the lowest temperature. In Ref. 10 we present a model based upon an effective decomposition via the frustration mechanism of the spin system into spin-glass-like and ferromagnetic networks. The ferromagnetic to spin glass crossover in $Eu_xSr_{1-x}S$ then is interpreted in terms of random field effects which arise when ferromagnetic and spin-glass order parameters are coupled together. The spin-glass freezing process will act to impose a random field on the ferromagnetic network, which destroys ferromagnetic order in any short-ranged coupled spin system with spatial dimensionality $D < 4$, according to recent theory and experiment /19/, hence the low-temperature phase is a pure spin-glass state, as really observed.

For intermediate temperatures, non-Lorentzian line profiles (eq. 2) in the "frustrated ferromagnetic" phase are observed. Just such an effect is expected for a random field /19/ which converts the ordinary delta-function-order parameter scattering in eq. 1, $B \cdot \delta(\vec{q})$, into the sum of a Lorentzian and its square:

$$B_m \cdot \delta(\vec{q}) \longrightarrow \frac{B''}{q^2 + \kappa^2} + \frac{B'}{(q^2 + \kappa^2)^2} \quad . \qquad (3)$$

For higher Eu concentrations ($x = 0.54$, compared to $x = 0.52$) within the crossover regime, the random-field induced scattering should include a larger Lorentzian squared component. Since κ is found experimentally to be very small, the scattering profile of eq. 3 may be indistinguishable from a power law form, q^{-y}, with an exponent $2 < y < 4$. Thus, this striking result on the spin correlations which was first observed in Ref. 6 may now properly be interpreted by recent theories of random-field effects.

IV. Conclusions

Our studies in $Eu_xSr_{1-x}S$ establish a ferromagnetic to spin glass crossover regime near $x_c = 0.51$ with novel magnetic order phenomena, which we tried to explain by the interplay between ferromagnetic and spin-glass type of ordering. As a consequence, a re-entrant ferromagnetic phase boundary for $x \geq x_c$ is observed.

First indications for such a crossover regime below about $x = 0.65$ are obtained in measurements of the specific heat and susceptibility at various concentrations. In several experiments the behavior around the Curie temperature T_c within the crossover regime is dominated by magnetic fluctuations, generated by the disorder and frustration, for example one observes a broad maximum in the magnetic specific heat and no pronounced critical scattering intensity of small-angle neutron scattering around T_c (see /1,6/). Nevertheless, characteristics of the ferromagnetic transition are still conserved within the crossover regime, for instance the ordinary value of the critical exponent $\gamma = 1.38$, as deduced by magnetization measurements (SQUID) in very small fields very near T_c, and the divergence of the ferromagnetic correlation length at T_c, as observed in neutron scattering experiments. The measured susceptibility just below T_c is also limited by the demagnetizating factor. Below $x_c = 0.51$, i.e. within the spin-glass regime, however, the χ-maximum gets reduced to a lower value which depends on x; we only observe a finite maximum of the ferromagnetic correlation length /6/ around the freezing temperature T_f in neutron scattering experiments (near x_c), and the temperature of the specific heat maximum does not coincide with the T_f-value determined by χ-measurements /4/.

Concentrated spin glasses reveal a complex high-frequency dependence of the susceptibility which can be understood due to the proximity to the re-entrant boundary. Deviations from the de Almeida-Thouless critical line are observed for spin-glass freezing in an applied field at high concentrations.

In neutron scattering experiments with $x \geqslant x_c$, no evidence is found for coexistence of ferromagnetic order with freezing of the transverse spin components at the lowest temperatures as predicted by Gabay and Toulouse /18/ for the infinite-range Heisenberg model. Campbell et al. /17/ arrived at the opposite conclusion for AuFe alloys by means of magnetic and Mössbauer experiments, however confirmation by neutron diffraction technique with single crystals would be valuable. These results would imply a fundamental difference of magnetic ordering within the crossover regime between metallic systems and short-ranged coupled insulating systems.

The re-entrant behavior observed in $Eu_xSr_{1-x}S$ can be explained in terms of random-field effects which arise when ferromagnetic and spin-glass order parameters are coupled together. Strong evidence for such mechanism is obtained by the striking observation of non-Lorentzian line profiles of the neutron spectra within the "frustrated ferromagnetic" regime, before pure spin-glass behavior is measured at the lowest temperature.

This work was supported in part by the NATO Research Grant No. 086.82. We thank H. Pink of Siemens AG, München and H. Lauterbach of KFA Jülich for preparing the single crystals.

References

1. H. Maletta, J. Appl. Phys. 53, 2185 (1982).
2. K. Binder, W. Kinzel, H. Maletta, and D. Stauffer, J. Magn.Magn.Mat. 15-18, 189 (1980); K. Binder and W. Kinzel in: Lecture Notes in Physics 149, 124 (1981, Springer Verlag).
3. A. Scherzberg, H. Maletta, and W. Zinn, J. Magn.Magn.Mat. 24, 186 (1981).
4. D. Meschede, F. Steglich, W. Felsch, H. Maletta, and W. Zinn, Phys. Rev. Lett. 44, 102 (1980).

5. K. Binder, W. Kinzel, and D. Stauffer, Z. Physik B36, 161 (1979).
6. H. Maletta and W. Felsch, Z. Physik B37, 55 (1980).
7. K. Binder, Z. Physik B48, 319 (1982).
8. G.P. Singh, M. v. Schickfus, and H. Maletta, submitted for publication.
9. J.A. Hamida, C. Paulsen, S.J. Williamson, and H. Maletta, submitted for publication.
10. H. Maletta, G. Aeppli, and S.M. Shapiro, J. Magn. Magn. Mat. 31-34, 1367 (1983).
11. H. Maletta and W. Felsch, Phys. Rev. B20, 1245 (1979); H. Maletta, J. Magn. Magn. Mat. 24, 179 (1981); H. Maletta in: NATO Institute on "Excitations in Disordered Systems", p. 431 (Plenum Press, 1982).
12. J.R.L. de Almeida and D.J. Thouless, J. Phys. A11, 983 (1978).
13. P. Monod and H. Bouchiat, J. Physique Lettres 43, L-45 (1982); R.V. Chamberlin, M. Hardiman, L.A. Turkevich, and R. Orbach, Phys. Rev. B25, 6720 (1982); Y. Yeshurun and H. Sompolinsky, Phys. Rev. B26, 1487 (1982).
14. A.P. Young, Phys. Rev. Lett. 50, 917 (1983); W. Kinzel and K. Binder, Phys. Rev. Lett. 50, 1509 (1983).
15. N. Bontemps, J. Rajchenbach, and R. Orbach, J. Physique Lettres 44, L-47 (1983).
16. M. Gabay and T. Garel, J. Phys. A14, 3411 (1981).
17. G.J. Nieuwenhuys, B.H. Verbeek, and J.A. Mydosh, J. Appl. Phys. 50, 1685 (1979); D.W. Carnegie and H. Claus, Phys. Rev. B20, 1280 (1979); S. Crane, D.W. Carnegie and H. Claus, J. Appl. Phys. 53, 2179 (1982); B.V.B. Sarkissian, J. Phys. F11, 2191 (1981); Y. Yeshurun, M.B. Salomon, K.V. Rao and H.S. Chen, Phys. Rev. Lett. 45, 1366 (1981); H. Maletta and P. Convert, Phys. Rev. Lett. 42, 108 (1979); I.A. Campbell, S. Senoussi, F. Varret, J. Teillet, and A. Hamzic, Phys. Rev. Lett. 50, 1615 (1983).
18. D. Sherrington and S. Kirkpatrick, Phys. Rev. Lett. 35, 1792 (1975); M. Gabay and G. Toulouse, Phys. Rev. Lett. 47, 201 (1981).
19. G. Parisi and N. Sourlas, Phys. Rev. Lett. 43, 744 (1979); Y. Imry and S.-K. Ma, Phys. Rev. Lett. 35, 1399 (1975); R.J. Birgeneau, H. Yoshizawa, R.A. Cowley, G. Shirane, and H. Ikeda, Phys. Rev. B28, 1438 (1983).

BROKEN SYMMETRY IN THE MEAN-FIELD THEORY OF THE ISING SPIN GLASS : REPLICA WAY AND NO REPLICA WAY

C. De Dominicis

SPh-T, CEN Saclay, 91191 Gif-sur-Yvette Cedex, France

- CONTENT -

Foreword and Summary

What follows is the response to the request of the Heidelberg conference organizers for a paper on replica symmetry breaking. The reader should be warned that this is no attempt at an exhaustive review but only a rather personal (and perhaps coherent) collection of aspects of it with which the author has had to deal with. Several items of interest are thus omitted, for example there is no discussion of the relevance of the order of limits involved (number of sites $N\to\infty$, number of replicas $n\to 0$)[1]. Likewise Heisenberg[2-4] or anisotropic systems[5-7] are left aside despite the fact that they have been recently examined from the point of view of replica symmetry breaking and have given rise to interesting results. For sheer reasons of size.

Thus we shall confine ourselves to the Sherrington-Kirkpatrick[8] (S.K) long range model described by the Hamiltonian

$$H = -\sum_{j<\ell} J_{j\ell}\,\sigma_j\sigma_\ell - \sum_j h_j\sigma_j \qquad (1)$$

with $\sigma_j = \pm 1$, Ising spin at site j (j=1,2...N). The couplings $J_{j\ell}$ are independent random variables with gaussian distribution (zero mean, width $J/\sqrt{N}$). This system, with quenched randomness, has a transition, in the thermodynamic limit, at $T_c=J$, and a spin glass phase below T_c. Its infinite (N-1) number of neighbours simulate high dimensionality and the solution of (1) can be considered as the first step (mean field or tree approximation) of a loop expansion for the corresponding short range system. This usually trivial step is transmuted here into what revealed itself a difficult problem.

By quenched it is meant that the bond averaging is taken over observables i.e. over lnZ rather than $Z = \mathrm{tr}\,\exp{-\beta H}$ itself. This is performed by the replica trick as shown by Edwards and Anderson[9], based on the relationship

$$\overline{\ln Z} \underset{n\to 0}{=} n^{-1}\,[\overline{(Z)^n}-1] \quad . \qquad (2)$$

On Z^n the bond averaging, symbolized by a bar, is easily performed with the result

$$\overline{Z^n} = \exp[n(N-n)(\beta J)^2/4] \int \prod_{(\alpha\beta)} \left(\frac{N}{2\pi}\right)^{1/2} dy_{\alpha\beta} \qquad (3)$$

$$\exp\left\{-\frac{N}{2}\sum_{(\alpha\beta)} y^2_{\alpha\beta} + N\ln(\mathrm{tr}_\sigma\, e^{L_H(y)})\right\}$$

$$L_H(y) = \beta J \sum_{(\alpha\beta)} y_{\alpha\beta}\sigma_\alpha\sigma_\beta + h\sum_\alpha \sigma_\alpha \tag{4}$$

with the saddle point condition

$$(\beta J)^{-1}\, y^c_{\alpha\beta} \equiv q_{\alpha\beta} = \langle\sigma_\alpha\sigma_\beta\rangle \tag{5}$$

$$= \mathrm{tr}_\sigma e^{L_H(\beta Jq)}\sigma_\alpha\sigma_\beta / \mathrm{Tr}_\sigma e^{L_H(\beta Jq)} \tag{6}$$

Sherrington and Kirkpatrick tried the ansatz $q=q_{\alpha\beta}$, the replica symmetric solution, on grounds that no replica should be preferred. But it was soon realized that that solution had unacceptable features (negative entropy) and was unstable under fluctuations[9,10]. The conclusion was unescapable, the solution of (5) had to be looked for with some sort of broken symmetry in replica space.

In section I we review the type of symmetry breaking involved in the solution discovered by Parisi[12] and in the static derivation of the solution first introduced via dynamics by Sompolinsky[13]. In section II we turn to a formulation of the problem due to Thouless, Anderson and Palmer[14](TAP) that replaces (3-6) by a set of equations for the magnetization $m_j=\langle\sigma_j\rangle_T$. A probability law $\mathcal{P}\{m_j\}$ for the magnetization is then built, but since TAP equations have many solutions, one may build different $\mathcal{P}$'s by attributing different weights to these solutions. We consider two cases : (i) a canonical distribution which is shown[15] to give identical results to the Hamiltonian formulation of (3-6) under a weak and physical assumption and (ii) a white distribution characterized by two matrices $q_{\alpha\beta}$ and a response $g_{\alpha\beta}$. We show[16] what symmetry breaking is necessary for q and g to recover Sompolinsky free energy. In section III we supplement replica indices in the Hamiltonian approach by "time" indices and show in particular that the analytic continuation involved in Sompolinsky's equilibrium derivation, is trying to mimick a translational symmetry breaking in "time" that incorporates Sompolinsky's ansatz of a long time scale sequence. In section IV we apply the same treatment to the white average approach and show[16] that, in that case, replicas can be altogether discorded and replaced by "time". Finally, mostly in last section, we briefly discuss recent work of Parisi[17,18] Houghton, Jain and Young[19] and De Dominicis and Young [20] and the attribution of distinct answers for the standard spin glass order parameter depending on the physical situation : equilibrium or non equilibrium associated with canonical or white (non canonical) initial conditions and density matrices.

I. THE REPLICA WAY : HAMILTONIAN APPROACH

(i) In the search for broken symmetry, away from the SK solution ($q_{\alpha\beta}=q, \alpha\neq\beta$) Bray and Moore[31] made a first attempt but, so distasteful was the idea of preferring some replicas to others, that they took their breaking infinitesimal (via a final limiting process). It was only with Blandin[22] and Blandin, Gabay, Garel[23] that a true symmetry breaking was attempted, together with the introduction of the idea of a real (or physical) replica. Unfortunately these schemes and others(*) all turned out to lead to unstable solutions. Then came Parisi[12]. By a bold generalization of Blandin et al. scheme he produced the first solution to withstand all tests.

(ii) Parisi : His solution is generated from the SK matrix by the following algorithm

Step (i)

$$q_{\alpha\beta}^{(o)} \equiv \overbrace{\boxed{\quad q_o \quad}}^{n} \longrightarrow q_{\alpha\beta}^{(1)} \equiv \overbrace{\begin{array}{|c|c|c|}\hline q_1 & q_0 & q_0 \\ \hline q_0 & q_1 & q_0 \\ \hline q_0 & q_0 & q_1 \\ \hline \end{array}}^{n} \;\} \, m_1 \tag{1.1}$$

Here the n×n, $q_{\alpha\beta}^{(o)}$ matrix is built with $(\frac{n}{m_1})\times(\frac{n}{m_1})$ constant blocks of size $m_1\times m_1$. The off diagonal q_o blocks are left untouched. The diagonal q_1 blocks are now submitted to

Step (ii)

$$\overbrace{\boxed{\quad q_1 \quad}}^{m_1} \longrightarrow \overbrace{\begin{array}{|c|c|c|}\hline q_2 & q_1 & q_1 \\ \hline q_1 & q_2 & q_1 \\ \hline q_1 & q_1 & q_2 \\ \hline \end{array}}^{m_1} \;\} \, m_2 \tag{1.2}$$

to constitute $q_{\alpha\beta}^{(2)}$ etc... up to the last and smallest diagonal block q_R of size $m_R\times m_R$. Here the successive sizes of the blocks are

$$n \geq m_1 \geq m_2 \dots \geq m_R \geq 1 \tag{1.3}$$

which, when the $n \to 0$ limit is taken, become

$$0 \leq m_1 \leq m_2 \dots \leq m_R \leq 1 \quad . \tag{1.4}$$

In the $R \to \infty$ limit

$$\frac{m_k}{m_{k+1}} \to 1 - \frac{dx}{x} \tag{1.5}$$

(*)About the same time Sommers[24] offered a new solution derived without replicas and that did not suffer from entropy catastrophe. It was later shown[25,26] to correspond to an infinitesimal symmetry breaking, and to be also unstable. It is the starting point for the iteration presented below that builds Sompolinsky equilibrium solution.

and $q_k \rightarrow q(x)$, $0 \leq x \leq 1$. That is, the order parameters q_k become the, by now well known, Parisi function $q(x)$. Note that one has a genuine broken symmetry. In the absence of a hypothetical field removing the (permutational) symmetry of the problem, the symmetry can be restored by combining solutions pointing to various directions in replica space.

There is a technical difficulty with Parisi's solution, namely that one is unable to write an explicit form for the stationary free energy in terms of the $q_k (k=1,2...R)$. Near T_c, the order parameter function $q(x)$ being small, one can work out explicit expansions. Far from T_c only partial differential equations are known[27].

(iii) Sompolinsky : This difficulty is no longer present in the solution proposed by Sompolinsky[13]. Inferred heuristically from a time dependent approach(*) (and more recently from TAP equations[29]) the solution produced by Sompolinsky displays, in particular, a stationary free energy, free of replicas (and hence free of the cumbersome limiting process to be taken in Parisi expressions). An equilibrium derivation, in the replica framework was then given[30] that runs parallel to Parisi's. Here the starting point, instead of being the SK constant matrix $q_{\alpha\beta}^{(o)}=q_o$, is the matrix that gives rise[25,26] to Sommers solution, i.e.

$$q_{\alpha\beta}^{(o)} \equiv \overbrace{\begin{array}{|c|c|c|} \hline q_0 & r_0 & r_0 \\ \hline r_0 & q_0 & r_0 \\ \hline r_0 & r_0 & q_0 \\ \hline \end{array}}^{n} \;\}\, p_o$$

$q_{\alpha\beta}^{(o)}$ is here built with $\frac{n}{p_o} \times \frac{n}{p_o}$ block matrices q_o (diagonal) and r_o (off diagonal) of size $p_o \times p_o$.

Step (i) :

$$\overbrace{\begin{array}{|c|} \hline \\ \quad q_0 \quad \\ \\ \hline \end{array}}^{p_o} \longrightarrow \overbrace{\begin{array}{|c|c|c|} \hline q_1 & q_0 & q_0 \\ \hline q_0 & q_1 & q_0 \\ \hline q_0 & q_1 & q_1 \\ \hline \end{array}}^{p_o} \;\}\, p_1$$

$$\overbrace{\begin{array}{|c|} \hline \\ \quad r_0 \quad \\ \\ \hline \end{array}}^{} \longrightarrow \overbrace{\begin{array}{|c|c|c|} \hline r_1 & r_0 & r_0 \\ \hline r_0 & r_1 & r_0 \\ \hline r_0 & r_0 & r_1 \\ \hline \end{array}}^{p_o} \;\}\, p_1 \qquad (1.6)$$

(*)An explicit dynamical derivation, using expansion-resummation techniques, has been since given by H.J. Sommers [28].

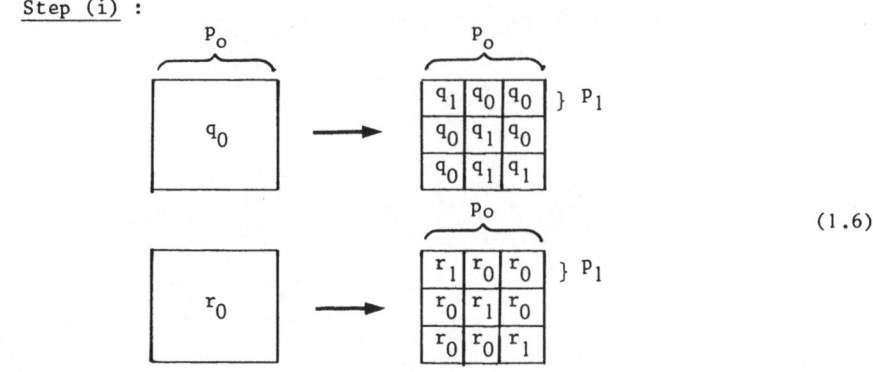

Parisi's step one is here effected on both q_o and r_o. Again off diagonal blocks (q_o, r_o) are untouched, and the diagonal blocks (q_1, r_1) are submitted to :

Step (ii) : It repeats the operations (1.6) on (q_1, r_1) etc ... up to the last and smallest diagonal blocks (q_R, r_R) of size $p_R \times p_R$. There are two main differences with Parisi procedure :

Here one works at given n, and $p_o p_1 \ldots p_R$. Analytic continuation is taken with

$$p_o >> p_1 >> \ldots >> p_R >> 1 \tag{1.7}$$

all the p's going to infinity in that order, and in the end $n \to 0$. An unnatural limiting process on which we shall return later.

We have now two functions, q_k as above and the "anomaly"

$$-\Delta'_k = p_k (q_k - r_k) \quad , \tag{1.8}$$

which are determined from Sompolinsky free energy functional,

$$-\beta f = \frac{\beta^2 J^2}{4} \left[(1-q_R)^2 + 2 \sum_{k=0}^{R} q_k \Delta'_k\right] + \int \prod_{k=0}^{R} \left(\frac{dz_k}{\sqrt{2\pi}} e^{-\frac{1}{2} z_k^2}\right) . \tag{1.9}$$

$$\cdot \left[\frac{\beta^2 J^2}{2} \sum_{k=0}^{R} \mu_k^2 \Delta'_k + \ln 2\cosh \beta h_s\right]$$

$$\beta h_s = \beta h + \sum_{k=0}^{R} \left[\beta J z_k (q_k - q_{k-1})^{1/2} - \beta^2 J^2 \mu_k \Delta'_k\right] . \tag{1.10}$$

Stationarity with respect to the magnetizations μ_k function of the effective local fields $z_o, z_1, \ldots, z_k$ yields

$$\mu_k(z_o, z_1, \ldots z_k) = \int \prod_{\ell=k+1}^{R} \left(\frac{dz_\ell}{\sqrt{2\pi}} e^{-\frac{1}{2} z_\ell^2}\right) \text{th } \beta h_s . \tag{1.11}$$

Stationarity with respect to Δ'_k and $q_k - q_{k-1}$ defines the order parameters q_k and Δ_k respectively

$$q_k = \int \prod_{j=0}^{k} \left(\frac{dz_j}{\sqrt{2\pi}} e^{-\frac{1}{2} z_j^2}\right) \mu_k^2 \tag{1.12}$$

$$\beta^{-1} \tilde{\chi}_k \equiv 1 - q_R + \sum_{j=k}^{R} \Delta'_j = \int \prod_{j=0}^{k} \left(\frac{dz_j}{\sqrt{2\pi}} e^{-\frac{1}{2} z_j^2}\right) \left[\beta J (q_k - q_{k-1})^{1/2}\right]^{-1} \frac{\partial}{\partial z_k} \mu_k . \tag{1.13}$$

The free energy (1.9,10) although still a functional (or R-multiple) integral over the effective field variables is free from the replica limiting process.

As $R \to \infty$ $(k/R+1 \to x)$, we are left with two functions $q(x)$ and $\Delta(x)$. The anomaly $\Delta(x)$ vanishes beyond the Almeida-Thouless[10] line where $q(x)$ becomes a constant. In zero field both $\Delta(x)$ and $q(x)$ vanish at and above T_c. We return later on the meaning of the x variable.

It is remarkable that (1.12,13) do not fully determine $q(x)$ and $\Delta(x)$ but leave the arbitrariness of a gauge choice[13]. Boundary values q(o), q(1) are fully determined (in terms of field and temperature) but otherwise one may choose any (monotonous) function for q(x), the gauge relation determines then $\Delta(x)$[13]. A closed form, in terms of spin correlations, for the gauge relation, has been derived by H.J. Sommers[47].

(iv) An immediate question is what is the relationship between these two solutions ? It is easily checked on fully explicit expressions obtained near T_c that Sompolinsky solution identifies with Parisi's provided[13]

$$\Delta'(x) = - xq'(x) \tag{1.14}$$

more generally it can be shown[31] that Sompolinsky solution satisfies a partial differential equation that reduces to Parisi's[27] provided a reparametrization $x \rightarrow u$ is allowed with

$$u(x) = -\Delta'(x)/q'(x) \quad , \tag{1.15}$$

so far no proof exists that u(x) is monotonous. However very recently H.J. Sommers[32] has been able to show that for any finite R, equations defining q_k in Parisi and Sompolinsky schemes become identical if one chooses constant Δ_k' and q_k' (i.e. a trapeze approcah to the continuum, rather than a rectangle (block) approach).

Another difference is of less consequence than it was hoped for. The Parisi function q(x) is monotonously increasing between x_o and x_1 (x_o,x_1 functions of h and β). Below and above, q(x) is a plateau at values q(o) ($x \leq x_o$) and q(1) ($x \geq x_1$). In Sompolinsky's formulation (functional of q' and Δ') the flat portions do not exist. This feature raised the hope that zero mass excitations associated with the flat portions in Parisi's fluctuations might be absent in Sompolinsky's solution.

(v) The stability of these solutions is discussed at this conference by I.Kondor. Here we just need mention results obtained near T_c (i) that the mass spectrum (i.e. eigenvalues of the quadratic form in $R_{\alpha\beta}$ obtained by replacing $y_{\alpha\beta}$ by $q_{\alpha\beta}+R_{\alpha\beta}$ in (4)) is identical for Parisi[33] and Sompolinsky[34] solutions, (ii) that it is semi definite : all masses are positive or zero. These features and others are in agreement with the extensive results obtained by Sompolinsky and Zippelius[35] for dynamic fluctuations.

(vi) One unresolved question is unicity. Are there other (semi) stable ansätze to the stationarity equations (4-6)? One class of solutions generalizing Sompolinsky's is obtained by starting e.g. from its first iteration that involves blocks q_1,q_0 and r_1,r_0. One applies then the iteration defined in (1.6) to q_1,q_0,r_1,r_0 yielding four functions $q_1^{(k)} q_\theta^{(k)} r_1^{(k)} r_0^{(k)}$. It is easy to convince oneself that these functions are related to one another and that the solution is again identical to Sompolinsky's. Likewise one could have started from the second iteration (or any fixed iteration) and apply (1.6) to $q_2q_1q_0,r_2r_1,r_0$ with the same result. More generally the game is to decompose $\sum_{\alpha\beta} q_{\alpha\beta}\sigma_\alpha\sigma_\beta$ into $\sum_b q_b(\Sigma\sigma)_b^2$, with $q_b > 0$, a necessary condition for stability. What we have explored however is far from exhausting the space of all admissible trial matrices, and unicity remains an open question.

II. THE REPLICA WAY : TAP APPROACH

(i) Instead of working directly with the spin Hamiltonian (1) of section I, and use a mean field procedure, one may as well directly introduce a self consistent local field u_j+h_j (where the external field h_j is singled out) and replace (1) by

$$H = -\sum_j \sigma_j(u_j+h_j) \tag{2.1}$$

$$u_j = \sum_\ell J_{j\ell}\, m_\ell(\not{j}) \tag{2.2}$$

where the magnetization m_ℓ is

$$m_\ell = \langle\sigma_\ell\rangle_T = \text{th}\beta(h_\ell+u_\ell) \tag{2.3}$$

and $m_\ell(\not{j})$ is the magnetization computed at site ℓ in the absence of spin at site j. With (2.2,3) we have

$$\begin{aligned} m_\ell(\not{j}) &= \text{th}\left[\beta h_\ell + \beta \sum_k J_{\ell k} m_k(\not{\ell},\not{j})\right] \\ &= \text{th}\left[\beta(h_\ell+u_\ell) - \beta J_{\ell j} m_j(\not{\ell})\right] \end{aligned} \tag{2.4}$$

i.e. to leading order,

$$m_\ell(\not{j}) = m_\ell - \beta J_{\ell j}(1-m_\ell^2)m_j \tag{2.5}$$

displaying the Onsager term. Using (2.2,3,5) we get the famous Thouless Anderson and Palmer[14] (TAP) equations(*)

$$\text{th}^{-1} m_j = \beta h_j + \sum_\ell \beta J_{j\ell} m_\ell - \sum_\ell \beta^2 J_{j\ell}^2 (1-m_\ell^2) m_j \quad . \tag{2.6}$$

It is useful to look at them as stationarity conditions

$$\partial F/\partial m_j = 0 \tag{2.7}$$

for the TAP free energy F{m}

$$\begin{aligned} -\beta F\{m\} = \frac{1}{2}\sum_{j,\ell} \beta J_{j\ell} m_j m_\ell + \frac{1}{4}\sum_{j,\ell} \beta^2 J_{j\ell}^2 (1-m_\ell^2)(1-m_j^2) + \sum_j \beta h_j m_j \\ - \frac{1}{2}\sum_j \left[(1+m_j)\ln\frac{1}{2}(1+m_j) + (1-m_j)\ln\frac{1}{2}(1-m_j)\right] \quad . \end{aligned} \tag{2.8}$$

(*) The above derivations is the simplest known to the author and was surely in the knowledge of TAP themselves. Other derivations abound [24,35].

From (2.8), it is convenient to build a probability law for the magnetization taking the value m_j (one may also work with the local field u_j)

$$\overline{\mathcal{P}\{m\}} = \overline{\prod_j \delta(\partial F/\partial m_j)\ \Delta\{m\}} \qquad (2.9)$$

where the bar stands for bond averaging and $\Delta\{m\}$ is the Jacobian

$$\Delta\{m\} = |\det \partial^2 F/\partial m_j \partial m_\ell| \qquad (2.10)$$

normalizing the delta functions and thus $\mathcal{P}\{m\}$ itself. If this were so, the bond averaging could triviallybe taken on $\mathcal{P}\{m\}$. However the system of TAP equations (2.7) admits a large number solutions[37-39], therefore one has instead

$$\mathcal{P}\{m\} = \sum_s \left(\prod_j \delta(m_j - m_j^s\{J\}) \right). \qquad (2.11)$$

This feature makes it necessary to specify which weight P(s) to give to each solution. We shall examine two cases[37,38] : the canonical averaging

$$\mathcal{P}_c\{m\} = \mathcal{P}\{m\}\ e^{-\beta F\{m\}} \qquad (2.12)$$

$$Z = \int \prod_j dm_j\ \mathcal{P}_c\{m\} \qquad (2.13)$$

where each solution carries a weight

$$P(s) = Z^{-1} \exp -\beta F\{m^s\} \qquad (2.14)$$

and the white averaging

$$\mathcal{P}_w\{m\} \equiv \mathcal{P}\{m\} \qquad (2.15)$$

$$\mathcal{N} = \int \prod_j dm_j\ \mathcal{P}_w\{m\} \qquad (2.16)$$

where each solutions has a constant weight $\mathcal{N}^{-1}$. To carry out the bond averaging we thus have to introduce replicas, since the normalizations of $\mathcal{P}_c$ and $\mathcal{P}_w$ are bond dependent. In the following we want to briefly describe replica symmetry broken solutions in those two situations.

(ii) Canonical average[15,46] : This is, for our purpose the lesser interesting case since it can be shown that, after replication and bond averaging, it yields a density matrix identical to the one derived from the Hamiltonian approach. This has been shown with the conjecture that there is no entropy contribution due to solution degeneracy. It is worth examining what it implies for various observables. We rewrite (2.13) as*

---

* : Note that we drop the Δ absolute value. This leaves identical eigenvalues for maxima, minima and saddle points with an even number of negative masses (we assume N to be even). This should be satisfactory for asymptotic or long time limits (not so if one wanted to estimate relaxation times or leading saddle point values).

$$Z = \int \prod_j dm_j \frac{d\hat{m}_j}{2\pi} d\eta_j^* d\eta_j \exp\left\{-\sum_j i\hat{m}_j (\partial F/\partial m_j) + \sum_{j,\ell} \eta_j^* (\partial^2 F/\partial m_j \partial m_\ell)\eta_\ell - \beta F\right\} \tag{2.17}$$

where η^*,η are Grassman (anticommuting) variables. After replication and bond averaging, the usual one-site expression is obtained ($f \equiv F/N$)

$$N^{-1}\,\overline{\ln Z} \;\equiv\; -\beta f = \lim_{n\to 0} n^{-1}\left[A_c + \ln D_c\right] \tag{2.18}$$

where

$$A_c = \left[-\mathrm{Tr}\left(i(q\tilde{q}+q\tilde{\hat{q}}+g\tilde{g}+n\tilde{n}) - \frac{1}{2}\beta^2 J^2 (q\hat{q}+g^2-n^2)\right) + \beta^2 J^2 \sum_\alpha (n_{\alpha\alpha}-g_{\alpha\alpha})(1-q_{\alpha\alpha})\right] \tag{2.19}$$

$$+ \left[-\beta^2 J^2 \mathrm{Tr}\left(\frac{1}{4}q^2+gq\right)+\beta^2 J^2 \sum_\alpha \frac{1}{4}(1-q_{\alpha\alpha})^2\right]$$

$$D_c = \int \prod_\alpha dm_\alpha \frac{d\hat{m}_\alpha}{2\pi} \det\left[(1-m^2)^{-1}\mathbb{1} - i\tilde{n}\right] \exp L_c \tag{2.20}$$

$$L_c = i\,\mathrm{Tr}(\tilde{q}mm + \tilde{g}i\hat{m}m + \tilde{\hat{q}}\, i\hat{m}i\hat{m})$$

$$+ \sum_\alpha \left[(\beta h - \mathrm{th}^{-1} m_\alpha)(m_\alpha + i\hat{m}_\alpha) - \frac{1}{2}\ln(1-m_\alpha^2)\right]. \tag{2.21}$$

The Grassman variables have been explicitly integrated out resulting in the determinant of (2.20), conversely we may also leave them into an $L_c(m,\hat{m},\eta^*,\eta)$ when needed to compute $\eta^*\eta$ averages. We have inforced the definitions

$$\begin{aligned} q_{\alpha\beta} &= N^{-1}\sum_j m_j^\alpha m_j^\beta \\ g_{\alpha\beta} &= N^{-1}\sum_j i\hat{m}_j^\alpha m_j^\beta \\ \hat{q}_{\alpha\beta} &= N^{-1}\sum_j i\hat{m}_j^\alpha i\hat{m}_j^\beta \\ n_{\alpha\beta} &= N^{-1}\sum_j \eta_j^{*\alpha}\eta_j^\beta \end{aligned} \tag{2.22}$$

for the standard observables (a step needed for the one site reduction), via the use of conjugated variables, the corresponding tilded matrices $\tilde{q},\tilde{\hat{q}},\tilde{g},\tilde{n}$. Stationarity of (2.18) with respect to tilded variables yields

$$\begin{aligned} q_{\alpha\beta} &= \langle m_\alpha m_\beta\rangle_c & g_{\alpha\beta} &= \langle i\hat{m}_\alpha m_\beta\rangle_c \\ \hat{q}_{\alpha\beta} &= \langle i\hat{m}_\alpha i\hat{m}_\beta\rangle_c & n_{\alpha\beta} &= \langle \eta_\alpha^* \eta_\beta\rangle_c \end{aligned} \tag{2.23}$$

where averages are computed with the density matrix D_c (2.20).

The fact that no extensive entropy arises due to the degeneracy of solutions implies for consistency that $\langle i\hat{m}_j\rangle_c = 0$ since h_j couples here to $(m_j+i\hat{m}_j)$ instead of m_j alone. Expressing that $\langle i\hat{m}_j\rangle_c$ is itself field and temperature independent yields,

$$g_{\alpha\beta} = -\hat{q}_{\alpha\beta} = n_{\alpha\beta} = \delta_{\alpha\beta}(1-q_{\alpha\alpha}) \ . \tag{2.24}$$

Together with stationarity with respect to $q,\hat{q},g,n$, we obtain then

$$i\tilde{q}_{\alpha\beta} = i\tilde{\hat{q}}_{\alpha\beta} = i\tilde{g}_{\alpha\beta}/2 = (\beta^2 J^2/2) q_{\alpha\beta} \ , \quad \tilde{n}_{\alpha\beta} = 0 \ . \tag{2.25}$$

With these relations, the free energy (2.18) rewrites with

$$A_c = -\frac{\beta^2 J^2}{4} \left(\sum_{\alpha\neq\beta} q^2_{\alpha\beta} - n \right) \tag{2.26}$$

and the one-site density matrix

$$D_c = \int_{-\infty}^{+\infty} \prod_{\alpha} dX_\alpha \ \frac{d\hat{\mu}_\alpha}{2\pi} \exp L_c \tag{2.27}$$

$$L_c = \frac{1}{2} \beta^2 J^2 \sum_{\alpha\neq\beta} q_{\alpha\beta} \ i\hat{\mu}_\alpha i\hat{\mu}_\beta + \sum_\alpha \left[(\beta h - X_\alpha) i\hat{\mu}_\alpha + \ln 2\cosh X_\alpha \right] \ . \tag{2.28}$$

Here we have made the change of variables

$$i\hat{\mu}_\alpha = i\hat{m}_\alpha + m_\alpha \tag{2.29}$$

$$X_\alpha = \mathrm{th}^{-1} m_\alpha \ . \tag{2.30}$$

We have also used the fact that in (2.27,28), $\hat{\mu}_\alpha$ is constrained to be ± 1 to eliminate $q_{\alpha\alpha}$. We have therefore recovered the Hamiltonian replicated form of (3,4). From there, one may easily check that the conjecture made (no entropy from solution degeneracy) is self consistent. This has been done with the ansatz for $q_{\alpha\beta}$ described in section I.

(iii) White average[16,46]: This average was commonly believed to lead to unphysical results. It turns out to be in fact most interesting. Consider now (2.16) i.e.

$$\mathcal{N} = \prod_j \int_{-1}^{+1} dm_j \int_{+\infty}^{+\infty} \frac{d\hat{m}_j}{2\pi} \int d\eta_j^+ d\eta_j \exp\left\{ -\sum_j i\hat{m}_j \ \partial F/\partial m_j + \sum_{j\ell} \eta_j^+ (\partial^2 F/\partial m_j \partial m_\ell) \eta_\ell \right\} \tag{2.31}$$

After replication and bond averaging, we obtain

$$\overline{N^{-1} \ln \mathcal{N}} = \lim_{n\to 0} n^{-1} \left[A_w + \ln D_w \right] \tag{2.32}$$

where A_w reduces to the first bracket of A_c in (2.19), the one site density matrix D_w is as in (2.20) and

$$L_w = i \ \mathrm{Tr} \ \tilde{q}mm + \tilde{g}i\hat{m}m + \tilde{\hat{q}}i\hat{m}i\hat{m} + \sum_\alpha (\beta h - \mathrm{th}^{-1} m_\alpha) \ i\hat{m}_\alpha \ . \tag{2.33}$$

Stationarity with respect to the tilded variables leave us with the same (2.23) equations averaged now over D_w.

In order to recover the same physical averages as in the canonical case we require no extensive entropy due to solution degeneracy at the physical saddle point. I.e. we are actually taking only an appropriate window of the white average (the terminology is perhaps misleading) such that $\overline{\ln\mathcal{N}}$ is non extensive i.e. $N^{-1}\times \overline{\ln\mathcal{N}}=0$ as $N\to\infty$. Among the extremas of (2.32), the chosen saddle point will have to satisfy, in particular, $\langle i\hat{m}_j\rangle_w=0$ as above (expressing that $N^{-1}\overline{\ln\mathcal{N}}$ is h independent). Expressing further field (and temperature) independence yields

$$\hat{q}_{\alpha\beta} = 0 \tag{2.34}$$

$$g_{\alpha\beta} = n_{\alpha\beta} \tag{2.35}$$

instead of (2.24). Stationarity with respect to $q,\hat{q},g,n$, together with (2.34,35) yields

$$i\tilde{q}_{\alpha\beta} = 0 \tag{2.36}$$

$$i\tilde{\hat{q}}_{\alpha\beta} = \frac{1}{2}\beta^2J^2q_{\alpha\beta} \tag{2.37}$$

$$i\tilde{g}_{\alpha\beta} = -i\tilde{n}_{\alpha\beta} = \beta^2J^2(g_{\alpha\beta}-\delta_{\alpha\beta}(1-q_{\alpha\alpha})) \equiv \beta J^2\chi_{\alpha\beta} \tag{2.38}$$

together with $A_w=0$. The density matrix becomes

$$D_w = \int \prod_\alpha dm_\alpha \frac{d\hat{m}_\alpha}{2\pi} d\eta_\alpha^* d\eta_\alpha \exp\left\{\frac{1}{2}\beta^2J^2q_{\alpha\beta}i\hat{m}_\alpha i\hat{m}_\beta - i\hat{m}_\alpha\left[th^{-1}m_\alpha-\beta h-\sum_\beta(\beta J^2\chi_{\beta\alpha}m_\beta)\right] + \eta_\alpha^*\left[(1-m_\alpha^2)^{-1}\eta_\alpha-\sum_\beta(\beta J^2\chi_{\beta\alpha}\,\eta_\beta)\right]\right\}, \tag{2.39}$$

on which the self consistency of (2.34,35) is easily checked. If we introduce the effective field

$$Y_\alpha = th^{-1}m_\alpha-\beta J^2\sum_\beta \chi_{\beta\alpha}m_\beta \tag{2.40}$$

the one site density matrix reduces to

$$D_w = \int \prod_\alpha dY_\alpha \frac{d\hat{m}_\alpha}{2\pi} \exp\left\{\frac{1}{2}\beta^2J^2\sum_{\alpha,\beta} q_{\alpha\beta}\, i\hat{m}_\alpha i\hat{m}_\beta + i\hat{m}_\alpha\left[\beta h-Y_\alpha\right]\right\} \tag{2.41}$$

obviously normalized to unity. We still have to show that (2.40-41) give rise to the same observable values.

(iv) Sompolinsky ansatz for white average : We exhibit now the ansatz that will reproduce Sompolinsky's result of section I for the above white average. The parameters q_k and Δ_k are to arise from the two $(n\times n)$ matrices involved, the correlation $q_{\alpha\beta}=\langle m_\alpha m_\beta\rangle_w$ and the response $g_{\alpha\beta}=\langle i\hat{m}_\alpha m_\beta\rangle_w$ (or $\chi_{\alpha\beta}$ as in 2.38).

The correlation matrix $q_{\alpha\beta}$ is built with $(n/p_o)^2$ identical block matrices of size $p_o\times p_o$, let q_{ab} $(a,b=1,2,\dots,p_o)$. The response matrix $\chi_{\alpha\beta}$ is built with (n/p_o) identical block matrices $p_o\times p_o$, along the diagonal, let χ_{ab}, and zero elsewhere.

$$q_{\alpha\beta} = \overbrace{\begin{array}{|c|c|c|}\hline q & q & q \\ \hline q & q & q \\ \hline q & q & q \\ \hline\end{array}}^{p_o} \left.\vphantom{\begin{array}{c}q\\q\\q\end{array}}\right\} n \qquad \chi_{\alpha\beta} = \overbrace{\begin{array}{|c|c|c|}\hline \chi & 0 & 0 \\ \hline 0 & \chi & 0 \\ \hline 0 & 0 & \chi \\ \hline\end{array}}^{p_o} \left.\vphantom{\begin{array}{c}q\\q\\q\end{array}}\right\} n \quad . \tag{2.42}$$

The block matrices q_{ab} and χ_{ab} are now formally constructed like a Parisi matrix(*) with successive blocks of linear size $p_o >> p_1 >> \ldots p_R >> p_M \equiv 1$. For example after one iteration

$$q_{ab}^{(1)} = \overbrace{\begin{array}{|c|c|c|}\hline q_1 & q_0 & q_0 \\ \hline q_0 & q_1 & q_0 \\ \hline q_0 & q_0 & q_1 \\ \hline\end{array}}^{p_1} \left.\vphantom{\begin{array}{c}q\\q\\q\end{array}}\right\} p_o \qquad \chi_{ab}^{(1)} = \begin{array}{|c|c|c|}\hline \chi_1 & \chi_0 & \chi_0 \\ \hline \chi_0 & \chi_1 & \chi_0 \\ \hline \chi_0 & \chi_0 & \chi_1 \\ \hline\end{array} \quad . \tag{2.43}$$

The q_{ab} and χ_{ab} matrixes also possess a non vanishing diagonal element $q_{aa} \equiv q_M$ and $\chi_{aa} \equiv \chi_M$. Finally

$$\chi_k \equiv -\Delta'_k / p_k \tag{2.44}$$

i.e. all matrix elements of χ_{ab} are infinitesimal since (as in Sompolinsky ansatz of section I) one is taking the limit where (i) all p'_ks go to infinity in succession, and (ii) $n \to 0$ only thereafter.

In order to see how Sompolinsky solution emerges here, and for later use, we work out explicitly the zeroth step i.e. $q_{ab}^{(o)}$ a constant q_o everywhere except on the diagonal q_M, and $\chi_{ab}^{(o)}$ likewise i.e. $\chi_o \equiv -\Delta'_o/p_o$ and $\chi_M \equiv -\Delta'_M$. Let us detail the α indices as $\alpha \equiv (\gamma, a)$ with $\gamma = 1,2,\ldots,n/p_o$ indexes the blocks of (2.42) and the index a reduces here to $j_o = 1,2,\ldots,p_o$ (but on further iterations is parametrized by $j_o j_1 \ldots$). Writing

$$\sum_{\alpha,\beta} q_{\alpha\beta}\, i\hat{m}_\alpha i\hat{m}_\beta = q_o \Big(\sum_{\gamma, j_o} i\hat{m}_{\gamma j_o} \Big)^2 + (q_M - q_o) \sum_{j_o} \Big(\sum_{\gamma} i\hat{m}_{\gamma j_o} \Big)^2 \tag{2.45}$$

$$\sum_{\alpha,\beta} \chi_{\beta\alpha}\, i\hat{m}_\alpha m_\beta = \chi_o \sum_{\gamma} \Big(\sum_{j_o} i\hat{m}_{\gamma j_o} \Big) \Big(\sum_{\ell_o} m_{\gamma \ell_o} \Big) + (\chi_M - \chi_o) \sum_{\gamma j_o} i\hat{m}_{\gamma j_o} m_{\gamma j_o} ,$$

we linearize the squares using auxiliary variables z_o, z_{jo}, and we define

$$M_\gamma = p_o^{-1} \sum_{j_o} m_{\gamma j_o} \tag{2.46}$$

via an associated constraint variable $i\widetilde{M}_\gamma$. We then obtain

---

(*) Note that if this $q_{\alpha\beta}$ matrix were inserted in the density (2.27,28) i.e. with no coupling to the $\chi_{\alpha\beta}$ matrix, one would only recover the trivial SK solution.

$$D_w = \int \frac{dz_o}{\sqrt{2\pi}} e^{-\frac{1}{2} z_o^2} \int \prod_\gamma \left(p_o dM_\gamma \frac{d\tilde{M}_\gamma}{2\pi} e^{-p_o M_\gamma i\tilde{M}_\gamma} \right)$$

$$\int \prod_{j_o} \left(\frac{dz_{j_o}}{\sqrt{2\pi}} e^{-\frac{1}{2} z_{j_o}^2} \right) . \int \prod_{\gamma j_o} \left(dm_{\gamma j_o} \frac{d\hat{m}_{\gamma j_o}}{2\pi} e^{i\tilde{M}_\gamma m_{\gamma j_o} - \xi_{\gamma j_o} i\hat{m}_{\gamma j_o}} [(1-m_{\gamma j_o}^2)^{-1} + \Delta'_M] \right) \quad (2.47)$$

$$\cdot \left[1 + \frac{1}{p_o} \sum_{\gamma j_o} \Delta'_o \frac{1}{(1-m_{\gamma j_o}^2)^{-1} + \Delta'_M} \right]$$

with

$$\xi_{\gamma j_o} = th^{-1} m_{\gamma j_o} - \left[\sqrt{q_o}\, z_o + \sqrt{q_M - q_o}\, z_{j_o} - \Delta'_o M_\gamma - \Delta'_M m_{\gamma j_o} \right] . \quad (2.48)$$

Here the determinant (Grassman integral) has been written as the product of diagonal terms $\prod_{\gamma j} (d\xi_{\gamma j_o}/dm_{\gamma j_o})$ and the last bracket. Leaving aside this last bracket, we may integrate over $i\hat{m}$ and ξ to obtain

$$D_W = \int \frac{dz_o}{\sqrt{2\pi}} e^{-\frac{1}{2} z_o^2} \left[\int \prod_\gamma \left(p_o dM_\gamma \frac{d\tilde{M}_\gamma}{2\pi} \right) \exp\left\{ -p_o \sum_\gamma M_\gamma i\tilde{M}_\gamma + \sum_{j_o} \ln \int \frac{dz_{j_o}}{\sqrt{2\pi}} e^{-\frac{1}{2} z_{j_o}^2 + \sum i\tilde{M}_\gamma m_{\gamma j_o}} \right\} \right] \quad (2.49)$$

Since $p_o \to \infty$ we may now evaluate M_γ, $i\tilde{M}_\gamma$ by a saddle point condition. We get

$$\tilde{M}_\gamma^c \equiv 0 \quad (2.50)$$

$$M_\gamma^c = p_o^{-1} \sum_{j_o} \int \frac{dz_{j_o}}{\sqrt{2\pi}} e^{-\frac{1}{2} z_{jo}^2} m_{\gamma j_o}(z_o, z_{j_o}; M_\gamma^c) \quad (2.51)$$

where $m_{\gamma j_o}$ is determined by (2.48) for $\xi=0$. This immediately results into

$$M_\gamma^c \equiv \mu_o(z_o) = \int \frac{dz_1}{\sqrt{2\pi}} e^{-\frac{1}{2} z_1^2} \mu_1(z_o, z_1; \mu_o) \quad (2.52)$$

$$m_{\gamma j_o}^c \equiv \mu_1(z_o, z_1) = th \left\{ \sqrt{q_o} z_o + \sqrt{q_M - q_o} z_1 - \Delta'_o \mu_o(z_o) - \Delta'_M \mu_1(z_o, z_1) \right\} . \quad (2.53)$$

Fluctuations around the saddle point do not contribute since they are to the power $\sum_\gamma \equiv n/p_o$. Likewise the discarded term in the last bracket of (2.47) behaves like n. As we shall see later this features are a clear advantage of the replica way. Evaluation of q_k, Δ_k, or the free energy would lead to the same expressions as given in [1,11-13]. In other words both density matrices (2.26-29) and (2.39-40) give rise to identical values for the magnetizations $\mu_k(z_o, \ldots, z_k)$, and the resulting q_k, Δ_k. The free energy has also the same value (but a distinct functional form). It is interesting to notice that this is not true of all observables. It has been argued[15] that the standard spin glass order parameter $\overline{\langle\sigma\rangle_T^2} \equiv Q$ should be given, in statistical mechanics, by an average over replicas (to take into account degenerate saddle points)

$$Q = n(n-1)^{-1} \sum_{\alpha\neq\beta} q_{\alpha\beta} \ . \tag{2.54}$$

If we use the ansatz described in section I, we obtain

$$Q = \frac{1}{n-1}\ [p_0(q_0-r_0)+\ldots+p_R(q_R-r_R)-q_R+\frac{n}{p_0}(p_0 r_0+p_1(r_1-r_0)+\ldots)] \tag{2.55}$$

which, as $p_0 \gg p_1 \gg \ldots \gg p_R \to \infty$, and $n \to 0$ yields

$$Q = \sum_{k=0}^{R} \Delta'_k + q_R \to \int_0^1 dx\, q(x)\ . \tag{2.56}$$

The white average ansatz, described by (2.42,43) can be obtained from (1.6) for what concerns $q_{\alpha\beta}$ by setting $q_k \equiv r_k$ (except for $q_M \equiv q_{\alpha\alpha}$ absent in (1.6)) Applied to (2.54) would result into $Q=q_M$, a surprising result.

III. ON REPLICAS AND "TIME" : CANONICAL AVERAGE

(i) In order to understand the occurrence of two distinct density matrices, we have to look beyond statistical mechanics. It is the merit of Houghton, Jain, and Young [19] to have pointed out the role of initial conditions in the time evolution towards equilibrium. In particular they have shown that, if initial conditions at t_o are governed by a canonical distribution, then at long times, one has

$$q_{\alpha\beta}(t') = \langle\sigma_\alpha(t)\sigma_\beta(t+t')\rangle \xrightarrow[t'\to\infty]{} q_{\alpha\beta} \tag{3.1}$$

observables being computed with the canonical density matrix that appeared in section I (and in section II for the canonical average) :

$$D_c = \int \prod_\alpha \left(dX_\alpha \frac{d\hat\mu_\alpha}{2\pi}\right) \exp\left\{\frac{\beta^2 J^2}{2} \sum_{\alpha\neq\beta} q_{\alpha\beta}\, i\hat\mu_\alpha i\hat\mu_\beta + \sum_\alpha (\beta h - X_\alpha)\hat\mu_\alpha + \sum_\alpha \ln 2\mathrm{ch} X_\alpha\right\} \tag{2.27}$$

$$X_\alpha = \mathrm{th}^{-1} m_\alpha \tag{2.30}$$

$$q_{\alpha\beta} = \langle \mathrm{th}(X_\alpha+\beta h)\,\mathrm{th}(X_\beta+\beta h)\rangle_c \ . \tag{3.2}$$

In other words, the structure that one might have expected in time (as postulated in the dynamical approach of Sompolinsky and reflected in the hierarchical form of its free energy (1.9-13)) is being relegated, via (3.1), in the replica structure of $q_{\alpha\beta}$. We have seen in (I) the bizarre analytic continuation needed to exhibit the postulated structure.

(ii) One may perhaps get some insight in it by doing the following which is only meant to have a pedagogical value. Instead of taking $\exp-\beta H(\sigma)$ as a weight one may consider the time evolution of the system from the initial time t_o to t_1. The Fokker-Planck conditional distribution

$\rho(\sigma_1;\sigma_o)$ is then of the form

$$\rho(\sigma_1;\sigma_o) \simeq \exp\left\{-\beta[H(\sigma_1)+H(\sigma_o)+\mathcal{M}(\sigma_1,\sigma_o)\ ]\right\} \tag{3.3}$$

$$\mathcal{M}(\sigma_1;\sigma_o) = \delta h(\sigma_1+\sigma_o)+u\sigma_o\sigma_o + v\sigma_1\sigma_1 + w\sigma_o\sigma_1 + \dots \tag{3.4}$$

where the coupling terms $\mathcal{M}$ between σ_1,σ_o all decay as t_1-t_o increases. Suppose we are in a situation where all correction terms (3.4) are small (in that case (3.3,4) is a realization of Blandin et al.[22,23] real copies with small coupling). One may iterate the procedure and introduce a large number of discrete times (or copies) $\theta=1,2,\dots T_o$ with an effective Hamiltonian

$$\sum_{\theta=0}^{T_o} H(\sigma_\theta) + \sum_{\theta=1}^{T_o} \mathcal{M}(\sigma_\theta,\sigma_{\theta-1}) \quad . \tag{3.5}$$

In the limit $\mathcal{M}\to 0$, (3.5) is a generalization of Blandin real copies approach, it comes formally identical to Orland[40] point of view. Usual physical average values involve now a time (or copy) average, e.g.

$$m = \frac{1}{T_o} < \sum_\theta \sigma_\theta > \tag{3.6}$$

for the magnetization. After bond averaging we obtain the same results as in section I with now a matrix $q_{\alpha\beta}^{\theta\theta'}$ with both replicas (α,β) and time indices. One may then write the associated standard Sompolinsky or Parisi solutions in two different fashions.

a) Leave all the <u>structure in replicas</u> :

$q_{\alpha\beta}^{\theta\theta'}$ is built with T_o^2 identical n×n matrices $q_{\alpha\beta}$, except for the diagonal terms $(q_{\alpha\alpha}^{\theta\theta}=0,\ q_{\alpha\alpha}^{\theta\neq\theta'}=q_{\alpha\alpha})$. By reshuffling indices it is easy to see that it amounts to take the standard e.g. Sompolinsky solution of section I and do the replacements

$$p_k \to p_k\, T_o \tag{3.7}$$

$$n \to n\, T_o \tag{3.8}$$

everywhere.

b) Bring all the <u>structure in time</u> :

$q_{\alpha\beta}^{\theta\theta'}$ is now built by giving structure to the $T_o\times T_o$ matrix $q^{\theta\theta'}$, i.e. introducing a sequence $T_o>>T_1>>\dots T_R>>1$ of times that are a measure of Sompolinsky time scales, with corresponding hierarchical values $\mathfrak{q}_k$ (constant in $T_k\times T_k$ blocks). To go to $q_{\alpha\beta}^{\theta\theta'}$ one associates to each $q^{\theta\theta'}$ a n×n matrix which now has a minimal structure i.e.

$\mathfrak{q}_k$ = 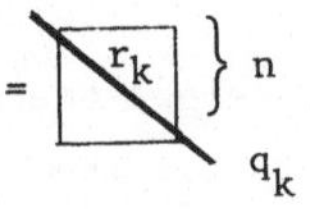

(3.9)

i.e. $q_{\alpha\alpha} = q_k$ (or zero for the n×n matrices associated with $q^{\theta\theta}$) and $q_{\alpha\beta} = r_k (\alpha \neq \beta)$. (En passant, let us remark that this is the structure tried by Blandin et al. for the case $T_o=2$). Again by reshuffling indices one easily realizes that what we have done is transform (a) into (b) via

$$p_k T_o \rightarrow T_k \quad . \tag{3.10}$$

Thus we see here that the bizarre analytic continuation involved in Sompolinsky statistical mechanics solution as derived in section I is a way of mimicking the time structure involved in the dynamical scale hypothesis. This time structure that precisely reflects itself into broken symmetry in replica space.

Although on physical grounds one may prefer the above introduction of time indices (and time structure) to having "copies" (like in Orland) for which the introduction of structure is harder to comprehend, one is still left perplex with the conditions to impose on the length of time intervals that allow neglecting the couplings $\mathcal{K}$. Perhaps this means that one ought to think here in terms of TAP solution indices rather than "time" indices. (Remember that after bond averaging the Hamiltonian route can be recovered from TAP probability density with a canonical weight ; we return later on the relationship between time and TAP solutions).

IV. ON REPLICAS OR TIME : WHITE AVERAGE

(i) It has been pointed out recently[19,20] that when initial conditions at t_o are not equilibrium ones (canonical), long time effects are resulting due to persistent memory of the initial state. The difference in structure between canonical (D_c) and white (D_w) in the density matrices can be understood from the difference in the initial distribution. The effective local field (that adds to the external field in D_c)

$$X_\alpha = th^{-1} m_\alpha \tag{2.30}$$

becomes (in D_w)

$$Y_\alpha = th^{-1} m_\alpha - \beta J^2 \sum_\beta \chi_{\beta\alpha} m_\beta \quad . \tag{2.40}$$

Both D_c and D_w give identical results for the free energy (and derivatives, parameters q_k, Δ_k thereof). In D_c, both components of Sompolinsky effective field $h_s = J\sum_k (z_k \sqrt{q_k'} - \beta J \Delta_k' \mu_k)$ come from the $q_{\alpha\beta}$ ansatz. In D_w, $q_{\alpha\beta}$ provides the regular part $\sum_k z_k \sqrt{q_k'}$ and $\chi_{\alpha\beta}$ the anomalous part. When going from D_c to D_w the number of parameters of the ansatz for $q_{\alpha\beta}$ is halved (the other half being taken care of by $\chi_{\alpha\beta}$). We saw in (III) that introducing "time" and transferring the $q_{\alpha\beta}$ block structure to the "time" dependence $q^{\theta\theta'}$ was leaving a trivial (two-parameter) structure (3.9) with the n×n replica matrix. Proceeding in the same way here should altogether allow us to drop all replicas.

(ii) This is indeed the case[16]. Let us assume that we can write a long time evolution equation that reduces to TAP equations at infinite time and yields a unique solution (for given initial conditions) for $m_{j,\theta}$. Memory terms and/or noise that allow for lifting the degeneracy are supposed to be small(and neglected).This can only make sense if we show that without replicas we can recover the result previously obtained with replicas. Let us look at the main features. We build a probability law (now normalized to one) with, instead of (2.15,16),

$$P_W \equiv \prod_{\theta=1}^{To} \mathcal{P}_w\{m_\theta\} \tag{4.1}$$

$$1 = \int \prod_{j,\theta} dm_j^\theta \; P_W \; . \tag{4.2}$$

After bond averaging one recovers the density matrix D_W

$$D_W = \int \prod_\theta dy_\theta \frac{d\hat{m}_\theta}{2\pi} \exp \frac{1}{2}\beta^2 J^2 \sum_{\theta,\theta'} q_{\theta\theta'} \, i\hat{m}_\theta i\hat{m}_{\theta'} + i\hat{m}_\theta(\beta h - Y_\theta) \tag{4.3}$$

$$Y_\theta = th^{-1} m_\theta - \beta J^2 \sum_{\theta'} \chi_{\theta'\theta} \, m_{\theta'} \tag{4.4}$$

where $\theta=1,2,\ldots,T_o$ and no $n\to 0$ is involved, in contrast to (2.40,41). The ansatz $q_{\theta\theta'}$, $\chi_{\theta\theta'}$ are given by (2.43) where instead of the sequence $p_o \gg p_1 \gg \ldots \gg p_R$ we have the time scales

$$T_o \gg T_1 \gg \ldots \gg T_R \gg T_M \equiv 1 \quad . \tag{4.5}$$

The result is Sompolinsky's[13] dynamic description with magnetizations $\mu_k(z_o z_1 \ldots z_k)$ that are coarse grained magnetizations over time T_k , still dependent on the effective local fields z averaged over longer time intervals ($T_\ell > T_k, \ell<k$). A full account of the system needs magnetizations coarse grained on all scales, this because the quenched averaging introduces time correlation over a strictly infinite range of time leaving fluctuations survive at all scales. With the μ_k's one builds as in (1.10-13), the correlation and response that become (within an obvious dilation of z)

$$q[T_k] = \left\langle \left(\frac{1}{T_k}\sum_\theta m_\theta\right)^2 \right\rangle_W = \langle \mu_k^2 \rangle_{z_o \ldots z_k} = q_k \tag{4.6}$$

$$\beta^{-1}\chi[T_k] = \frac{1}{T_k}\left\langle \frac{\partial}{\partial z_k}\mu_k - \frac{\partial}{\partial z_{k+1}}\mu_{k+1}\right\rangle_{z_o \ldots z_k} = -\Delta_k'/T_k \; . \tag{4.7}$$

The factor T_k^{-1} is easily understood. The response is here the time response. In (1.13) it is the Fourier transform $\tilde{\chi}(\omega_k \sim T_k^{-1})$, which in terms of (4.7) then writes

$$\tilde{\chi}(\omega_k) \simeq \sum_{j=k}^{R} T_j \chi[T_j] \simeq \sum_{\theta=0}^{T_k} \chi[\theta] \tag{4.8}$$

a natural relation if one recalls that the Fourier weight $\exp -i\theta/T_k$ cuts off for θ in the time scales $T_j \gg T_k$, $j<k$.

(iii) The derivation of Sompolinsky's results with time indices instead of replicas is however more difficult. Indeed now fluctuations around saddle point contributions are no longer affected with a power proportional to n, nor are the off diagonal contributions of the Jacobian (as appearing in the last bracket of (2.47)) given a weight n since those helpful replicas are gone. We only quote a few salient points

a) Fluctuations around saddle points : writing $M=\mu_o(z_o)+\delta M$ and $\widetilde{M}=\delta\widetilde{M}$, one may notice that in (2.47) the $(\delta M)^2$ is proportional to $\langle i\hat{m}i\hat{m}\rangle_W = 0$. The fluctuations is then just the inverse coefficient of $i\delta\widetilde{M}\delta M$ i.e.

$$\left[1+\frac{\Delta'_o}{T_o}\left\langle \sum_{\ell_o} i\hat{m}_{\ell_o} \sum_{j_o} m_{j_o}\right\rangle_W\right]^{-1} = \left[1+\Delta'_o\left\langle\frac{\partial}{\partial z_o}\mu_o(z_o)\right\rangle_{z_o}\right]^{-1} \tag{4.9}$$

b) Jacobian : the off diagonal Jacobian piece reads now

$$\left[1+\frac{\Delta'_o}{T_o}\left\langle\sum_{j_o}\left(\partial\xi_{j_o}/\partial m_{j_o}\right)^{-1}\right\rangle_W\right] = 1+\Delta'_o\left\langle\frac{\partial}{\partial z_o}\mu_o(z_o)\right\rangle_{z_o} \tag{4.10}$$

and thus exactly cancels (4.9).

c) Corresponding cancellations occur at each step of the procedure. Fluctuations around the very first saddle point ($N\to\infty$) in $q,\hat{q},g,n$ and corresponding tilded variables concur with a contribution of order N^{-1} of the Jacobian to impose a relationship between correlations involving four elements of m or $\hat{m}$.

d) A point whose meaning escapes us : in order to have genuine saddle points one needs to have a number of saddle variables much smaller than the large parameter. On (2.47) the large parameter is (here) T_o, and there is one saddle variable, $1 \ll T_o$. In general a sufficient condition is

$$T_o/T_k \ll T_k \quad , \qquad (k=0,1,\ldots,R) \tag{4.11}$$

to get satisfied together with (4.5).

e) An unsatisfactory point is the fact that in (4.4) the index (here time) summation is unrestricted whereas a response in time should be a triangular matrix (due to retardation). In some sense $\chi[T_k]$ relates to triangular properties, since in (4.5) the matrix element is a difference. In fact, it can be shown[41], that with appropriate coarse graining taken on the time dependent TAP equations themselves, one recovers a triangular matrix (on the average).

f) The connection with Sompolinsky and Zippelius[48,13,35] approach is easy to comprehend in broad terms. Indeed, leaving aside the part of their Lagrangian relevant for short time dynamics, one is left with a part that contributes to long times ($T_R \le t \le T_o$) i.e.

$$\int dt\, dt' \left(\frac{1}{2}\, q(t;t')\, i\hat{m}_t i\hat{m}_{t'} + \beta^{-1}\chi(t;t')\, i\hat{m}_t m_{t'}\right)$$

as in (4.3,4). Here q and χ, correlation and response, depend on both variables t and t' (reducing to a single variable t-t' for canonical initial conditions). The block ansatz is then made on $q(t;t')$ and the Fourier transform $\widetilde{\chi}(\omega;\omega')$.

Note that if we ask here what is the standard order parameter Q, we can only define it now via time average

$$N^{-1}\sum_j \left\langle\left(\frac{1}{T_o}\sum_{\theta=1}^{T_o} m_{j,\theta}\right)^2\right\rangle_W = T_o^{-2}\sum_{\theta,\theta'} q_{\theta\theta'} = q(o) \;, \tag{4.6}$$

which is the answer proposed by Sompolinsky[13] and Sommers[42], distinct from (2.54-56). And not surprisingly since we are asking here a distinct question. We return on that in the end.

V. ON ORDER PARAMETERS

(i) Up until recently the only available interpretation for the order parameter $q(x)$ was Sompolinsky's proposal ($q(x)\equiv q[\tau_x], 1-q(1)+\Delta(x)\equiv\beta^{-1}\tilde{\chi}(\omega_x\sim\tau_x^{-1})$). Although a connection between state overlap i.e.

$$q^{ss'}\equiv N^{-1}\sum_j \overline{m_j^s m_j^{s'}} \tag{5.1}$$

and $q(x)$ was considered physically reasonable and heuristically used [29] it was only recently established by Parisi[17]. This author showed that the probability $W(q)$ for a state overlap equal to q

$$W(q)\equiv \sum_{s,s'} P(s)P(s')\delta(q-q^{ss'}) \tag{5.2}$$

where $P(s)$ is the canonical weight of (2.14), is related to $q(x)$, by the relationship

$$\frac{dx(q)}{dq} = W(q) \ . \tag{5.3}$$

Houghton, Jain and Young [19] had also suspected such a relation by noticing its validity on the first moment of (5.3). An independent derivation was given by Orland [40].

From there, the standard spin glass order parameter writes

$$Q = \int_{q(o)}^{q(1)} dq\, W(q)q = \int_0^1 dx\, q(x) \tag{5.4}$$

as in (2.56).

(ii) More recently Parisi[18] has conjectured a form for the (canonical) time evolution of the systems. It assumes that at a (large) time t, the system has visited all states with an overlap larger than $\hat{q}(t)$ (the minimum overlap). As t becomes infinite, $\hat{q}[t]\to q(o)$, and $Q[t]\to\int_0^1 dx\, q(x)$ with

$$Q[t] = t^{-2}N^{-1}\int_0^t dt\int_0^t dt' \sum_j \overline{m_j(t)m_j(t')} \tag{5.5}$$

The proposal is clearly meant to retain at a time t the fraction of saddle points associated with the corresponding time scales. In our terms this would mean retaining in the sum $\sum_{\alpha\neq\beta}$ contributing to Q (2.54) contributions of blocks up to size p_k (for a scale T_k) i.e. instead of (2.55)

$$Q[T_k] = \sum_{j=k}^{R}\Delta_j' + q_R \Rightarrow q(x)x + \int_x^1 dy\, q(y) \tag{5.6}$$

Rewriting in terms of $W(q)$ and $\hat{q}(t)$, we have

$$Q[t] = \hat{q}[t]\int_{q(o)}^{\hat{q}[t]} W(q)dq + \int_{\hat{q}[t]}^{q(1)} W(q)q\,dq \tag{5.7}$$

Parisi proposal retains last term of (5.7) (normalized by the coefficient of $\hat{q}[t]$ in the first term).

It has been further proposed[20] (but perhaps this was already meant in Parisi's) that $\hat{q}[t]$ be identified with Sompolinsky's correlation function i.e., using results of section (IV),

$$q[T_k] = < \left(\frac{1}{T_k} \sum_{\theta=1}^{T_k} m_\theta\right)^2 >_W = \frac{1}{T_k^2} \sum_{\theta,\theta'=1}^{T_k} q_{\theta\theta'} = q_k \quad . \tag{5.8}$$

Finally, taking the time derivative of (5.7), one obtains

$$\frac{dQ[t]}{dt} = \frac{dq[t]}{dt} \int_{q(x=0)}^{q[t]} W(q)dq = \frac{dq[t]}{dt} x[t] \tag{5.9}$$

This last expression is what replaces $dq[t]/dt$ in the fluctuation dissipation theorem at large times[35] (of the order of Sompolinsky's time scales) i.e. $Q[t]$ is built in to remove violations of the fluctuation dissipation theorem, which $Q \equiv Q[\infty]$ is already doing at equilibrium[15,43-45]. The picture then seems to be the following:

For a system away from equilibrium (here "white average") the evolution is to visit states as different as possible with at t an overlap, which remains minimal, $q(t)$ (no returns). For a system starting from equilibrium (canonical average) the evolution is to wander around exploring all states, and building $Q[t]$.

In the off equilibrium system the only questions that may be asked are time averages. In the equilibrium system, sums over (dominant) saddle points.

This situation leads to distinct answers for the standard order parameter : $Q=\int dx q(x)$ for the average over <u>all</u> blocks, $q[T_o]=q(x=0)$ for the average over the corresponding first $(p_o \times p_o)$ block with a weight favouring the largest block scanned (likewise on shorter time scales : $Q[T_k]$ for the average over all $(p_k \times p_k)$ blocks and $q[T_k]$ for the average over the corresponding first $(p_k \times p_k)$ block as it is clear in (5.8) with weight favouring the scale T_k).

It is clear that the off equilibrium description $q[T_k]$ at all time scales contains all the informations to build the equilibrium result Q by appropriate scanning and weighting of all saddle points. Conversely Parisi[17,18] has pointed out that the presence of a magnetic field varying with the copy would select among saddle points and bring a crossover from $Q = \int dx\, q(x)$ to $q(o)$. A time dependent magnetic field could precisely play this role. However, in the academic problem of strict statistical mechanics, there seem to be little room for such an effect.

* * *

The author gratefully acknowledges discussions with M. Gabay, T. Garel, I.Kondor, G. Parisi, H.J. Sommers and A.P. Young which in the last period helped him to develop and clarify the material presented here. He is thankful to H.J.Sommers for a critical reading of the manuscript. Finally he has come to share D.Sherrington's point that this contribution ought to be properly renamed "Replicas as a substitute for Dynamics".

REFERENCES

[1] van Hemmen J.L. and Palmer R.G., J.Phys. A12 (1979) 563
[2] Cragg D.M., Sherrington D. and Gabay M., Phys. Rev. 49 (1982) 158
[3] Elderfield D. and Sherrington D., J.Phys. A15 (1982) L513
[4] Gabay M., Garel T. and De Dominicis C., J.Phys. C15 (1982) 7165
[5] Elderfield D. and Sherrington D., J.Phys. A15 (1982) L437
[6] Elderfield D. and Sherrington D., Imperial College preprint (1983)
[7] Sherrington D., Proceedings of NATO ASI on Multicritial Phenomena (Geilo, Norway 1980)
[8] Sherrington D. and Kirkpatrick S., Phys.Rev.Lett. 35 (1975) 1792
[9] Edwards S.F. and Anderson P.W., J.Phys. F 5 (1975) 965
[10] De Almeida J.R.L. and Thouless D.J., J.Phys. A11 (1978) 983
[11] Pytte E. and Rudnick J., Phys.Rev. B19 (1979) 3603
[12] Parisi G., Phys.Rev.Lett. 43 (1979) 1754
[13] Parisi G., J.Phys. A13 (1980) 1101, 1887
Phil.Mag. 41 (1980) 677
[13] Sompolinsky H., Phys.Rev.Lett. 47 (1981) 935
[14] Thouless D.J., Anderson P.W. and Palmer R.G., Phil.Mag. 35 (1977) 593
[15] De Dominicis C. and A.P. Young, J.Phys. A16 (1983) 2063
[16] De Dominicis C., Gabay M., and Sommers H.J., in preparation
[17] Parisi G., Phys.Rev.Lett. 50 (1983) 1946
[18] Parisi G., Rome preprint 1983
[19] Houghton A., Jain S. and Young A.P., J.Phys. A16 (1983) L 375
Houghton A., Jain S. and Young A.P., Phys.Rev. B28 (1983) 2630
[20] De Dominicis C. and Young A.P., J.Phys. A16 (1983) L 641
[21] Bray A.J. and Moore M.A., Phys.Rev.Lett 41 (1978) 1068
J.Phys. C12 (1979) 79
[22] Blandin A., J.Physique 39 C6 (1978) 1499
[23] Blandin A., Gabay M. and Garel T., J.Phys. C13 (1980) 403
[24] Sommers H.J., Z.Phys. B31 (1978) 301
[25] Bray A.J. and Moore M.A., J.Phys. C13 (1980) 419
[26] De Dominicis C. and Garel T., J.Physique Lett. 41 (1980) L575
[27] Parisi G., J.Phys. A13 (1980) L115
Phys.Rep. 67 (1980) 97
[28] Sommers H.J., Z.Phys. B50 (1983) 97
[29] Dasgupta C. and Sompolinsky H., Phys.Rev. B27 (1983) 4511
[30] De Dominicis C., Gabay M. and Orland H., J.Physique 42 (1981) L523
[31] De Dominicis C., Gabay M. and Duplantier B., J.Phys. A15 (1982) L47
[32] Sommers H.J., private communication
[33] De Dominicis C. and Kondor I., Phys.Rev. B27 (1983) 606
[34] Kondor I. and De Dominicis C., J.Phys. A16 (1983) 73
[35] Sompolinsky H. and Zippelius A., Phys.Rev.Lett 50 (1983) 1297
[36] De Dominicis C., Phys. Rep. 67 (1980) 36
[37] De Dominicis C., Gabay M., Garel T. and Orland H., J.Physique 41 (1980) 923
[38] Bray A.J. and Moore M.A., J.Phys. C13 (1980) L469
[39] Tanaka F. and Edwards S.F., J.Phys. F10 (1980) 2471
[40] Orland H., J.Physique Lett. 44 (1983) L 673
[41] Sommers H.J., De Dominicis C. and Gabay M., Saclay preprint 1983
[42] Sommers H.J., J. Physique 43 (1982) L719
[43] Young A.P. and Kirkpatrick S., Phys.Rev. B25 (1982) 440
[44] Hertz J., J.Phys. C16 (1983) 1233
[45] Fisher K. and Hertz J., J.Phys. C16 (1983) to appear
[46] Bray A.J. and Moore M.A., J.Phys. A14 (1981) L371
[47] Sommers H.J., J.Phys. A16 (1983) 447
[48] Sompolinsky H., and Zippelius A., Phys.Rev.Lett 47 (1981) 359
Phys.Rev. B25 (1982) 6860.

THE INFINITE-RANGED m-VECTOR SPIN GLASS

by: David Sherrington
Physics Dept., Imperial College, London SW7 2BZ, U.K.

Abstract

The principal features of the mean field theory of a vector spin glass as characterised by an infinite-range model are discussed by analogy with the modern theory of the Ising case.

In 1975 Edwards and Anderson (EA) wrote a paper [1] which revolutionized the statistical mechanics of disordered spin systems. Stimulated by Cannella and Mydosh's observation of cusps in the a.c. magnetic susceptibilities of alloys such as AuFe and CuMn [2] and by the early mean field theory of Adkins and Rivier [3], EA identified the relevance of randomly competing exchange interactions, introduced a theoretically attractive model with symmetrically distributed bond disorder for which no conventional magnetic order is possible, employed a novel mathematical procedure (replication) for averaging the physically relevant but mathematically inconvenient free energy, introduced a new characteristic order parameter and, within a generalized mean field theory for the replicated and averaged system, used it to characterize a phase transition to a new type of ordered phase, the spin glass.

The infinite-range model of Sherrington and Kirkpatrick (SK) [4] was originally introduced as a model for which the mean field analysis of Edwards and Anderson would be exact. An Ising version was considered [4] since this was believed by the authors to contain the essential new physics and to be mathematically simpler. In fact, it led to the realization that the problem was more subtle than had first been anticipated since the "exact" solution gave a negative ground state entropy, impossible for an Ising system but the norm for continuous classical spins such as employed by EA. Subsequently there followed other demonstrations of the inadequacy of the detailed analyses of EA and SK [5,6], extendable to vector spins. The resolution of the inadequacies has proven to be both subtle and instructive. In this article it will be described briefly and qualitatively with particular regard to the case of general classical vector spins, but with discussion of the Ising situation first for orientation where appropriate.

The first point of note is that the subtleties alluded to above occur at the level of mean field theory as epitomized by the infinite-range model and are independent of fluctuations beyond mean field theory, which are being discussed in other articles in this book [7]. An adequate mean field theory of spin glasses is highly non-trivial, but is now largely understood, at least for the model discussed here which is characterized by the Hamiltonian

$$\mathcal{H} = - \sum_{(ij)} J_{ij} \underline{S}_i \cdot \underline{S}_j - \sum_i \underline{H} \cdot \underline{S}_i , \tag{1}$$

where the $\underline{S}$ are classical vector spins of dimension m and length $\sqrt{m}$, located at sites i,j, (ij) denoting a pair of such sites, $\underline{H}$ is an external magnetic field, and the J_{ij} are quenched exchange interactions distributed randomly with mean J_0/N and variance J^2/N, where N is the number of spins in the system, to be allowed to tend to infinity in the final thermodynamic limit. The scaling with N^{-1} ensures an exact mean field theory, suppressing thermodynamic fluctuations, while the offset J_0 allows an extension to include ferromagnetism [8]. Since all spins interact equivalently space dimensionality is irrelevant.

Let us first note that a high temperature series analysis demonstrates that a system described by (1) will exhibit phase transitions - the infinite-range and N scaling make such series summable to thermodynamic relevancy, each site summation yielding a contribution N, each single bond $(\beta J_o/N)$, each double bond $(\beta J)^2/N$, any higher multiple bond being irrelevant [9,10]. Ferromagnetic transitions, which are course possible only for H=o, are signalled, as usual, by a divergence of the conventional susceptibility

$$\chi^{(1)}_{\mu\nu} = \beta N^{-1} \sum_{ij} \overline{\langle S_{i\mu} S_{j\nu} \rangle} \tag{2}$$

where μ,ν label Cartesian coordinates and the bar designates disorder averaging (not strictly necessary in the thermodynamic limit). The more interesting spin glass transition is signalled only by a more complicated susceptibility function. For simplicity consider first the Ising case with H=o and in the absence of prior ferromagnetic instability, for which the relevant susceptibility is

$$\chi^{(2)} = \beta^2 N^{-1} \sum_{ij} \overline{\langle S_i S_j \rangle^2} \tag{3}$$

which describes both the response of the average square of the local magnetization to random local fields and the third order average magnetization response to an infinitesmal uniform field. For H=o $\chi^{(1)}$ diverges at $\beta J_o=1$, $\chi^{(2)}$ at $\beta J=1$, with only the lower $\beta(=(kT)^{-1})$ being relevant. For the m-vector case with H=o $\chi^{(2)}$ is generalized to carry coordinate labels as in (3) but the transition temperatures are unaltered. For H≠o $\chi^{(2)}$, as given by equation (3), is insufficient, the relevant susceptibility signalling spin-glass onset being

$$\chi^{(3)}_{\mu\nu} = \beta^{-2} N^{-1} \{\langle M_\mu M_\nu \rangle^2 - 2\langle M_\mu M_\nu \rangle \langle M_\mu \rangle \langle M_\nu \rangle + 3\langle M_\mu \rangle^2 \langle M_\nu \rangle^2\} \tag{4}$$

where $M_\mu = \sum_i S_{i\mu}$. To the best of the author's knowledge no explicit high temperature series analysis has been performed for this general case but from other analyses (to be reported below) it is known that H leads to a reduction in the temperature of onset of spin glass behaviour [6,11,12] and that divergence of $\chi^{(3)}$ is the signal [13 and unpublished].

The principal method of analysis of a spin-glass model such as (1) is the replica procedure [1,4] in which the free energy is averaged and analyzed as follows:

$$-\beta F = \overline{\ln Z} = \lim_{n\to o} (Z^n - 1)/n = \lim_{n\to o} (-\beta F(n)) \tag{5}$$

where

$$\exp(-n\beta F(n)) = \int [d\underline{\underline{Q}}] [d\underline{M}] \exp(-N(n\beta\hat{F}(n)/N) \quad (6)$$

$$(n\beta\hat{F}(n)/N) = \tfrac{1}{4}(\beta J)^2 \sum_{\alpha,\beta} (\underline{\underline{Q}}^{\alpha\beta})^2 + \tfrac{1}{2}(\beta J_o) \sum_{\alpha} (\underline{M}^{\alpha})^2 - \ln \mathrm{Tr} \exp(-\beta\tilde{H}) \quad (7)$$

$$-\beta\tilde{H} = \tfrac{1}{2}(\beta J)^2 \sum_{\alpha,\beta} \underline{S}^{\alpha}.\underline{\underline{Q}}^{\alpha\beta}.\underline{S}^{\beta} + \beta\sum_{\alpha}(J_o\underline{M}^{\alpha} + \underline{H}).\underline{S}^{\alpha} \quad (8)$$

and the explicit spins and trace are single site (but replicated). Assuming commutation of the limits $n\to o$, $N\to\infty$ [14], (7) is intensive and the integral in (6) is extremally dominated, so that

$$F(n) = \hat{F}(n) \quad (9)$$

where $\hat{F}(n)$ is now interpreted as its extremal. $\underline{\underline{Q}},\underline{M}$ are determined by the extremal relations

$$\underline{M}^{\alpha} = \mathrm{Tr} \exp(\underline{S}^{\alpha} \exp(-\beta\tilde{H}))/ \mathrm{Tr} \exp(-\beta\tilde{H}), \quad (10)$$

$$\underline{\underline{Q}}^{\alpha\beta} = \mathrm{Tr} \exp(\underline{S}^{\alpha}\underline{S}^{\beta}\exp(-\beta\tilde{H}))/\mathrm{Tr} \exp(-\beta\tilde{H}). \quad (11)$$

This defines the mean field theory for the model, believed to be exact. Further utilizing the symmetry of (1) in spin-space it is natural to parametrise $\underline{\underline{Q}}$,M in the form

$$\underline{\underline{Q}}^{\alpha\beta} = \delta^{\alpha\beta} \left[\begin{array}{c|ccc} 1+(m-1)y^{\alpha} & & o & \\ \hline & 1-y^{\alpha} & & o \\ o & & \ddots & \\ & o & & 1-y^{\alpha} \end{array}\right] + (1-\delta^{\alpha\beta}) \left[\begin{array}{cccc} q_{\parallel}^{\alpha\beta} & & & o \\ & q_{\perp}^{\alpha\beta} & & \\ & & \ddots & \\ o & & & q_{\perp}^{\alpha\beta} \end{array}\right] \quad (12)$$

$$\underline{M}^{\alpha} = [\tilde{m}^{\alpha}, o, \ldots, o] \quad (13)$$

where we have taken the external field direction as $\mu=1$. The parameters $q_{\parallel}^{\alpha\beta}$, $q_{\perp}^{\alpha\beta}$, y^{α}, $\tilde{m}^{\alpha}$ are associated respectively with longitudinal spin glass, transverse spin glass, quadrupolar and longitudinal ferromagnetic order.

The replica symmetric (RS) approximation solves the above equations subject to the ansatz that all replicas are equivalent, so that $\tilde{m}^{\alpha}$, y^{α} are independent of α while the $q^{\alpha\beta}$ are independent of the (different) pair ($\alpha\beta$). This is the analogue of the procedure originally followed by EA and SK for the ($m=3$, $J_o=H=o$) and $m=1$ cases respectively. Although inadequate for a description of the whole parameter space it does already indicate correctly a new feature, namely a transition associated with the transverse order parameter $q_{\perp}$ [11], even in the presence of finite $q_{\parallel}$ and $\tilde{m}$ induced by magnetic field or ferromagnetism. This feature is indicated by the surface

T_{GT} in Fig 1.

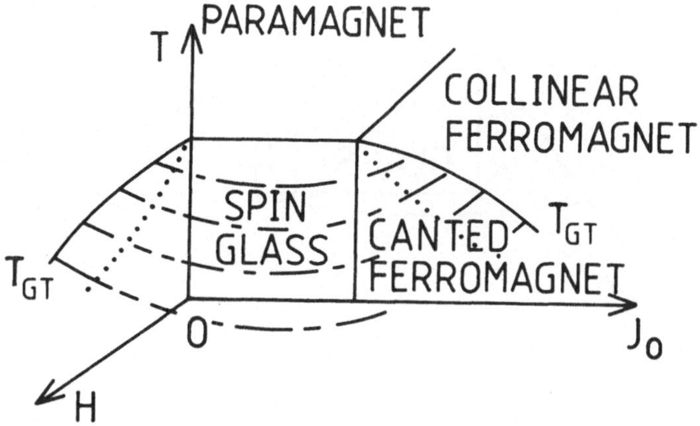

Fig 1. Schematic phase diagram for m>1 vector model. Phase transition lines are indicated by solid curves, surface by chains. Dotted lines designate crossovers.

Generally, the stability of a phase can be ascertained by examination of the matrix $S^{\alpha\beta,\gamma\delta}$ defined by

$$S^{\alpha\beta,\gamma\delta} = \beta N^{-1} \lim_{n\to o} \partial^2 (nF(n))/\partial X^{\alpha\beta}\partial X^{\gamma\delta} \tag{14}$$

where $X^{\alpha\beta} = (q_{\parallel}^{\alpha\beta}, q_{\perp}^{\alpha\beta}, y^{\alpha}\delta^{\alpha\beta}, \tilde{m}^{\alpha}\delta^{\alpha\beta})$. Zero eigenvalues signal transitions, changes of sign instabilities. Within the RS subspace the solutions mentioned above are stable (although the S-matrix could be employed as their indicator). More interesting are modes which break replica-symmetry. It is straightforward to demonstrate that above the surface T_{GT} all RS-breaking modes are stable. However, below this surface, which is only appropriate for m>1, the system not only acquires a finite $q_{\perp}$ order paramter but also is RS-unstable. For the Ising case (m=1) the only order parameters are $q_{\parallel}$ and $\tilde{m}$ but there is a RS-breaking transition associated with a surface connecting lines qualitatively like those indicated by dots in Fig 1, the phase diagram now being given in Fig 2.

When RS is unstable a new ansatz is required. Such an ansatz was provided for the Ising model by Parisi [15] and has been shown to be stable with respect to the fluctuations in replica-space characterized by the reduced S-matrix appropriate to that case [16]. The corresponding extension to the vector case is mathematically more complicated [17,18] but the characteristic features and their consequences can be indicated qualitatively.

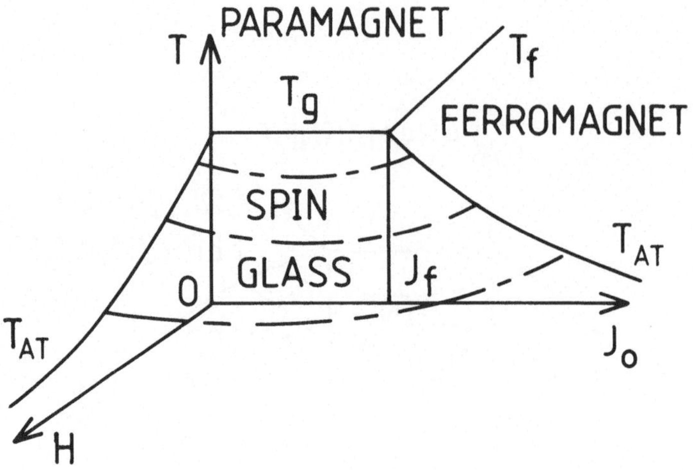

Fig 2. Schematic phase diagram for Ising case.

For orientation let us first consider the Ising case and take $J_o = o$. In replica-symmetric approximation all the off-diagonal $q^{\alpha\beta}$ ($q_{\parallel}^{\alpha\beta}$ above) are taken as equal to a single order parameter q. The Parisi ansatz [15] consists of taking a fractal decomposition to a continuous limit to produce instead an order parameter function q(x), where x lies in the interval (o,1); for a pictorial illustration see [19] or the author's other contribution to this volume [20]. Although originally mysterious, a simple physical interpretation of q(x), or strictly its inverse, has recently been demonstrated [21,22,23]. This is that dx/dq is the probability that the overlap between the physical extremal states (often called metastable states but strictly stable in the thermodynamic limit [24]) of the unaveraged system is q; the overlap $\tilde{q}$ is defined by

$$\tilde{q} = \sum_{ss'} P(S)P(S')N^{-1} \sum_i M_i^s M_i^{s'} \tag{15}$$

where the S label the states, P(S) their probability, and M_i^s the magnetization at site i in state S. Thus the local "equilibrium" susceptibility, corresponding to a Gibbs average over all states, is given by

$$\chi^{\ell}(\text{equil}) = \beta\,(1 - \int_o^1 q(x)dx) \tag{16}$$

while that for a system restricted to a single state S (but averaged over choices) is given by

$$\chi^{\ell}(\text{restricted}) = \beta(1-q(1)) \; . \tag{17}$$

χ(equil) and χ (restricted) are often identified with the experimental field-cooled (FC) and zero-field-cooled (ZFC) susceptibilities and below we use the appellations interchangeably.

Although it is possible to write a closed form expression for a free energy functional from which q(x) is in principle obtainable extremally, thereby yielding the physical free energy and other observables, it is more instructive for our present purposes to consider simply the general shape of q(x) which results from lowest non-trivial perturbation order for small reduced temperature $\tau=(T_g-T)/T_g$ where $T_g=J/k_B$ is the spin-glass transition temperature in zero field. This is illustrated in Fig 3 for small τ,H.

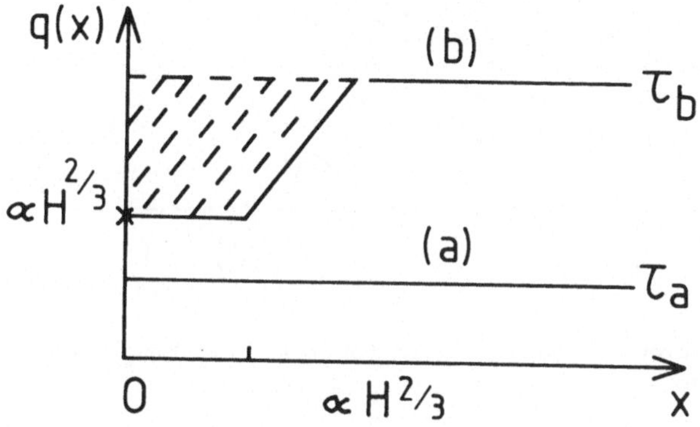

Fig. 3 Parisi function for Ising spin glass for (a) $T>T_{AT}$ (b) $T<T_{AT}$. The hatched area determines the anomaly Δ.

Curve (a) corresponds to a value of τ less than a critical value $\alpha H^{2/3}$. On (a) q(x) is constant and consequently the field-cooled and zero-field-cooled susceptibilities are equal. The constant value, τ_a, is given by

$$\tau_a=\tfrac{1}{2}\,\{\tau + (\tau^2 + 2H^2)^{\frac{1}{2}} + \ldots\ldots\}\tag{18}$$

so that for $\tau>>|H|$, $\tau_a \simeq \tau$, although for $|\tau|<<|H|$, $\tau_a \simeq |H|/\sqrt{2}$. By contrast curve (b) corresponds to $\tau>\tau_{AT}=\alpha H^{2/3}$, the Almeida-Thouless [6] reduced temperature at which the S-matrix becomes unstable. It is clear that in this case the two susceptibilities are unequal, with the hatched region, of area Δ, giving a measure of the difference or anomaly. For the purpose of qualitative extension to the vector case the transition from curve (a) to curve (b) should be visualized as arising in the following manner: As the temperature is lowered τ_a increases but q(o) is subject to a constraint to be no greater than $\alpha H^{2/3}$. As τ_a is increased past this value q(o) becomes pinned and with it q(x), small x, while q(x), large x, continues to have a rising plateau. In lowest order perturbation theory there is also a small x plateau and a region of unit slope separating them.

Let us now turn to the case of vector spins of dimension m>1, but again initially taking $J_o=0$. The quadrupolar order parameter y, as also $\tilde{m}$, remains replica-symmetric but Parisi extension is now required for both $q_{\parallel}(x)$, $q_{\perp}(x)$. Let us consider the behaviour of these functions as the temperature is reduced through the regions indicated in Fig. 4, again restricting discussion to small H. In this figure the solid

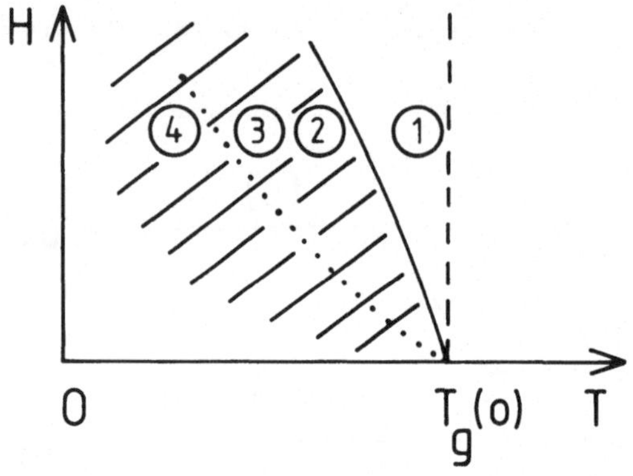

Fig 4. Schematic section of phase diagram for m>1. Hatching indicates the anomalous region. Numbers refer to regions discussed in the text.

line between regions 1 and 2 is the Gabay-Toulouse [11] line on which $q_{\perp}$ becomes non-zero - it is given by $\tau \sim H^2$. In region 1 $q_{\perp}(x)$ is everywhere zero and $q_{\parallel}(x)$ is constant and given by τ_a of eqn.(18). Region 2 refers to $\tau_{GT}<\tau<<H$ and its q(x) behaviour is illustrated in Fig 5(a). $q_{\perp}(x)$ behaves much as q(x) in an Ising model in zero-field (or more strictly as q(x) in an (m-1) dimensional system in zero field [25]); that is, it has essentially unit slope from x=o up to a plateau at $q=\tau'=\tau-\tau_{GT}$, and correspondingly an anomaly $\Delta_{\perp}\sim(\tau')^2$. $q_{\parallel}(x)$ becomes replica-symmetry broken simultaneously with $q_{\perp}(x)$ but its anomaly is weaker, $\Delta_{\parallel}\sim(\tau')^3$. The plateau value is essentially as in region 1, given by eqn(18), and of magnitude $\sim|H|/\sqrt{2}$. Region 3 refers to $H<<\tau<<H^{2/3}$ and is illustrated in Fig 5(b). It differs from region 2 principally in that now the plateau values of $q_{\parallel}(x)$, $q_{\perp}(x)$ are comparable, of order τ, each having risen relative to the values shown in Fig 5(a). The next interesting feature occurs for $\tau=\tau^*\sim H^{2/3}$, as indicated by the dotted curve in Fig 4. This arises because, as for the simpler Ising case, $q_{\parallel}(o)$ is constrained to a maximum value proportional to $H^{2/3}$. τ^* is the reduced temperature at which this maximum is attained. As the temperature is lowered further, τ increased, the large x plateau values of q(x) continue to rise but $q_{\parallel}(o)$ is pinned, as illustrated in Fig 5(c) which corresponds to region 4 of Fig 4. Thus, although the transverse anomaly continues to scale as $(\tau')^2$, or τ^2 to leading order for $\tau'>>H^2$, the longitudinal anomaly experiences a crossover from $(\tau')^3$ to $(\tau')^2$ as τ^* is traversed.

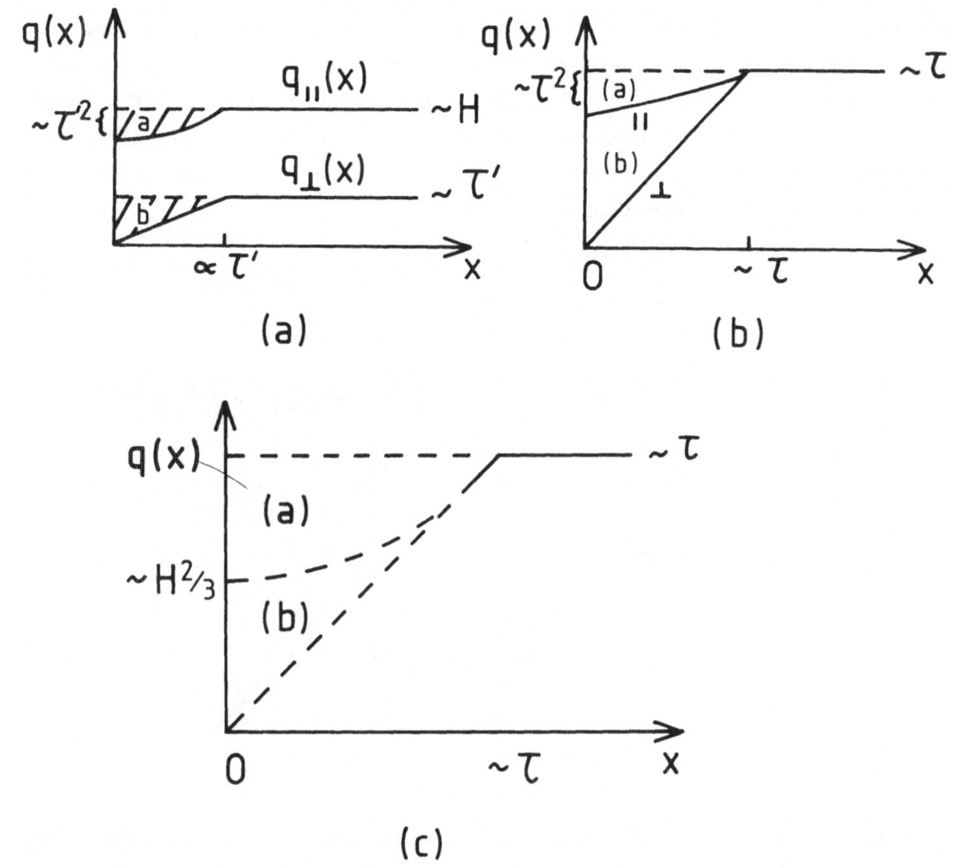

Fig 5 Schematic representation of the Parisi functions for a vector spin glass in an applied field. Figures (a), (b), (c) correspond to regions 2, 3, 4 of Fig 4. Regions (a), (b) indicate longitudinal and transverse anomalies.

The local susceptibilities follow directly from

$$\chi_{\lambda}^{\ell}(\text{equil}) = \beta^{-1}(Q_{\lambda}^{\alpha\alpha} - \int_{0}^{1} q_{\lambda}(x)dx), \tag{19}$$

$$\chi_{\lambda}^{\ell}(\text{restricted}) = \beta^{-1}(Q_{\lambda}^{\alpha\alpha} - q_{\lambda}(1)) \tag{20}$$

where $\lambda = \parallel$ or $\perp$ and, as given in eqn (12),

$$Q_{\parallel}^{\alpha\alpha} = 1 + (m-1)y, \tag{21}$$

$$Q_{\perp}^{\alpha\alpha} = 1 - y \tag{22}$$

Global susceptibilities for $H \neq 0$ depend upon non-local as well as local correlations. However, equilibrium values follow directly from $\tilde{m}(H)$. No direct evaluations of restricted global susceptibilities have been performed but an estimate for the longitudinal response may be made by assuming that the global and local longitudinal anomalies are identical. This leads to the predictions of Figs 6 and 7 for the Ising [26] and m>1 vector [27, 28] cases respectively.

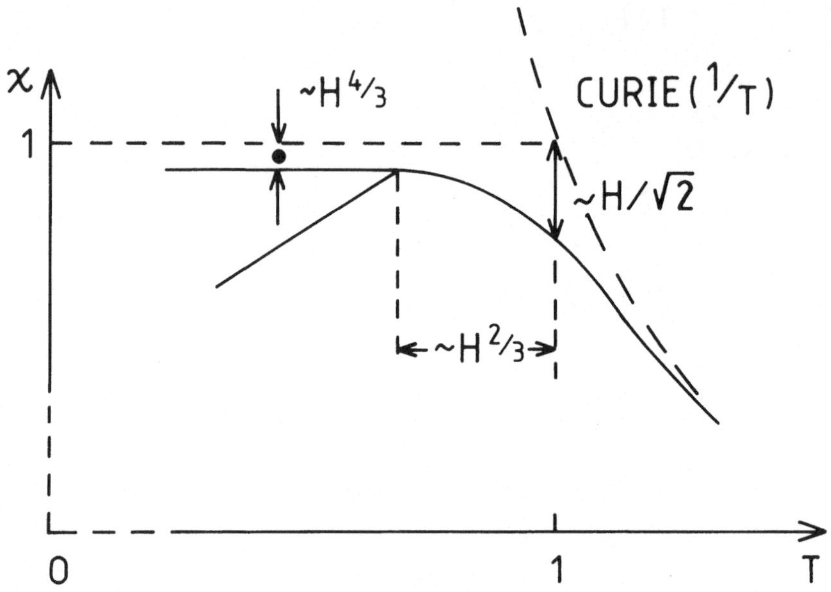

Fig 6 Schematic plot of global susceptibilities of an Ising spin glass. The upper curve corresponds to field-cooled, the lower to zero-field-cooled.

The behaviour when $J_o \neq 0$ can be deduced directly from the above by appropriate mapping [29]. The results are illustrated in Figs 1 and 2. The phases referred to as ferromagnets have $\tilde{m} \neq 0$ for $H \to 0$. A collinear ferromagnet has $q_\perp = 0$, a canted ferromagnet has $q_\perp \neq 0$. The surfaces T_{GT} and T_{AT} indicate the onset of replica-symmetry breaking anomalous behaviour. The line (isolated for H=0 only) separating the spin glass and ferromagnetic phases is determined by

$$J_o \chi^{\ell}(\text{equil}) = 1 \tag{23}$$

and is vertical since it can be proven rigorously [30] that the Parisi ansatz gives

$$\chi^{\ell}(\text{equil}) = J^{-1}; \text{ all } T<T_g \tag{24}$$

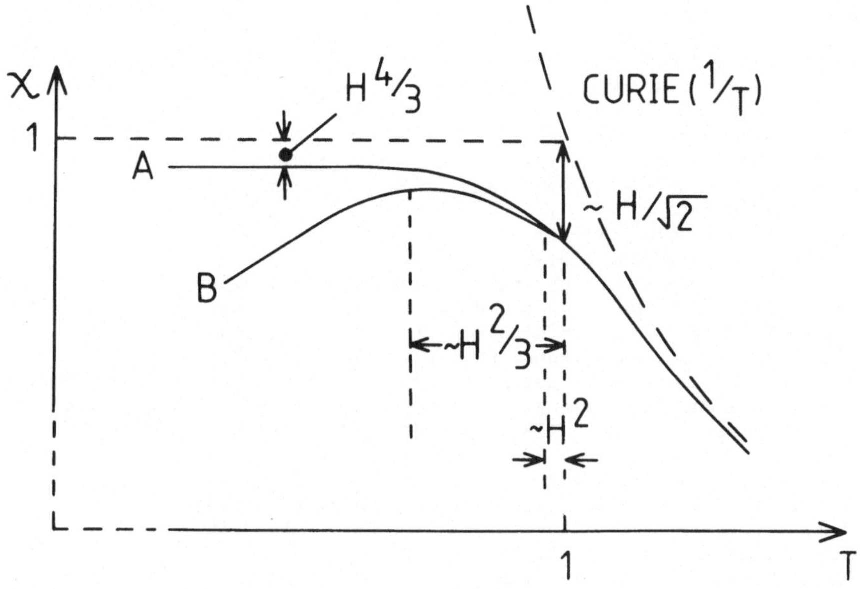

Fig 7 Corresponding schematic plot of global susceptibility of a vector spin glass.

Related analyses can be performed for anisotropic models but will not be discussed here - see instead references [31-33, 13].

Finally, we note that the above analysis is all for the infinite-range model, which corresponds to non-trivial mean field theory. Its theoretical inadequacy for short-range models is the subject of other authors in these proceedings [7] and will not be dwelt upon here. However, it is interesting to note that its predictions bear an uncanny resemblence to many experimental features of real systems [13, 28] .

References

1. S.F. Edwards and P.W. Anderson, J. Phys. F5, 965 (1975)
2. V. Cannella and J.A. Mydosh, Phys. Rev. B6, 4220 (1972)
3. K. Adkins and N.Y. Rivier, J. Physique C4-237 (1974)
4. D. Sherrington and S. Kirkpatrick, Phys. Rev. Lett. 35, 1792 (1975)
5. E. Pytte and J. Rudnick, Phys. Rev. B19, 3603 (1979)
6. R.J. de Almeida and D.J. Thouless, J. Phys. A11, 983 (1978)
7. See in particular:
   K. Binder, this volume (1983)
   A.P. Young, this volume (1983)
8. D. Sherrington and B.W. Southern, J. Phys F5, L49 (1975)
9. D.J. Thouless, P.W. Anderson and R.G. Palmer, Phil. Mag. 35, 593 (1977)
10. S. Kirkpatrick and D. Sherrington, Phys. Rev. B17, 4384 (1978)
11. M. Gabay and G. Toulouse, Phys. Rev. Lett. 47, 201 (1981)
12. D.M. Cragg, D. Sherrington and M. Gabay, Phys. Rev. Lett. 49 158 (1982)
13. D. Elderfield and D. Sherrington, to be published in J. Phys. C (1983)
14. J.L. van Hemmen and R.G. Palmer, J. Phys. A12, 563 (1979)
15. G. Parisi, Phys. Rev. Lett. 23, 1754 (1979)
16. C. de Dominicis and I. Kondor, Phys. Rev. B27, 606 (1983)
17. D. Elderfield and D. Sherrington, J. Phys. A15, L513 (1982)
18. M. Gabay, T. Garel and C. de Dominicis, J. Phys. C15, 7165 (1982)
19. B. Duplantier, J. Phys. A14, 283 (1981)
20. D. Sherrington, this volume (1983)
21. G. Parisi, Phys. Rev. Lett. 50, 1946 (1983)
22. A. Houghton, S. Jain and A.P. Young, J. Phys. C16, L375 (1983)
23. A.P. Young, this volume (1983)
24. N.D. MacKenzie and A.P. Young; Phys. Rev. Lett. 49, 301 (1982)
25. A.J. Bray and M.A. Moore, J. Phys C15, L301 (1982)
26. G. Parisi and G. Toulouse, J. Physique Lettr. 41, L361 (1980)
27. D. Elderfield and D. Sherrington, J. Phys. C15, L783 (1982)
28. D. Sherrington, D.M. Cragg, D. Elderfield and M. Gabay, J. Phys.Soc. Jpn. 52, 229 (1983)
29. G. Toulouse, J. Physique Letts. 41, L447 (1980)
30. H.J. Sommers, J. Phys. A16, 447 (1983)
31. D.M. Cragg and D. Sherrington, Phys. Rev. Lett. 16, 1190 (1982)
32. S.A. Roberts and A.J. Bray, J. Phys. C15, L527 (1982)
33. D. Elderfield and D. Sherrington, J. Phys. A15, L437 (1982)

Long-range Heisenberg Spin Glasses in a Magnetic Field: Theory and Experiment

U. Krey

Naturwiss. Fak. II - Physik, Universität Regensburg,
D-8400 Regensburg, F.R.G.

Abstract:

A survey is given of recent experimental and theoretical results concerning long-range Heisenberg spin glasses in a magnetic field, together with a tentative interpretation of certain formal aspects of the theory in physical terms.

I. Introduction

The last few years have seen considerable progress in the field of spin glass magnetism, both in theory, and in experiment /1/. Theoretically, the main problem is still to understand the strange properties of the mean-field, i.e. infinite-range, model /2/, particularly in the spin glass regime. This model, which definitely has a phase transition at a finite temperature T_c (whereas short-range spin glasses have $T_c = 0$, at least in zero magnetic field $\overline{H}$, see /3/, is certainly not a realistic description of the canonical spin glasses as Cu Mn where the exchange interaction $J_{\ell m}$ is $\sim \cos(2k_F r_{\ell m})/r_{\ell m}^3$, i.e. oscillatory and decaying. For spin-glass systems with a truly finite range interaction, e.g. the $Eu_xSr_{1-x}S$ system /4/, where the interaction reaches only first- and second-nearest neighbours /5/, it is still more far from reality. However, in spite of this fact there is an astonishing agreement, as we will see below, of many of the theoretical predictions for the infinite range model with experiments, particularly with experiments in a strong magnetic field after (field-) cooling down to the apparent freezing transition and below.

On the other hand, the spin glass properties are exciting already in the paramagnetic phase, where one observes nonlinear susceptibilities and nontrivial scaling behaviour, deviating strongly from the predictions of the infinite range model (see below). Thus, having an appparent agreement with mean field theory below the apparent transition, but strong deviations above, one would like to know how this is compatible.

In the present survey, spin glass properties in a magnetic field will be considered mainly for Heisenberg spin glasses, however for comparíson , sometimes also the Ising case must be considered.

Hithertoo, the theoretical results for the infinite-range model of a Heisenberg spin glass in a magnetic field have almost exclusively used the replica trick; the most comprehensive recent series of papers comes from D. Sherrington and coworkers /6/, who use both the replica-symmetric scheme, which is unstable below the transition /7/, and also the replica-breaking scheme of Parisi /8/, which is stable /9/, but leads to a continuous set of order parameters $q(x,T)$.

There, of course, the questions arise, how the formal aspects of these theories can be interpreted, particularly

(i) what replica-breaking means physically, and what is the meaning of the family of order parameters $q(x,T)$;

(ii) whether linear response theory does hold or not;

(iii) whether one is in thermodynamic equilibrium or not, or under which conditions;

(iv) whether spin glasses are ergodic or not, and how an apparent nonergodicity is related e.g. to $q(x,T)$; and finally, of course:

(v) how the apparent success of the mean field theory at low temperatures can be understood in view of the discrepancies above the apparent transition, and in view of the almost certain nonexistence /3/, of an equilibrium transition for spin glasses with short range interactions (and on the mean field level, as will be shown below, even RKKY spin glasses are short ranged).

These questions are intimately related to Sompolinsky's recent dymamical approach /10/, which introduces <u>two</u> continuous order parameter functions $q(\nu)$ and $\Delta(\nu)$, depending on a reciprocal time scale ν . Of these two functions, $\Delta(\nu)$ describes an <u>anomalous response.</u>

The existence of this anomalous response is sometimes believed to be typical for spin glasses. However, for an understanding of the spin glass phenomena it may be important to note, as will be shown below, that in the limit $\nu \to 0$ even for pure ferromagnets there is an anomalous response, which is related to the breaking of ergodicity in the low temperature phase, and which is given by $T \cdot (\chi_T - \chi_{is})$, where T is the temperature and χ_T and χ_{is} are the isothermal and isolated susceptibilities, respectively. Furthermore, it turns out that the anomalous response, i.e. $T \cdot (\chi_T - \chi_{is})$, vanishes if $M(T)$ is flat $(\partial M/\partial T = 0)$, which is just the case in the field-cooled state at low T.

The outline of the following chapters is as follows: In chapter II we present experimental results on Heisenberg spin glasses, with the intention, (i) to locate the facts which have to be explained by the theory, e.g. the existence of two apparent high-field spin glass phases instead of the usual single one, which would be observed in the Ising case; (ii) to stress the above-mentioned problem of apparent agreement (below the apparent transition) and striking disagreement (above T_f) between experiment and mean field theory; and (iii) to discuss the question to which extent the experimental observations, i.e. the field-dependence of the different transition lines, and on the other hand the nontrivial scaling behaviour observed in the paramagnetic phase, prove or disprove the existence of one or more sharp phase-transitions.

Then in chapter III, the theoretical aspects will be outlined. To explain the experiments we use a direct molecular field approach for quantum spins (i.e. we do not use the classical approximation, $s \gg 1$, for the spin quantum number s). Also the replica trick /2/ will be avoided, as well as the complications of introducing Onsager's reaction fields /11/. The reasons for this are (i) to keep things as simple as possible, and (ii) to point out a possible doorway leading from microscopic equations beyond the infinite range model, by including dynamics, and also spatial fluctuations of the order parameters and fields. This direct approach, which for the case of infinite range interaction and classical spins would correspond to the replica-symmetric solution of /2/, is exact for that case above the freezing temperature, where already many interesting phenomena happen, and it gives a sufficient description of the important observations even below T_f, see /6/ for a comparison.

Finally, in chapter IV certain formal aspects of improved theories as that of Parisi /8/, which is used in /2/ and /9/, and that of

Sompolinsky /11/, are interpreted in terms of jumps between metastable states, and the anomalous response appearing in these theories is related to the difference between isolated, isothermal and adiabatic susceptibilities appearing already in ferromagnetic systems.

Finally, it should be noted that we do not consider Heisenberg systems with anisotropies: A certain amount of anisotropy is probably necessary to stabilize the spin glass phase at $H = 0$, see /12/, but as long as the anisotropies are small compared to the isotropic exchange, the behaviour in high fields is <u>not</u> changed, see /6/.

II. Experimental Observations

II. 1) One or more Spin-Glass Phases?

In a number of Heisenberg spin glasses, in a high magnetic field, not one, but two apparent spin glass phases have been observed: One particular example is the $Cd_{1-x}Mn_xTe$ - system studied at the department of the present author /13/, for x = 0.54 and 0.63. In this system one has antiferromagnetic interactions between Mn neighbours, which have as usual the spin quantum number s = 5/2; however, since the spins are placed on a fcc - lattice, one has strong frustration effects, which together with the disorder leads to spin glass behaviour, as observed in a number of experiments /14-16/. However, whereas in /14-16/ the conventional spin glass characteristica, cusps of the susceptibility in low fields, rounded maxima of the specific heat, and irreversibilities, have been studied, the measurements in /13/ concentrated on the magnetization M(T,H) induced by strong magnetic fields, from 0.3 to 6.5 Tesla, at temperatures between 4 and 45 K. Actually, M(T,H) has been measured by Faraday rotation, but for comparison some results have also been obtained by a SQUID magnetometer.

In Fig.1 results are presented for a Mn - concentration of x = 0.54 for fields of H = 1 Tesla (Fig.1a) and 3 Tesla (Fig.1b), respectively: Note that in these two figures there are two different curves: The "virginal" (or non-field cooled) one, which has lower magnetization and where the temperature is increasing from low values, and the higher curve which is obtained by cooling the system down from high temperatures in the field. The results for virginal curve are of course irreversible, whereas the field cooled results are reversible, once the system has been cooled down in the field.

The striking observation in Fig.1 is the existence of two apparent transitions; an upper one at a temperature $T_2(H)$ separating the paramagnetic phase from a region, where M(T), for given H ist almost flat, and the hysteresis is small, and a lower transition, at $T_1(H)$, where the deviation between the virginal and the field-cooled magnetization becomes pronounced. In any case, however, I would like to stress at this place: In Fig.1 there is not only this pronounced hysteresis below T_1, but also a certain amount of hysteresis between T_1 and T_2, and some hysteresis exists even above T_2, if T_2 is defined as that temperature, where the field cooled magnetization has its local maximum, see Fig.1a.

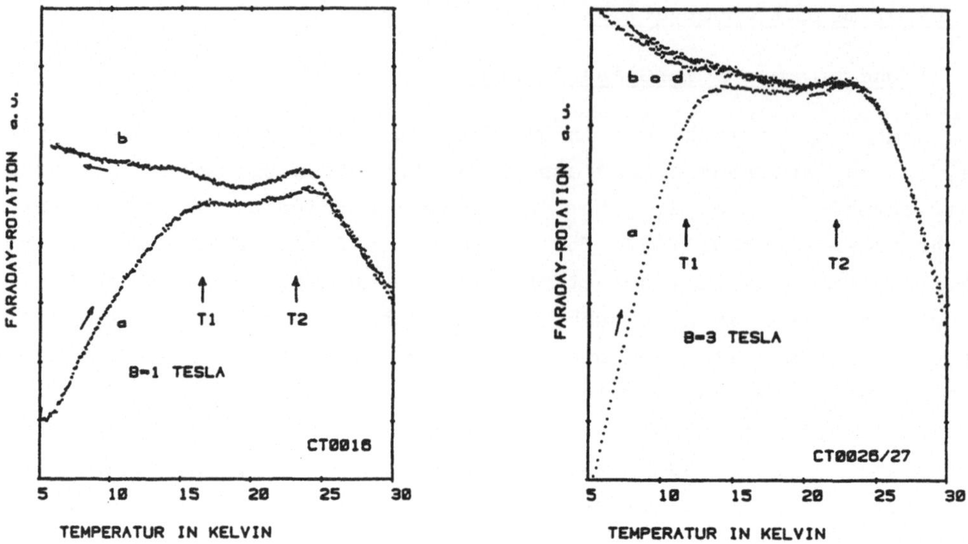

Fig. 1 For $Cd_{0.46}Mn_{0.54}Te$, the magnetization (in arbitrary units) as measured by the Faraday rotation is plotted over the temperature for fields of H = 1 Tesla and 3 Tesla, respectively. (From ref. /13/)

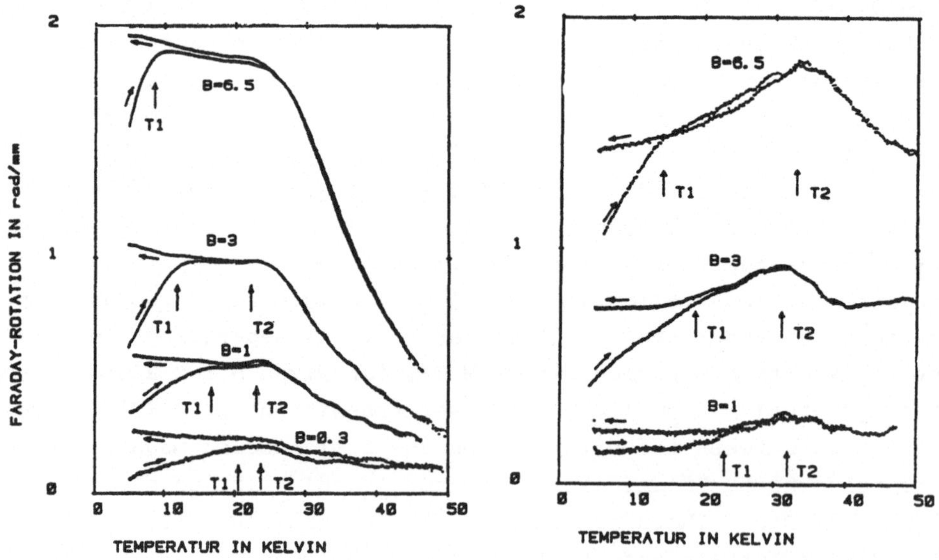

Fig. 2 Overview of results for different fields . Left part: $Cd_{0.46}Mn_{0.54}Te$. Right part: $Cd_{0.37}Mn_{0.63}Te$.

These results are confirmed by SQUID measurements, which I don't present, see /13/. Instead, Fig.2a,b gives an overview of the results obtained for the two different samples with $x = 0.54$ and $x = 0.63$, for $H = 0.3$, 1, 3, and 6.5 Tesla. Here the following observations are important: For $x = 0.63$ the magnetization is flat only below the lower temperature $T_1(H)$, whereas between $T_1(H)$ and $T_2(H)$, $M(T)$ decreases with decreasing temperature in contrast to the case of $x = 0.54$, where $M(T)$ is essentially flat already below the upper temperature $T_2(H)$. This different behaviour may be due to a partial coexistence and competition between spin glass behaviour and clusters, where antiferromagnetic behaviour dominates. In any case, however, one should keep in mind that in the region below $T_1(H)$, where the hysteresis becomes very strong, $M(T)$ is essentially flat in both cases.

Another notable fact is that in both systems $T_1(H)$ and $T_2(H)$ extrapolate to a common value T_o $(= T_1(o) = T_2(o))$, namely that temperature, where according to /14-16/ for $H = 0$ the spin glass freezing starts. However, from the lowest curve in Fig.2a, for $H = 0.3$ Tesla, it is quite obvious that neither $T_1(H)$ nor $T_2(H)$ represent sharp phase transitions, since there hysteresis is visible even up to $T \cong 2T_2$.

Of course, the experimentalists always like to give precise definitions of transition temperatures even in the case of rounded transitions; e.g. in the present case, see /13/, $T_2(H)$ was simply fixed by taking the local maximum of the field-cooled magnetization in the transition region, while in case of T_1 the extrapolation procedure is visualized in Fig.3, where the difference between the field-cooled and virginal magnetization has been plotted over T and taking $T_1(H)$ as that temperature, where the two linearly extrapolated branches of the curve cross each other.

Having defined $T_1(H)$ and $T_2(H)$ in this way, one obtains for $x = 0.54$ and $x = 0.63$ the "phase diagrams" presented in Fig.4a,b. In these figures, the black lines are not just a guide for the eye, but quantitative results of a theory presented in chapter III, see also /6/, /17/ and /18/. This theory predicts that in the mean field limit there should be a true transition at $T_2(H)$, given in the classical limit $s \gg 1$ by /19/

$$T_2(H) = T_o \cdot \{1 - \frac{1}{4} \frac{m^2+4m+2}{(m+2)^2} (\frac{g\mu_B s \cdot H}{k_B \sqrt{m}\, T_o})^2\} . \quad (1a)$$

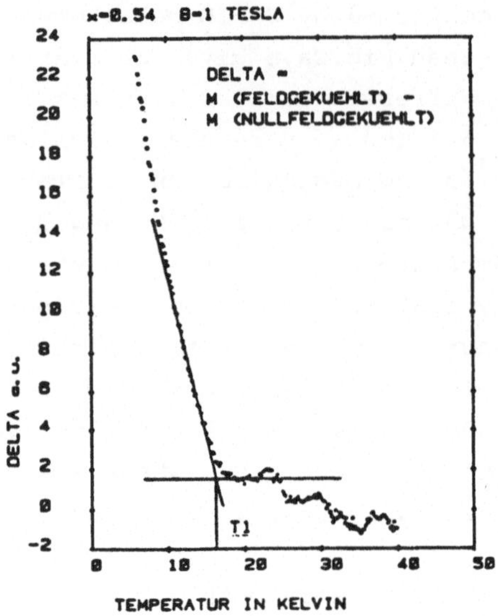

<u>Fig. 3</u> For $Cd_{0.46}Mn_{0.54}Te$ in a field of 1 Tesla it is shown, how the temperature $T_1(H)$ is determined from a plot of the difference of the field-cooled and the virginal magnetization over the temperature T .

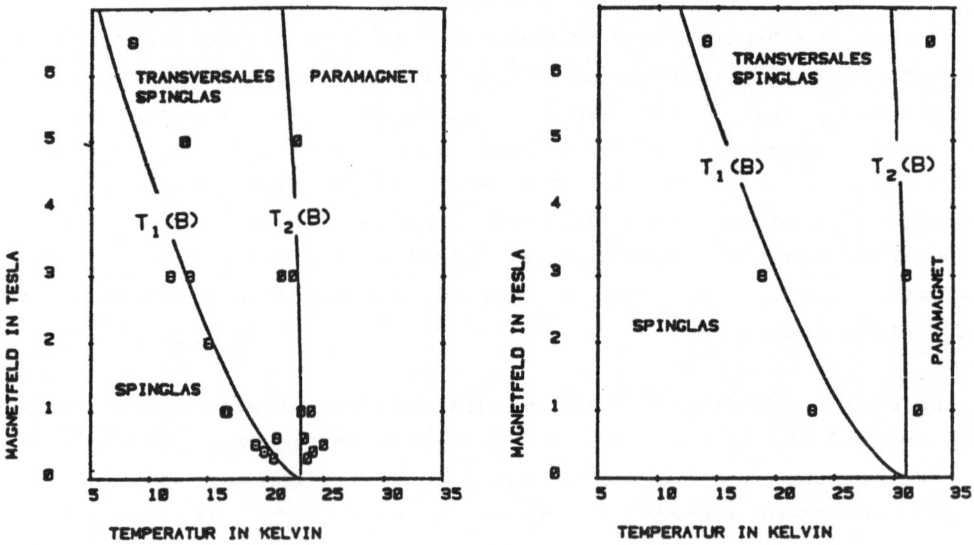

<u>Fig. 4</u> For $Cd_{0.46}Mn_{0.54}Te$ (left part) and $Cd_{0.37}Mn_{0.63}Te$, respectively the three apparent phases derived from Fig. 2 are presented. The lines $T_1(H)$ and $T_2(H)$ are theoretical results (see the text).

Here m (= 3) is the number of spin components, μ_B Bohr's magneton, $s(\gg 1)$ the spin quantum number, and k_B Boltzmann's constant.

Evaluating (1a) with $m = 3$ and $s = 5/2$
one has

$$T_2(H) = T_0 \cdot (1 - \frac{0.8646}{T_0^2} \cdot H^2) , \tag{1b}$$

where T is measured in Kelvins and H in units of 1 Tesla.

Below $T_2(H)$, according to the theory, there is transverse - spin glass freezing, i.e. if $\vec{\sigma}_\ell := <\vec{S}_\ell>_T$ is the thermal expectation value of the spin at site ℓ, and σ_ℓ^z and σ_ℓ^t are the longitudinal and transverse components, respectively, of $\vec{\sigma}_\ell$, referred to the direction of the external field $\vec{H}$ (= $H \cdot \hat{z}$), then above $T_2(H)$, σ_ℓ^t vanishes, while below $T_2(H)$ $\sigma_\ell^t \neq 0$. On averaging over the sites ℓ, one has then $<\sigma_\ell^t>_\ell = 0$, since there is no preferred average direction, but averaging, instead, the square one obtains a finite order-parameter $q^t := <(S_\ell^t)^2>_\ell \neq 0$.

According to the theory, see /6/, /17/ and /18/, hysteresis starts already as soon as $T<T_2$, in agreement with the observation; however the hysteresis is small, involving mainly the transverse components, until with further decreasing temperature around $T_1(H)$ a crossover to strong hysteresis of the longitudinal components happens. For $s \gg 1$, $T_1(H)$ would be given by

$$T_1(H) = T_0 \cdot \{1 - (\frac{m+2}{4})^{1/3} (\frac{g\mu_B \, s \, H}{k_B \sqrt{m} \, T_0})^{2/3} \} , \tag{2a}$$

or, with $m = 3$, with H in Tesla, T in Kelvin:

$$T_1(H) = T_0 \cdot \{1 - \frac{1.675}{T_0^{2/3}} \cdot H^{2/3}\} . \tag{2b}$$

Note that the $H^{2/3}$- behaviour is similar to the Almeida - Thouless instability in the Ising case, see /7/; there the line $T_2(H)$ is not present, whereas $T_1(H)$ is a sharp transition. In the Heisenberg case, however, $T_1(H)$ plays the role of a crossover line where the longitudinal susceptibility becomes strongly negative, see chapter III; thus it plays a similar role as the spinodal line in undercooled first-order phase transitions.

Let us finally stress again that in Fig.4 the agreement of $T_1(H)$ and $T_2(H)$ with the theoretical expressions (1b) and (2b) is quantitative, although experimentally not only T_1 but also T_2 (where in the mean field theory a sharp transition happens) represent only crossover temperatures (see above).

These observations have been confirmed by H. Kett in his thesis by a computer simulation for the $Cd_{0.45}\,Mn_{0.55}\,Te$ - system /13/; moreover: simular quantitative agreement with the theory has been found even for the canonical spin glasses Ag Mn with 2.6 % and 4 % of Mn by Chamberlin et al. /20/; and the $H^{2/3}$ - behaviour is also found in static observations on Pt Mn spin glasses /21/and in the work of Bouchiat and Monod /22/. Finally we mention in this context that in a number of recent dynamical experiments /23-25/ or computer-experiments /26/, a $H^{2/3}$ - behaviour of an apparent transition line has been found, where however the prefactors depended on the time scale of the measurement and partially differed considerably from the theoretical prediction (2a), in contrast to the above-mentioned results. We will return to this fact in chapter IV.

II. 2) Scaling Behaviour

Whereas the mean field-theory fits so well to the experiments at the freezing temperature $T_2(H)$ and below, this is not the case in the paramagnetic region: There several authors have observed scaling behaviour and strongly diverging nonlinear susceptibilities, which could be descripted by nontrivial critical exponents β and γ , deviating strongly from the mean field predictions, see below. Particulary surprising, at least on a first glance, is also the wide range of the scaling behaviour.

Perhaps the most impressing recent work in this connection is that of R. Omari et al. /26/. These authors have measured the magnetization M(T,H) of a canonical Cu Mn 1 % spin glass between 0 and 7 Teslas in a temperature range between 11 and 40 K (i.e. from 1.1 T_0 to 4 T_0, with T_0 = 10.05 ±0.05 K). In this work the authors concentrate on the singular part $M^{sing}(T,H)$ of the magnetization, which is given by

$$M^{sing}(T,H) = M(T,H) - \chi_0(T)\cdot H \,, \qquad (3)$$

where $\chi_0(T) \sim 1/T$ is the linear part of the susceptibility

(Curie law). From the mean-field theory one would predict (see chapter III) that $M^{sing}(T,H)$ follows a scaling behaviour given by

$$\frac{M^{sing}(T,H)}{H} = (\frac{T-T_o}{T})^{\beta} f[(\frac{g\mu_B H}{kT})^2 / (\frac{T-T_o}{T})^{\beta+\gamma}] \quad (4)$$

Here μ_B is Bohr's magnetic moment, g is the Landé factor and T_o the critical temperature for $H = 0$, which according to the mean field theory is given by

$$k_B T_o = \frac{s\cdot(s+1)}{3} \{<\sum_m J_{\ell m}{}^2>_\ell\}^{1/2}, \quad (5)$$

see chapter III, where $J_{\ell m}$ is the exchange integral appearing in the Heisenberg quantum Spin Hamiltionian

$$\mathbb{H} = -\frac{1}{2}\sum_{\ell,m} J_{\ell m}\vec{S}_\ell\cdot\vec{S}_m - g\mu_B H \sum_\ell S_\ell{}^z . \quad (6)$$

β and γ are critical exponents, which in the mean-field theory are given by $\gamma = \beta = 1$, and $f[x]$ is a scaling function, which according to the mean field theory should be universal, allowing for $x<<1$ an expansion of the form

$$f[x] = \sum_{n=1}^{\infty} C_n\cdot x^{2n} , \quad (7)$$

with $C_1 \neq 0$, while for $x>>1$ $f[x]$ should be $\sim x^\delta$, with $\delta = 1+\frac{\beta}{\gamma}$, such that for $x>>1$, or $T \to T_o$, the behaviour of $M^{sing}(H,T)$ could also be described by

$$\frac{M^{sing}(T,H)}{H} = H^{2/\delta}\cdot g\,[(\frac{T-T_o}{T}) / (\frac{g\mu_B H}{kT})^{2/\beta\delta}] , \quad (8)$$

where $g[y]$ should have a finite limit for $y \to 0$.

Here two remarks are in order:

(i) In contrast to conventional ferromagnetic systems, the scaling field is not H, but H^2;

(ii) The expansion is always in $(T-T_o)/T$ instead of $(T-T_o)/T_o$, which makes an important difference if one is not in the limit $(T-T_o)/T <<1$, as is the case in the present context.

In fact, both (i) and also (ii) are natural consequences of the molecular field theory of chapter III, and will be discussed later . At this place, however, we concentrate on the results, presented in Fig.5a,b.

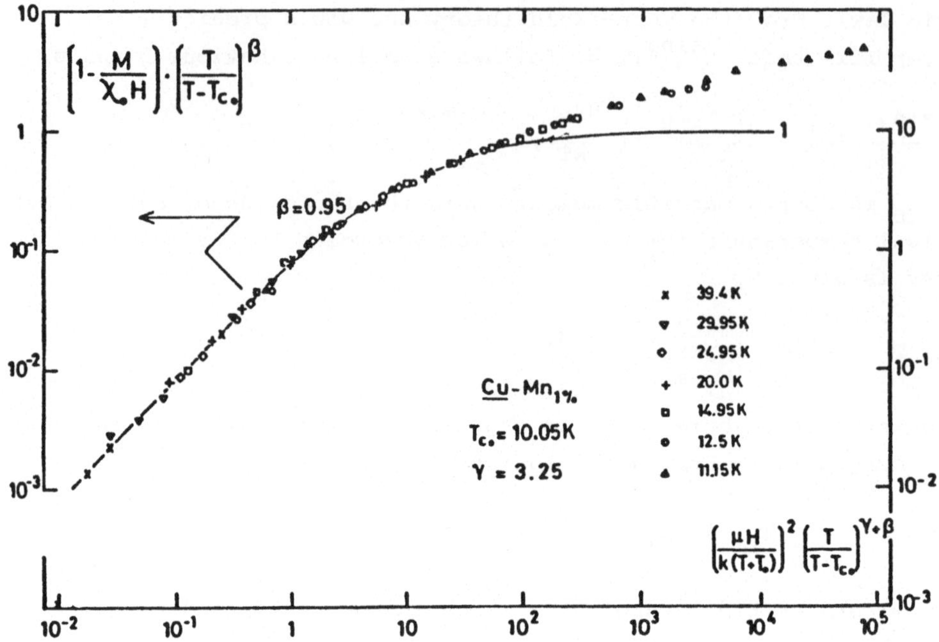

Fig. 5a For a Cu Mn 1% - spinglass the results for the magnetization M(T, H) for T between 11.15 K and 39.4 K ($1.1\ T_0 < T < 4\ T_0$) collapse onto one single curve (for H between 0 and 7 Tesla), if $(1-(M/\chi_0 H))(T/(T-T_0))^{\beta}$ is plotted over $(\mu H/k_B(T+T_b))^2(T/(T-T_0))^{\beta+\gamma}$, with $\beta = 0.95$, $\gamma = 3.25$, $T_0 = 10.05$ K and $T_b = -0.8$ K. (Taken from /27/). See the text for further details.

In Fig.5a, all the data obtained for $M(T,H)$ with H from 0 to 7 Tesla and T from $1.1\ T_o$ to $4\ T_o$ are collapsed onto one single curve, representing
$(-M^{sing}/H) \cdot (\frac{T}{T-T_o})^{\beta}$ over $(g\mu_B H/(T-T_p))^2 \cdot (T/(T-T_o))^{\beta+\gamma}$; here $T_p = -0.8$ K is a small "paramagnetic Curie temperature", which might reflect a small negative value of the average $\langle J_{\ell m}\rangle_\ell$, and $\beta = 0.95$ (± 20 %) , $\gamma = 3.25$ (± 3 %) are the best values, with respect to an optimal data collapsing.

With this result, expanding in Equ. (4) according to (7), one would obtain for $(-M^{sing}/H)$:

$$-M^{sing}/H = C_3 \left(\frac{T}{T-T_o}\right)^{\gamma} \left(\frac{g\mu_B H}{k_B(T-T_b)}\right)^2 + C_5 \cdot \left(\frac{T}{T-T_o}\right)^{\gamma+2\beta} \cdot \left(\frac{g\mu_B H}{k_B(T-T_b)}\right)^4 + \ldots \quad . \tag{9}$$

Thus, according to (9) one would obtain strongly divergent nonlinear susceptibilities, diverging $\sim (T-T_o)^{-\gamma}$ and $(T-T_o)^{-(\beta+2\gamma)}$, respectively; however, the formula (9) would also fit exactly to the high temperature Curie behaviour, if one assumes $T \gg T_o$, or $T/(T-T_o) \cong 1$, and $C_3 = (1/45)[(s+1/2)^4-(1/2)^4]$, $C_5 = -(2/945)[(s+1/2)^6-(1/2)^6]$,which are the expansion coefficients appearing in the Brillouin function for spin quantum number s .

In fact, assuming (somewhat unnecessarily) $s \gg 1$, and then replacing $g \cdot s \cdot \mu_B$ by $5.45\ \mu_B$ (which yielded the best fit for the linear susceptibility $\chi_o(T)$) , the authors obtained from (9) a second and independent fit leading to the same value of γ , but a smaller value of β , namely $\beta = 0.75$ or $\beta+2\gamma = 7.25$; instead of $\beta+2\gamma = 7.4$. However for this set, where the two-term-expression (9) worked best giving a fit of all the data, although $C_3 \cdot (\frac{T}{T-T_o})^{\gamma}$ varied over more than three orders of magnitude, and
$C_5 \cdot (\frac{T}{T-T_o})^{\gamma+2\beta}$ even over six magnitudes, the data collapsing according to (4) was worse; therefore in Fig.5a, the fit with $\beta = 0.75$, which can be found in /27/, has been omitted for greater clarity, leaving only the best fit with $\beta = 0.95$.

The authors of /27/ have also studied the field-dependence of $M(T,H)$ for $T = T_o$, for $Cu_{1-x}Mn_x$ samples with $x = 0.01$, 0.05, and 0.08 .

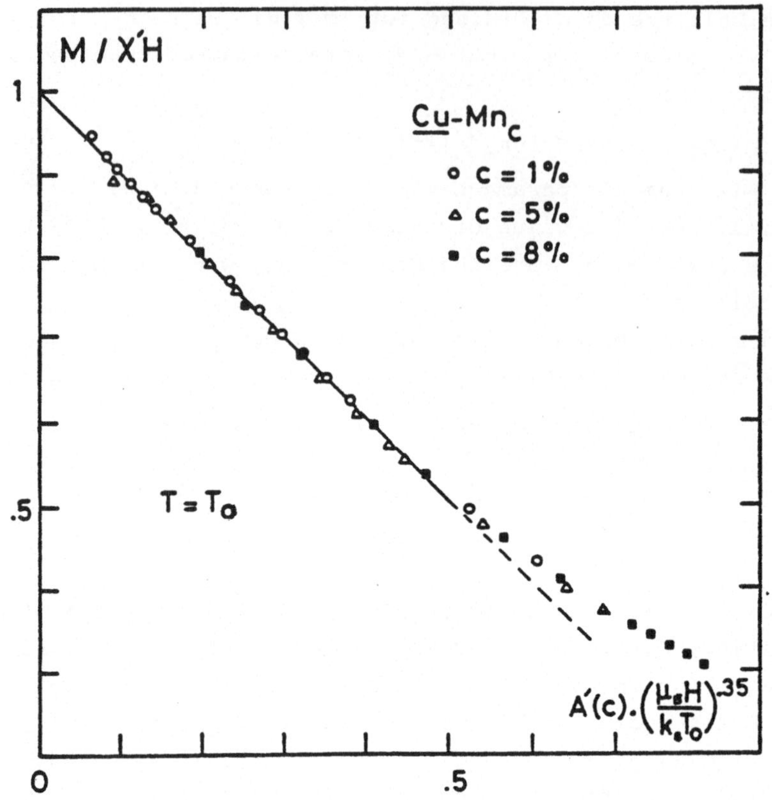

Fig. 5 b A power law $1 - (M/\chi_0 H) \sim H^{2/\delta}$, with $2/\delta \cong 0.35$, describes the magnetization M(H) at the temperature T_0 for three different Cu Mn alloys with 1%, 5%, and 8% Mn over a wide range of magnetic fields. In large fields there are some systematic deviations from the $H^{2/\delta}$ - behaviour. (Taken from /27/ .)

Results are presented in Fig.5b, which shows a power-law behaviour for not too large fields, with an effective exponent 0.35. According to (7) and (8), one would expect the exponent 0.35 appearing in Fig.5b to be identified with $2/\delta = 2\beta/(\beta+\gamma)$. However the best values of Fig.5a ($\gamma = 3.25$, $\beta = 0.95$) would lead instead to $2/\delta = 0.45$ ($\delta = 4.4$) . (The second set $\gamma = 3.25$, $\delta = 0.75$, determined directly from the nonlinear susceptibilities would better fit to Fig.5b, since then $2/\delta = 0.375$, on the other hand with this set the overall scaling along Fig.5a would be worse).

Actually, with the exponent $2/\delta = 0.45$ one can also get a good data-collapsing for $M(T_o,H)$ for all three Cu Mn samples, but instead of a power law one has in that case $M(T_o,H) / \chi_o H = \exp(-A\cdot H^{2/\delta})$, see Fig.12 in /27/. Thus there remains some ambiguity concerning the determination of β , and particularly, concerning the compatibility of exponents obtained above T_f and at T_o , respectively.

II. 3) Discussion of II.2 and II.1

In the following, the implications of the experimental results of the last two paragraphs will be discussed.

The first astonishing point , besides the fact of nontrivial scaling itself, is of course the wide range of temperatures and fields over which this behaviour can be observed. However, this can be explained quite naturally as follows:

According to Equ. (5), in a spin glass with a typical number Z of interacting neighbours of a given site, and with a typical magnitude $|J|$ of the exchange interactions, the freezing temperature should be given by

$$k_B T_o \cong \frac{s\cdot(s+1)}{3} \sqrt{Z}\cdot |J| \ , \tag{10a}$$

while for a ferromagnet (where $J>0$) the corresponding value would be given by

$$k_B T_c \cong \frac{s\cdot(s+1)}{3} Z\cdot |J| \ . \tag{10b}$$

Therefore one expects a relation $T_c \cong T_o^2/K$ between these temperatures.

The important consequence of this rough relation (which, besides, is well fullfilled in some experiments, where one has a double transition with a ferromagnetic phase lying above a spin glass phase, e.g. /28/, is that not just in the vicinity of T_o but already below $\sim T_o^2/K$ (i.e. below ~ 100 K (!) in the case of the experiments of /27/ presented in Fig.5) one would expect the strongly cooperative behaviour which usually leads to a description in terms of droplets, correlation lengths and scaling. This means theoretically that in this region renormalization would be a useful concept, even if the renormalization flows would not lead ultimately to a fixed point, or if there would be a crossover to a different fixed point as that which one would extrapolate from the apparent critical behaviour e.g. in the region between 1.1 T_o and 4 T_o of the experiments in /27/.

In fact such apparent transitions are well known from other fields; e.g. in the Anderson - localization problem in two dimension, a power law behaviour $r_o(W) \sim (W-W_o)^{-\nu}$ with $\nu \cong 0.8$ has been found /29/ for the localization lengths r_o as a function of the strength of disorder W for a wide range of r_o - values, although in this case there is an ultimate crossover, i.e. all states are localized even for $W < W_o$, and W_o is just a "crossover - value", where a crossover from exponential localization to weak localization takes place /30/.

Other examples in this context are apparent ferroelastic phase transitions at low temperatures, which are ultimately suppressed by quantum fluctuations /31/, or even many amorphous ferromagnets, which show both an unexceptionally wide range of scaling above T_c and unusually large effective γ - exponents /32/. In fact, Fähnle et al. have found a phenomenological description, by a "correlated molecular theory", for many of the experimental features both of amorphous ferromagnets /32/, and of spin glasses /33/.

In this connection it should finally be mentioned that K. Binder in a recent computer experiment /31/ on an Ising spin glass on a square lattice, with the usual Gaussian distribution of nearest-neighbour interaction has found a nice scaling behaviour of $M(T,H)$ above the apparent critical temperature, although for this system there is convincing evidence (see Morgenstern and Binder, /3/) that in this case, for $H = 0$, there is no true transition.

If the interpretation of the experiments of /27/ as pseudo-transition would be correct then also the apparent contradiction between the value of β and γ as determined from Fig.5a and the value of

$2/\delta = 0.35$ from Fig.5b would no longer be so unnatural, since one would no longer expect the usual identity of exponents above, at, or below the transition if there is no true transition. Also one might well expect that in completely different systems such as the Cu Mn spin glass systems as contrasted to the Gd Al system the "fractality" of the droplet landscape in the correlated region between T_o and T_o^2/K could be different, leading to different effective exponents, which means nonuniversality.

This would explain the fact that the exponents which have been found for the canonical Cu Mn or Ag Mn spin glasses by various authors /35-37/, namely $\gamma \approx 3.3$, $\beta \approx 1$, lead to δ - values around 4.3, while in /35/ for the Gd Al spin glass $\delta \approx 6$ has been found above T_o.

The authors of /35/ have also performed some measurements for a limited range of temperatures below T_o. For the Cu Mn system the experiment could be described by a result $M^{sing}/H \sim H^{2/\delta'} \cdot (T_o - T)^{-\gamma'}$, with $2/\delta' = 0.43$ and $\gamma' = 0.07$. This result is only compatible with (8), if $g(x) \sim |x|^{-\gamma'}$, and $2/\delta' = (2/\delta)\cdot(1+\gamma'/\beta)$ $(>2/\delta)$. However with $\delta \approx 4.15$ or $2/\delta \approx 0.48$ as obtained in /35/ one would obtain a contradiction, since experimentally $2/\delta'$ is smaller as $2/\delta$. This contradiction would also remain if we take for δ and β the results of /27/ which are more accurate.

Taken all these experimental observations, according to the opinion of the present author there is presently reason enough to interprete the observed scaling phenomenon as representing a pseudo-transition only.

Looking back at all the results of the last two paragraphs, one therefore finds a lot of fundamental problems: (1) Absence, probably, of true phase transitions, but (ii) an obvious presence of two different pseudo-transitions, where (iii) the behaviour in the two different "spin glass phases" fits well into the features of mean field theories, e.g. (iv) the roughly constant behaviour of the field cooled magnetization M(T) and particularly (v) the $H^{2/3}$- behaviour of the lower transition temperature $T_1(H)$, while on the other hand, (vi) the scaling behaviour above the higher pseudo-transition reminds strongly to true critical phenomena and is obviously beyond mean field theory: All these points would have to be explained simultaneously and quantitativly by the theory, which is far beyond what theory does at present.

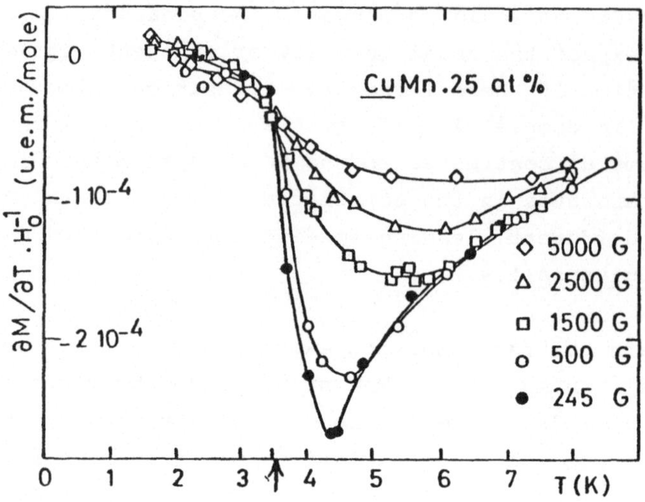

Fig. 6 a) (= upper part) : For Cu Mn .25% the magnetocaloric effect has been measured for various fields as a function of T

b) (= lower part) The characteristic temperatures derived from Fig. 6a are presented. H_c corresponds to the turning points, H_m to the minima in Fig. 6a. From /36/.

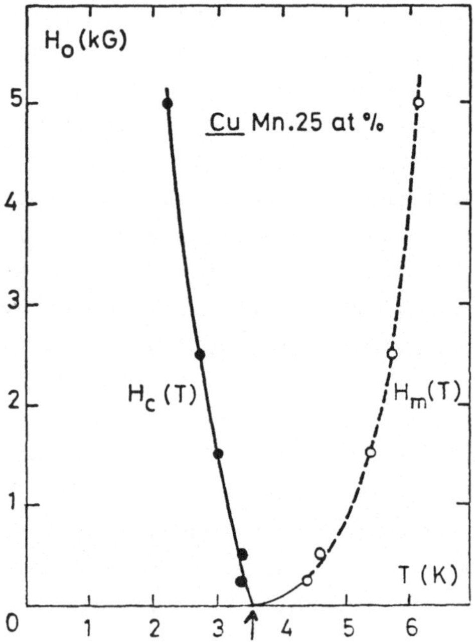

As a characteristic example which makes these points quite obvious, in Fig.6a and 6b results of Berton at al. /36/ are presented; these authors have studied the magnetocaloric effect as a function of T and H in the vicinity of the transition, for Cu Mn 0.25 %. In Fig.6a, the results for $(\partial M/\partial T)/H$ are presented over T, for H between 245 and 5000 Oersteds. From this figure, the authors derive two characteristic temperatures, a lower one, $T_c(H)$, given by the turning point of the curve, to the left of the minimum, and a higher characteristic temperature, $T_m(H)$, given by the minimum itself. Fig.6b presents $T_c(H)$ and $T_m(H)$ over H. Here $T_c(H)$ shows once more the Almeida-Thouless behaviour, $T_c(H) = T_o - \text{const.} \cdot H^{2/3}$, while $T_m(H)$ follows a relation $T_m(H) = T_o - \text{const.} \cdot H^{2/(\gamma+2)}$, with $\gamma \approx 3.5$.

III. Theory

III. 1) Basic Equations

The simplest theoretical approach for a magnetic system is usually a mean-field theory, which then should be extended to take into account fluctuations and critical phenomena /38/. In this sense, one might start with the following Equ. (11) for the expectation value $\vec{\sigma}_\ell$ of the the spin at site ℓ in a given field $\vec{h}_\ell = g\mu_B \cdot \vec{H}_\ell$, where $\vec{H}_\ell$ may comprise both the external field $H \cdot \hat{z}$ and random fields or thermal noise fields (however, unless explicitly stated, only the external field is taken into account below!). The starting point is:

$$\frac{1}{\Gamma_o} \frac{d\vec{\sigma}_\ell}{dt} + \vec{\sigma}_\ell = (\vec{h}_\ell + \sum_m J_\ell \vec{\sigma}_m) \cdot$$

$$\cdot \{ a_1(T) - a_3(T) (\vec{h}_\ell + \sum_m J_{\ell m} \vec{\sigma}_m)^2 + a_5(T) \cdot (\ldots)^4 - \ldots \} . \tag{11}$$

Here the coefficient functions

$$a_1(T) = \frac{s \cdot (s+1)}{3 k_B T} , \tag{12a}$$

$$a_3(T) = \frac{1}{45} \cdot \left(\frac{1}{k_B T}\right)^3 \cdot [(s+\tfrac{1}{2})^4 - (\tfrac{1}{2})^4] , \tag{12b}$$

$$a_5(T) = \frac{2}{945} \cdot \left(\frac{1}{k_B T}\right)^5 \cdot [(s+\tfrac{1}{2})^6 - (\tfrac{1}{2})^6] \tag{12c}$$

arise from the expansion of the Brillouin function for given spin quantum number s.

On the l.h.s. of (11) the term $\Gamma_o^{-1} \, d\vec{\sigma}_\ell/dt$ describes the approach to equilibrium, where precessional terms have been neglected for simplicity; Γ_o is a microscopic relaxation frequency.

Now in a spin glass one may assume that on average the bonds $J_{\ell m}$ have a Gaussian distribution with

$$\langle J_{\ell m} \rangle_\ell = 0 , \tag{13a}$$

$$\langle J_{\ell m_1} J_{\ell m_2} \rangle_\ell = \langle J_{\ell m_1}^2 \rangle_\ell \, \delta_{m_1, m_2} , \tag{13b}$$

$$<J_{\ell m_1} \; J_{\ell m_2} \; J_{\ell m_3} \; J_{\ell m_4} \; \sigma_{m_1}{}^{i} \; \sigma_{m_2}{}^{i} \; \sigma_{m_3}{}^{j} \; \sigma_{m_4}{}^{j}>_\ell$$

$$= <J_{\ell m_1}{}^2>_\ell \; \{\cdot <J_{\ell m_3}{}^2>_\ell \; \delta_{m_1,m_2} \; \delta_{m_3,m_4} + <J_{\ell m_2}{}^2>_\ell \cdot \delta_{ij} \cdot$$

$$\cdot \; [\delta_{m_1,m_3} \; \delta_{m_2,m_4} + \delta_{m_1,m_4} \circ \delta_{m_2,m_3}]\} \; q_i q_j \; , \tag{13c}$$

i.e. generally all cumulants with an odd number of exchange vanish (or are very small) while the even cumulants do not vanish. The "order parameters" q_i $(i = x,y,z)$ in (13c) are given by

$$q_i := <(\sigma_\ell{}^i)^2>_\ell \; . \tag{14}$$

Here $< \; >_\ell$ means an average over the sites ℓ ; note that we do not necessarily deal with a so-called random-bond problem, since the properties described by the Eqs. (13) can simply result from site-randomness, too. Also we do not assume that the range of interaction is infinite, instead, what we have in mind is a mean field approximation, with some sketched improvements, for real systems with interactions of finite range.

At a certain point below, we even need the cumulants with six instead of four exchange integrals, the corresponding equation (13d) would be rather lengthy and has therefore been omitted.

Now, by carefully separating out the longitudinal and transverse parts in (12), by squaring the equations and averaging afterwards, one would obtain equations for q_x, q_z, and for the longitudinal magnetization per site σ $(= <\sigma_\ell{}^z>_\ell)$, namely:

$$- \tau_o q_x + \frac{2a_3(T)}{a_1(T)} \left(\frac{T_o}{T}\right)^2 \cdot h^2 q_x$$

$$+ \frac{2a_3}{a_1{}^3} \cdot \left(\frac{T_o}{T}\right)^4 \cdot q_x \cdot (3q_x + q_y + q_z)$$

$$+ \xi_o{}^2 \nabla^2 q_x + \frac{2\dot{q}_x}{\Gamma_o} + \frac{\dot{q}_x{}^2}{\Gamma_o{}^2} = <(\delta h_x)^2> a_1{}^2 \; , \tag{15a}$$

$$- \tau_o q_z - h^2 a_1{}^2 + 2\frac{a_3}{a_1} \left(\frac{T_o}{T}\right)^2 \cdot h^2 (6q_z + q_x + q_y)$$

$$+ \frac{2a_3}{a_1^3} \cdot (\frac{T_o}{T})^4 \cdot q_z \cdot [3q_z + q_x + q_y]$$

$$+ \xi_o^2 \nabla^2 q_x + \frac{2\dot{q}_x}{\Gamma_o} + \frac{\dot{q}_x^2}{\Gamma_o^2} = <(\delta h_z)^2> a_1^2 , \tag{15b}$$

$$\frac{\dot{\sigma}}{\Gamma_o} + \sigma = ha_1 \cdot \{1 - \frac{a_3}{a_1^3} \cdot (\frac{T_o}{T})^2 \cdot (3q_z + q_x + q_y) + \ldots\} . \tag{15c}$$

We have omitted the equation for q_y which is similar to (15a). Furthermore in Eqs. (15), τ_o is given by

$$\tau_o := (\frac{T_o}{T})^2 - 1 , \tag{16a}$$

with T_o from Equ. (5), while

$$\xi_o^2 := \frac{1}{2} \cdot \frac{<\sum_m J_{\ell m}^2 (x_\ell - x_m)^2>_\ell}{<\sum_m J_{\ell m}^2>_\ell} . \tag{16b}$$

Finally, δh_x and δh_z are random fields, partially quenched fields and partially thermal fields, with zero average, but non-zero variance, while h corresponds to the constant external field.

In the following, at first the mean-field equilibrium solutions of (15) will be discussed, which means that time derivatives, gradient terms and random fields will be neglected.

One should note at this place that according to the equations (15), as long as the quantities $q_x = <(S_\ell^x)^2>$ etc. do not vary appreciably from site to site, even the canonical R K K Y spin glasses as Cu Mn can be considered as short ranged, since the relevant quantities in Eqs. (15) are the <u>squares</u> $J_{\ell m}^2$ of the exchange interactions, which decay $\propto 1/|\vec{r}_\ell - \vec{r}_m|^6$, and not $J_{\ell m}$ itself.

III. 2) Mean-Field Approximation for Heisenberg Spin Glasses with $H \neq 0$

Under the above-mentioned assumptions $\delta h_x = \delta h_y = \delta h_z \equiv 0$, $\nabla^2 \equiv 0$, $\partial/\partial t \equiv 0$, the Eqs.(15) can easily be solved: From (15b) it follows, that there will always be a field-induced finite value of q_z, which for $T \gg T_o$ would be given by $h^2 a_1^2/|\tau_o|$; on the other hand, for $q_x (=q_y)$ there exists always the trivial solution $q_x \equiv 0$ of (15a); a non-trivial solution can only arise if T is sufficiently small. To obtain the corresponding critical temperature, which will be called $T_2(h)$ (compare chapt. I.1), one has to solve at first (15b) for q_z, with $q_x = q_y \equiv 0$; then, substituting the result into (15a) one obtains, after a somewhat lengthy, but straightforward calculation:

$$\tau_o(h) = \left(\frac{T_o}{T}\right)^2 \cdot \left\{\frac{a_3}{2a_1} \cdot h^2 + \sqrt{\frac{a_3 h^2}{a_1} + \frac{a_3^{\,2}}{4a_1^{\,2}} h^4}\right\}. \tag{17a}$$

Note, that this formula leads to different results in the classical limit $(s >> 1)$ and in the quantum case:

In the classical case $a_3/a_1 \cong s^2/(15\ k_B T_o)^2$, and for $g \cdot s \cdot \mu_B \cdot H/k_B T_o >> 1$ $(s \to \infty)$ the second term under the square root dominates. Therefore, with $T_o \tau_o \simeq 2(T_o - T_2(H))$ one has in the classical case

$$T_2(H) = T_o \cdot \left\{1 - \frac{1}{30}\left(\frac{g\mu_B H}{k_B T_o} \cdot s\right)^2 + \ldots\right\}, \tag{17b}$$

while for general s:

$$T_2(H) = T_o \cdot \left\{1 - \frac{1}{2}\left(\frac{1}{15}\right)^{1/2} \frac{g\mu_B H}{k_B T_o} \cdot \left[\frac{(s+\frac{1}{2})^4 - (\frac{1}{2})^4}{s \cdot (s+1)}\right]^{1/2} + \ldots\right\}. \tag{17c}$$

Thus, for quantum spins, in contrast to the classical case treated in /6/ and /17,18/, the deviation of $T_2(H)$ from T_o starts with a linear term in H ; however the experiments are certainly not accurate enough, at present, to give a clear proof of this behaviour.

After having determined $T_2(H)$, the critical temperature for the onset of transverse ordering $(q_x = q_y \neq 0$ für $T < T_2(H))$, we discuss some properties of the mean field solution in greater detail.

In Fig.7 a qualitative phase diagram is given, where three phases are distinguished : The phase III denotes the paramagnetic phase, which is

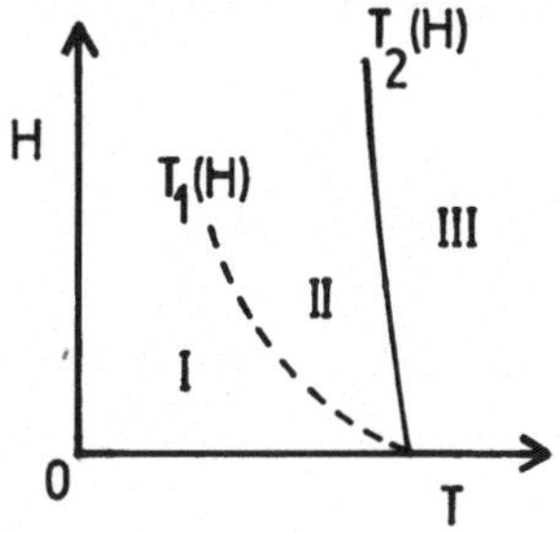

Fig. 7 Qualitative Phase Diagram

separated from the spin glass phase by the black line $T_2(H)$ which in the framework of the mean-field theory represents a true transition. The spin glass phase itself is separated into two different regions I and II by the dashed line $T_1(H)$, which has the meaning of a crossover temperature, see below!

In the paramagnetic region, according to the Eqs. (15), one has $q_x = q_y \equiv 0$, but $q_z \cong h^2 a_1{}^2/|\tau_0|$, leading with (15c) to

$$\sigma(T,H) = a_1(T) \cdot h \cdot \{ 1 - \frac{3a_3}{a_1} (\frac{T_0}{T})^2 \frac{h^2}{|\tau_0|} + \ldots\} . \qquad (18)$$

Here the first term yields the usual uncorrected Curie behaviour, while the second term describes a divergent nonlinear susceptibility diverging for $\tau_0 \to 0$ with a critical exponent $\gamma = 1$. Also one can see explicitly at this place those properties of the scaling Ansatz, which have been used in chapt. II, Eqs. (7-9).

On the other hand, for $\tau_0 = 0$ $(T=T_0)$ one obtains

$q_z \simeq (a_1{}^5/6a_3)^{1/2} |h|$, and

$$\sigma = a_1(T_0) \cdot h - (\frac{3}{2} a_3 a_1)^{1/2} h \cdot |h| + \ldots . \qquad (19)$$

If one enters the spin glass phase but stays in region II far enough from the dashed line one can neglect q_x and finds that σ is still increasing as T decreases, however with decreasing slope $|d\sigma/dT|$. On the other hand, in region III, near the axis $H = 0$, one has $q_z \cong q_x + \frac{h^2 a_1{}^2}{\tau_0} + \ldots$; as long as $h^2 a_1{}^2/\tau_0 \ll q_z$, q_x itself is given by

$$q_x = \frac{a_1{}^3}{10\, a_3} (\frac{T}{T_0})^4 \tau_0 \cdot [1 + \frac{7}{20} (\frac{T}{T_0})^2 \cdot (1 + \frac{2a_1 a_5}{a_3{}^2}) \cdot \tau_0 + \ldots] . \qquad (20)$$

Surprisingly it turns out that σ does not depend on a_5 to this order; namely

$$\sigma = h a_1(T) \{1 - \frac{T_0}{2} (\frac{T}{T_0})^2 (1 + \frac{7}{20} \tau_0) - \frac{3a_3}{a_1} (\frac{T_0}{T})^2 \cdot \frac{h^2}{\tau_0} + \ldots\} , \qquad (21a)$$

which by a careful expansion of τ_0, $a_1(T)$ and (T/T_0) in terms of the quantity $\varepsilon = (T_0 - T)/T_0$ yields

$$\sigma = h \cdot \frac{s\cdot(s+1)}{3k_BT_o} \cdot \{1 - \frac{\varepsilon^2}{5} - \frac{3a_3}{2a_1} \cdot \frac{h^2}{\varepsilon} + \ldots\} \,. \qquad (21b)$$

Therefore, σ should decrease for $T \to 0$, and should possess a maximum around

$$\varepsilon = \varepsilon^* = (\frac{15}{4}\frac{a_3}{a_1} \cdot h^2)^{1/3} \equiv \{\frac{1}{4} \cdot \frac{(s+\frac{1}{2})^4-(\frac{1}{2})^4}{s(s+1)} \cdot (\frac{g\mu_BH}{k_BT_o})^2\}^{1/3} \qquad (22a)$$

For $\varepsilon = \varepsilon^*$ the bracket in (21b) deviates from 1 by an amount $\sim h^{4/3}$.

In (22a) one has the notorious $H^{2/3}$ behaviour, for $T_o-T_1(H)$, which corresponds in case of the Ising model to the Almeida-Thouless instability line /7/, where the instability against breaking of the replica symmetry breaking happens. In fact, in the Ising case, the result corresponding to (21b) can simply be obtained by putting $s = 1/2$ and replacing $\varepsilon^2/5$ (= $\varepsilon^2/(m+2)$, where m is the number of spin components), by $\varepsilon^2/3$; then instead of (22b) one would obtain

$$\varepsilon^* = \{\frac{3}{16}(\frac{g\mu_BH}{k_BT_o})^2\}^{1/3}, \qquad (22b)$$

which is the exact result of (7), with $g = 2$.

In constrast to the Ising case, however, where ε^* defines a true phase transition, which is the only one for that case, in the present Heisenberg case the temperature $T_1(H) = T_o\cdot(1 - \varepsilon^*(H))$ is only a characteristic temperature corresponding to a crossover: In the vicinity of ε^* the longitudinal spin-glass susceptibility becomes strongly unstable. This is shown in the following section:

III. 3) The Instability Line $T_1(H)$

In the context of replica formalism, /2/, /6/, /8/, /17/, /18/, as already mentioned, the instability breaks the replica-symmetry, and it is not easy to see what this means. Fortunately the present direct approach, by avoiding the replica trick, sheds some light on the possible interpretation:

It is useful to start from the wellknown fact that for a ferromagnet, in a Gaussian or mean-field approximation, the local magnetic susceptibility χ_ℓ is given by

$$\chi_\ell = \frac{\delta\langle S_\ell^z\rangle_T}{\delta h_\ell} \equiv \frac{1}{k_B T} \cdot \frac{\langle (S_\ell^z)^2\rangle_T - \langle S_\ell^z\rangle_T^2}{1 - \frac{T_c}{T} \cdot \frac{[\langle (S_\ell^z)^2\rangle_T - \langle S_\ell^z\rangle_T^2]}{\langle (S_\ell^z)^2\rangle_T}} \qquad (23a)$$

Here the expectation values are taken in the mean-field approximation (independent sites) and T_c is the mean-field value for the Curie temperature.

Now according to the philosophy and derivation of Eqs. (15a-c) the corresponding relation for the local spin-glass susceptibility χ_ℓ^{SG} is

$$\chi_\ell^{SG} = \frac{\delta\langle\langle S_\ell^z\rangle_T^2\rangle_\ell}{\langle(\delta h_\ell^z)^2\rangle_\ell}$$

$$= \frac{1}{(k_B T)^2} \cdot \frac{\langle\langle (S_\ell^z - \langle S_\ell^z\rangle_T^2\rangle_\ell}{1 - (\frac{T_o}{T})^2 \cdot \frac{\langle\langle (S_\ell^z - \langle S_\ell^z\rangle_T)^2\rangle_T^2\rangle_\ell}{\langle\langle (S_\ell^z)^2\rangle_T^2\rangle_\ell}} \qquad (23b)$$

Thermodynamic stability requires $\chi^{SG} \geq 0$ which implies for the denominator of (23b):

$$(\frac{T_o}{T})^2 \; [1 - \frac{2q_z}{\frac{s(s+1)}{3}} + \frac{\langle\langle S_\ell^z\rangle_T^4\rangle_\ell}{\frac{s^2(s+1)^2}{9}}] < 1 \; . \qquad (24)$$

Here we have assumed that we are in region I of Fig,7, near the $H = 0$ axis.

But $\langle(\sigma_\ell^z)^4\rangle_\ell \equiv a_1^4 \langle (\sum_m J_{\ell m}\sigma_m)^4\rangle_\ell \equiv 5(\frac{T_o}{T})^4 \cdot q_z^2$, therefore the stability condition is
$(\frac{T_o}{T})^2 \cdot [1 - \tilde{q}_z]^2 + 4\tilde{q}_z^2 < 1$, where $\tilde{q}_z := q_z/[s(s+1)/3]$, and with our results for q_z one obtains after some calculation a critical temperature $\varepsilon^{**}(H) = 1 - \frac{T_1(H)}{T_o}$ for the instability, given by

$$\varepsilon^{**} = \{\frac{s(s+1)}{3} \cdot (\frac{g\mu_B H}{k_B T_o})^2\}^{1/3} \; / \; \{\frac{4\; s^4(s+1)^4}{[(s+\frac{1}{2})^4 - (\frac{1}{2})^4]^2} - \frac{2}{5}\}^{1/3} \; . \qquad (25)$$

For $s \gg 1$ ε^{**} simplifies to $(\frac{5}{54})^{1/3} (\frac{g\mu_B s}{k_B T_o})^{2/3}$, which is by a factor $\sim 5/7$ smaller than the expression given by Equ. (22a) ,

$$\varepsilon^* \simeq (\frac{1}{4})^{1/3} (\frac{q\mu_B H}{k_B T_0})^{2/3} .$$

In any case however one should stress at this place that in contrast to the ferromagnet the present results, which would correspond to the replica symmetric solution within the replica formalism, see /6/, are an approximation even for the case of infinite range interactions. The exact solution for this case is probably given by calculations within the Parisi replica breaking formalism , see /6/. In that calculation it is shown that $T_1(H)$ is only a crossover temperature; the reason is that replica symmetry will be broken already below $T_2(H)$. In fact, an instability for the transverse spin glass susceptibility can be derived in a similar way as for the longitudinal treated above, and one can easily show that the transverse spin glass susceptibility would become negative already below $T_2(H)$.

III. 4) Collection of Results

The results of the present chapter are visualized in Fig.8, which is taken from Sherrington et al. /6/. In this figure, which presents the susceptibility over the temperature, for a finite field, the following points are stressed:

(i) The susceptibility deviates from the Curie-behaviour already at rather high temperatures;

(ii) At T_0 the deviation is proportional to H ;

(ii) The critical temperature $T_2(H)$ deviates from T_0 by a small negative amount which would be proportional to H^2 for classical spins, $(g\mu_B sH/k_B T) >> 1$, but $\sim H$ in the quantum case;

(iv) below T_0 there is a difference between the virginal magnetization (B) and the field cooled magnetization; this difference becomes pronounced below a crossover temperature $T_1(H)$ which deviates from T_0 by an amount $\sim H^{2/3}$.

Finally from the Parisi replica breaking scheme it follows, see /6/, that at temperatures below $T_1(H)$ the field cooled magnetization $M(T,H)$ is essentially constant, and the susceptibility deviates from the value which it would have at T_0 according to the Curie behaviour by an amount $\sim H^{4/3}$, in agreement with Equ. (22a).

What is the reason for the difference between the field-cooled and the virginal magnetization? With this question we come to the last

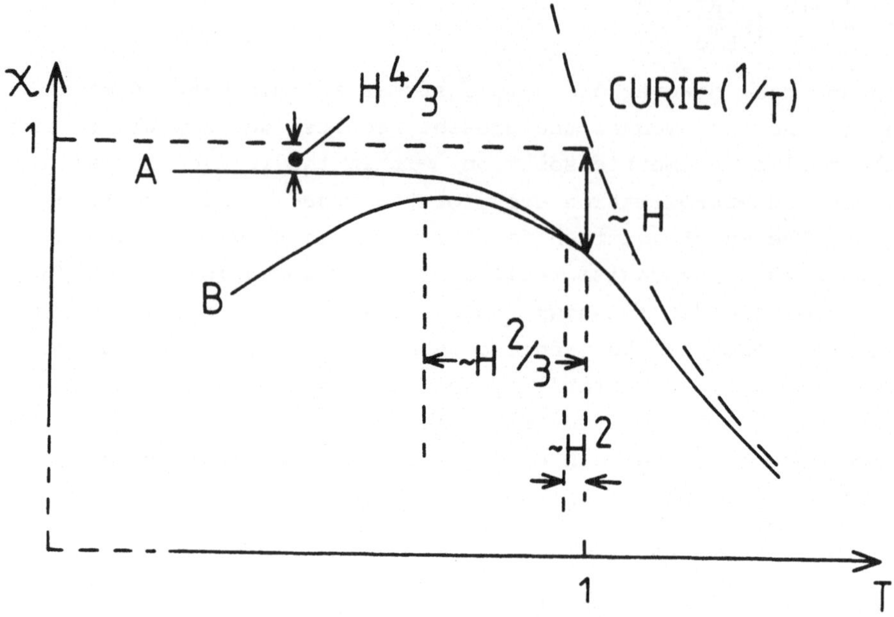

<u>Fig. 8</u> The susceptibility is presented over the reduced temperature T (= T/T_o , with T_o given by Equ. (5)). This qualitative sketch, which is taken from Sherrington et al., ref /6/, presents all the typical features derived by the theory , particularly the typical $H^{2/3}$- and $H^{4/3}$ - dependences, the H^2-decrease of the critical temperature and so on. For a detailed description see the text.
Curve A represents the field-cooled longitudinal susceptibility, while curve B corresponds to the virginal susceptibility.

chapter, where we try to interprete, as far as possible, some formal aspects of improved theories in terms of more pictorial concepts, and where we also try to stress a possible analogy between the anomalous response observed in spin glasses and the more conventional difference between adiabatic and isothermal susceptibilities.

IV. Formal Aspects of Improved Theories and an Interpretation

IV. 1) Introduction: Metastable States and Barkhausen Jumps

According to experience gained from experiments, particularly from computer experiments /34,39/, the essential point for an understanding of the spin glass phenomena seems to be the presence of an enormous number of metastable states with energy barriers depending on T which become extremely large below T_o /39/. These metastable states are more or less loosely related to the famous concepts of frustration /40/, two-level centers /41/, and spin clusters, as explained in somewhat greater detail by the present author in a former publication, see /1/. In fact, even for one-dimensional Ising-spin glass models with Gaussian nearest-neighbour interaction, where no true frustration exists, precise numerical simulations /42,43/ have found essentially the same dynamical behaviour as in conventional three-dimensional spin glasses. Particularly, in /43/ it was found that at low temperatures the decay of the remanent magnetization could be described by a formula

$$M(t) = M(o) \cdot \int_0^\infty dE\, \rho(E) \exp(-t/\tau(E)) \ , \tag{26}$$

where $\rho(E)$ was the distribution of energy barriers of the metastable states and $\tau(E) = \tau_o \cdot \exp(E/k_BT)$ the corresponding relaxation time. From the observed decay of $M(t)$, which for $T \ll T_o$ was logarithmic over a wide range of times, $\rho(E)$ could be calculated, and turned out to be <u>flat</u> over an extremely wide range of energies, and practically not to depend on T.

Now, if a free energy $f_\alpha(\sigma)$ can be assigned to the different metastable states α, then the corresponding quasi-equilibrium value of σ would be determined by the thermodynamic relation $\partial f_\alpha/\partial\sigma = h$, as presented in Fig.9a, and if h becomes larger than the local maximum of $\partial f/\partial\sigma$ next to the equilibrium point in Fig.9a, then (at the latest) there will be a "Barkhausen jump" leading to a new state with larger σ (see Fig.9b), since among attainable different states with the same h and roughly the same value of f_α the system prefers that one which minimizes the free enthalpy $g = f - h\sigma$.

These arguments are similar to those, which are used to derive the conventional spinodal instability for a ferromagnet below T_c, where after reversing the external field the magnetization must eventually

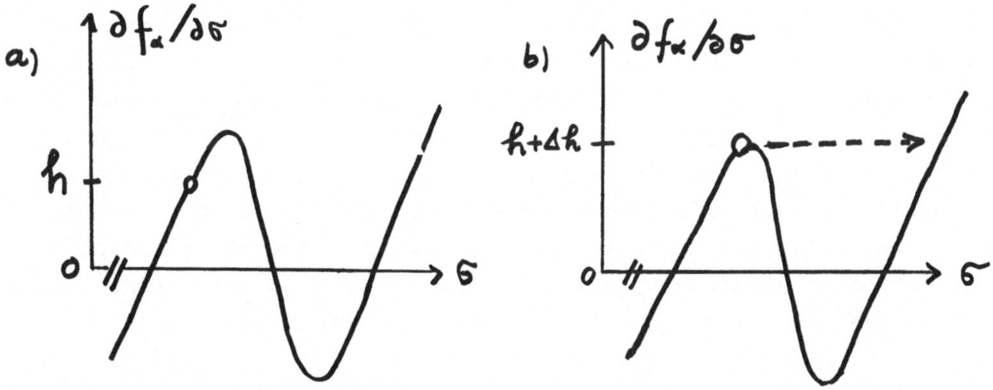

<u>Fig. 9a,b</u> In a metastable state α with free energy f_α the quasi-equilibrium value σ_α of the magnetization per spin is given by the condition $\partial f_\alpha / \partial \sigma = h$, where h is the reduced magnetic field. An enhancement of h leads to an irreversible Barkhausen jump.

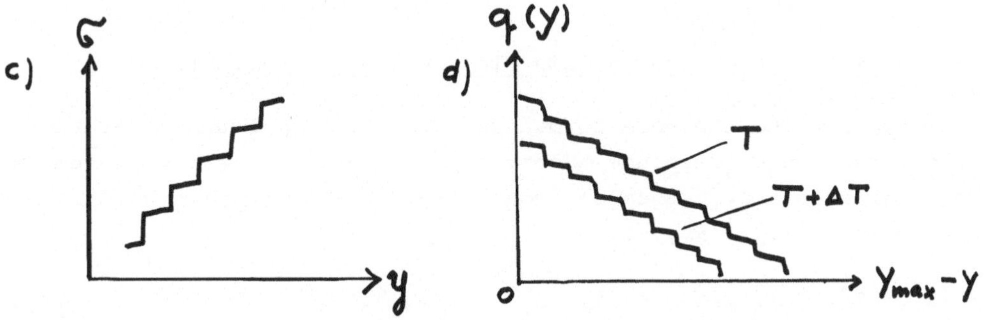

<u>Fig. 9c</u> Gross effects of Barkhausen jumps: The variable y can be interpreted (i) as the magnetic field, or (ii) as the temperature. In case (i) the figure explains the difference between the effective (e.g. virginal) susceptibility and the smaller single-valley result; in case (ii) it explains the difference between the steeper virginal M(T) and the almost flat field-cooled results.

<u>Fig. 9d</u> The order parameter q(y) is presented over $(y_{max}-y)$; the variable y can be interpreted as the variable x of Parisi, /8/, or as a logarithmic frequency scale. The area under the curve is the anomalous response Δ(0).

jump from $+\sigma$ to $-\sigma$ when $(-h)$ has become so large that in the metastable, undercooled state the susceptibility $\chi = \partial\sigma/\partial h$ vanishes. In the present case, however, the instability is probably signalled by the spin-glass susceptibility χ_ℓ^{SG} instead of χ, compare (23a,b).

This difference may also explain the recent numerical results of Mackenzie and Young /39/ for a finite-size approximation of the infinite range Ising model, where the relaxation to the field-cooled magnetization is governed by a broad spectrum of relaxation times $\tau \sim \exp(c\cdot N^{1/4})$, while for $H = 0$ there is an ultimate relaxation from the field-cooled state to the Curie behaviour governed by a relaxation time $\sim \exp(c\cdot N^{1/2})$; in the first case the instability would be given by χ_ℓ^{SG}, in the second case by χ.

In any case, the increase of σ through the jumps could explain qualitatively or even quantitatively the difference between the virginal and the field-cooled susceptibility, see Fig.9c; and the cross-over-temperature $T_1(H)$ from above, with the $H^{2/3}$-behaviour, would just correspond to that region, where the concurrent influences of the nonlinear susceptibility, which tends to reduce σ, and of the jumps, which would enhance σ, compensate each other.

IV. 2) Anomalous Response in Spin Glasses and Ferromagnets

Let us now come to the more formal aspects: If one assumes that the conventional linear response theory applies, as long as the system is in a metastable state, then one would have for the <u>total</u> susceptibility χ^{ij} (with i,j = x,y,z):

$$\chi^{ij} = \frac{\partial\sigma^i}{\partial h^j} = \frac{1}{k_B T} \langle \sum_m (\langle S_\ell^i S_m^j\rangle_T - \langle S_\ell^i\rangle_T \langle S_m^j\rangle_T)\rangle_\ell \ . \tag{27a}$$

For $H = 0$, by the symmetry of the exchange distribution the terms with $\ell \neq m$ vanish on average, which leads to the Fischer-relation /45/, namely the identity of (27a) with the <u>local</u> susceptibility

$$\chi_\ell^{ij} = \frac{\partial\sigma_\ell^i}{\partial h_\ell^j} = \frac{1}{k_B T} \langle\langle S_\ell^i S_\ell^j\rangle_T - \langle S_\ell^i\rangle_T \langle S_\ell^j\rangle_T\rangle_\ell \ . \tag{27b}$$

For $H \neq 0$, in the Heisenberg case, the identity of (27a) and (27b) would still be given for the transverse susceptibilities (e.g. i=j=x), but not for χ^{zz}, except for small enough fields H.

Now, according to the successful dynamical approach of Sompolinsky /10/, the relation (27) should be modified by an <u>anomalous response</u> $\Delta_i(\tau)$, depending on the time scale τ $(= 1/\nu)$ of the measurements: For $i=j$, for example (27b) would be replaced by

$$\chi_\ell^{ii}(\nu) = \frac{1}{k_B T} \{\langle\langle (S_\ell^{\,i})^2\rangle_T\rangle_\ell - q_i(\nu_{max}) + \Delta_i(\nu)\} , \qquad (27c)$$

where for short times $(\nu > \nu_{max})$ $\Delta_i(\nu)$ would vanish, i.e. conventional linear response theory would apply, which for $\nu < \nu_{max}$ $\Delta_i(\nu)$ would be a monotoneously increasing positive function attaining it's maximum at $\nu = O(\tau = \infty)$.

Theoretically, the anomalous response has been calculated for the Ising case by Sompolinsky /10/ and Parisi /8/, and for the Heisenberg case in a field by Sherrington et al. /6/, using the Parisi approach. That approach, which uses a special replica-symmetry breaking scheme, see /9/, leads to a continous order parameter function $q_i(x)$, which increases monotoneously between a minimum value $q_i(x_{min}) \sim H^{4/3}\,\delta_{i,3}$, /6, 46/, and a maximum $q_i(x_{max}) \equiv q_i(\nu_{max})$, /47/. Although the meaning of the quantity x and of the function $q_i(x)$ is obscured by the formalities of the replica trick and of the particular symmetry-breaking scheme,one expects that an interpretation in terms of Barkhausen jumps should be possible. Some temptative suggestions in this direction are made in the following:

According to Sompolinsky, /10/ (see also /9, 6, 46, 48/) it is

$$\Delta(\nu) = \int_{\nu}^{\nu_{max}} x(\tilde{\nu})dq(\tilde{\nu}) = \int_{t_{min}}^{t} x(\tau)\left(\frac{dq}{d\tau}\right)\cdot d\tau . \qquad (28)$$

Here we have omitted the dummy index i and substituted $\tilde{\nu}$ by $1/\tau$ and identified $x(1/\tau)$ by $\tilde{x}(\tau)$, omitting the tilde afterwards, for simplicity:

Now the Parisi-approach makes definite predictions for $q(x)$, and Sompolinsky makes definite predictions for $\Delta(x)$ $(= \Delta(\nu(x)))$, however the precise relation between ν and x is left open and may be system-dependent or subject to scale-transformations. However it is temptating to interprete (28) in a scale-invariant way, taking $x(\tau)$, apart from a factor, as a jump-rate decreasing with τ as $q(t_{min})/\tau$, while $(\frac{dq}{d\tau})\,d\tau$ would be proportional to the change induced on average by a single jump, with an essentially constant "velocity" $dq/d\tau$ $(\sim k_B T)$. In fact, this interpretation would be in agreement with an activated behaviour similar as in (26) : There, from $\tau(E) =$ $= \tau_o \cdot \exp(E/k_B T)$, one can replace dE by $k_B T\ d\{\ln(\tau/\tau_o)\}$, and $\exp(-t/\tau(E))$ roughly by $\Theta(\tau(E)-t)$, and since $\rho(E)$ is essentially flat beyond $E \approx E_{min}$, one gets a logarithmic decay over a wide range of times, see /43/.

A similar logarithmic increase would result for the anomalous response $\Delta(\tau)$ directly from the above-mentioned assumptions on $x(\tau)$ and $dq/d\tau$. Furthermore, if one extrapolates the corresponding logarithmic decrease of $q(\tau) = q(t_{min}) - \Delta(\tau)$ to zero, as presented schematically in Fig.9d, then from the fact that $q(t_{min}) \cong \varepsilon + \varepsilon^2 + \dots$, with $\varepsilon = 1-(T/T_o)$, see /10/, and assuming that $(dq/d\tau)d\tau$ does not depend drastically on T , one will get $\int_0^{q(t_{min})} x(q)dq \sim \varepsilon^2$.

The fact, that e.g. for the Ising case the result is not just $\sim \varepsilon^2$, but exactly $= \varepsilon^2 + \dots$, which leads to a constant $\chi(T)$ below T_o , does of course not follow from such simple considerations (see however below).

Interestingly, the phenomenon of anomalous response is not completely characteristic for spin glasses, contrarily to a wide-spread opinion; in fact, even in an ideal ferromagnet below T_c something similar happens: Namely, according to D.L. Huber /49/ the following relation applies between the total magnetization $M = \sum_\ell \langle S_\ell^z \rangle_T$ and the isothermal and isolated susceptibilities

$\chi_T = (\partial M/\partial h)_T$ and $\chi_{is} = \lim_{\omega \to 0} \{\chi'(\omega)\}$:

$$\frac{\hbar}{2\pi} \oint_{-\infty}^{+\infty} d\omega \cdot \coth\left(\frac{\hbar\omega}{2k_B T}\right) \cdot \chi''(\omega) = \langle M^2 \rangle_T - \langle M \rangle_T^{\,2}$$

$$+ k_B T \{ \chi_{is} - \chi_T \} . \qquad (29)$$

This relation, where the integral on the l.h.s. means the principal part, follows from the well known quantum fluctuation-dissipation theorem, where always a small, but finite field of order $k_B T/N \cdot \mu_B$ has to be applied to select a particular copy of the different ergodicity components /50/, which would exist for $H = 0$ in the thermodynamic limit below T_c . (E.g. in the Heisenberg ferromagnet the different ergodicity components can be transformed into each other by rotating all spins by a common angle); thus, as rule, the limit $N \to \infty$ has always to be taken first, before $H \to 0$ or $\omega \to 0$ or $t \to \infty$.

Finally, in (29), $\chi'(\omega)$ and $\chi''(\omega)$, mean the real and imaginary parts of the ac-susceptibility.

Now, according to R. Kubo /51/, the l.h.s. of (29) can also be expressed as the limit $t \to \infty$ of the expression $<(M(0) - M(t))\cdot(M(0) - <M>_T)>_T$. Therefore, using (29) and this identity one can define two different order-parameters of the Edwards-Anderson type, namely $\tilde{q} := \lim_{t \to \infty} <M(0)\cdot M(t)>_T$ and $q_T := <M>_T^2$, which are related to each other by (29) as follows:

$$\tilde{q} - q_T = k_B T \cdot \{\chi_T - \chi_{is}\} \ . \tag{30}$$

But in the spin glass case, $\tilde{q}$ would just correspond to $q(\nu_{max})$, the one-valley result, and χ_{is} would correspond to $\chi(\nu_{max})$, while q_T would correspond to the equilibrium, or field-cooled value $q_T = q(\nu_{max}) - \Delta(0)$. Therefore one has $\Delta(0) = k_B T \cdot (\chi_T - \chi_{is})$.

Now it is well known, see /49/ and references therein, that $\chi_T \geq \chi_S \geq \chi_{is}$, where χ_S is the adiabatic susceptibility, which generally, but not always, is identical with the isolated susceptibility. On the other hand, for ferromagnets below T_c , even in the limit $H \to 0$ the isothermal susceptibility χ_T will be larger than χ_S ; namely according to well-known Maxwell relations one has

$$\Delta(0) \geq k_B T \cdot (\chi_T - \chi_S) \equiv k_B T^2 \cdot (\frac{\partial M}{\partial T})^2 / C_H > 0 \ , \tag{31}$$

where C_H is the heat capacity.

This relation may be of some importance for the interpretation of the anomalous response in spin glasses.

The difference between the spin glass case and the ferromagnet is of course that in the spin-glass case $\Delta(\nu)$ is > 0 for all time scales $\tau = 1/\nu$ with $\nu < \nu_{max}$ whereas in the ferromagnet only $\Delta(0) > 0$, whereas $\Delta(\nu) = 0$ for all $\nu > 0$. However, considering the irreversible spin glass phenomena as a cascade of single events, one is temptated to rewrite (28) in the light of (31) as

$$\Delta(t) \cong k_B T^2 \int_{t_{min}}^{t} \frac{(\frac{\partial \sigma}{\partial T})^2(\tau)}{c_H(\tau)} \, d\,(\log \tau) \ , \tag{32}$$

where now Δ , σ , and c_H refer to one site, and $(\frac{\partial \sigma}{\partial T})$ and c_H are functions of the metastable state attained at time-scale τ , which (as has been discussed above) should be measured in logarithmic units:

The interesting point with (31) and (32) is that $\Delta(t)$ should stop to increase as soon as the metastable state, which has been obtained at time scale τ, would have $\partial\sigma/\partial T = 0$: This is just the flatness of the magnetization $\sigma(T)$ for the field-cooled state! (The fact that $\sigma(T)$ is not exactly constant in the Heisenberg case may be related to the small difference, which may still exist between χ_S and χ_{is}.)

V. Conclusions

We have discussed Heisenberg spin glasses in a magnetic field, putting equal emphasis on experimental and theoretical aspects. Among those points which have been stressed in this review, one should perhaps keep in mind the natural explanation which we have found for the wide range of the apparent scaling behaviour in the paramagnetic region, and the arguments which have been given for the compatibility of the apparent scaling behaviour with the probable absence of a true equilibrium phase transition. Also it has been stressed that there is not neccessarily a contradiction between the facts that the scaling behaviour cannot be described by the mean field theory while the behaviour below the upper freezing temperature $T_2(H)$ fits well into that theory. Namely to the opinion of the present author both transitions, particularly the lower one with the notorious $H^{2/3}$ - dependence, are more analogous to a spinodal crossover in an undercooled system than to an equilibrium phase transition .

Concerning the theoretical aspects discussed in chapter III, we have avoided the replica trick, using a direct mean-field approximation for systems with short range interaction and quantum spins. This approximation turned out to be sufficient for a qualitative and partially quantitative explanation of most phenomena, except for the non-trivial scaling in the paramagnetic region, but including e.g. the $H^{2/3}$ - behaviour of the crossover-line $T_1(H)$.

In principle the above-mentioned theoretical approach can be generalized beyond the mean-field approximation, to obtain a dynamic theory along the lines of /10/.

Finally in chapter IV we have tried a temptative interpretation of some formal aspects of improved theories as that of Sompolinsky /10/, Parisi /8/ and Sherrington /6/ in terms of Barkhausen jumps, and have related the anomalous response appearing in these theories to a difference between isothermal and adiabatic susceptibilities. By this approach we have found a somewhat speculative non-formal explanation for the flatness of the field-cooled magnetization $\sigma(T)$ at low temperatures.

Acknowledgements:

The author would like to thank K. Binder, W. Gebhardt, H. Kett and J. Souletie for stimulating discussions, and C. De Dominicis, J. Hertz, H. J. Sommers, D. Sherrington, and H. Sompolinsky for preprints.

Rererences:

/1/ For recent reviews see e.g.
R. Rammal, J. Souletie, in: Magnetism of Metals and Alloys (M. Cyrot, Ed.), North Holland, Amsterdam, 1982;
U. Krey, in: Berichte der Arbeitsgemeinschaft Magnetismus, Vol.I, (H. Mende, Ed.), Verlag Stahleisen, Düsseldorf (1983);
K. Fischer, phys.stat.sol. (b), 116,357, (1983)
/2/ D. Sherrington, S. Kirkpatrick, Phys. Rev. Lett. 35, 1792 (1975);
S. Kirkpatrick, D. Sherrington, Phys. Rev. B 17, 4384 (1978)
/3/ R. Fisch, A.B. Harris, Phys.Rev.Lett. 38, 785 (1977);
I. Morgenstern, K. Binder, Phys. Rev. Lett. 43, 1615 (1979);
Phys. Rev. B 22, 288 (1980); Z. Physik B 39, 227 (1980);
A. J. Bray, M. A. Moore, J. Phys. C 12, 79 (1979);
H. Sompolinsky, A. Zippelius, Phys. Rev. Lett. 50, 1297 (1983)
/4/ H. Maletta, in: Excitations in Disordered Systems (M. F. Thorpe, Ed.), Plenum Press, New York 1982
/5/ H. G. Bohn, W. Zinn, B. Dorner, A. Kollmar, J. Appl. Phys. 52, 2228 (1981)
/6/ D. Sherrington, D. M. Cragg, D. Elderfield, M. Gabay, J. Phys. Soc. Japan 52 (suppl.), 229 (1983);
D. M. Cragg, D. Sherrington, M. Gabay, Phys. Rev. Lett. 49, 158 (1982);
D. Elderfield, D. Sherrington, J. Phys. C, in press
/7/ J. R. L. de Almeida, D. J. Thouless, J. Phys. A 11, 983 (1978)
/8/ G. Parisi, Phys. Lett. 73 A, 203 (1979); Phys. Rev. Lett. 43, 1754 (1979); J. Phys. A 13, L 115; 1101; 1887 (1980); Phil. Mag. B 41, 677 (1980); Phys. Rep. 67, 25 (1980); preprint (1983)
/9/ C. de Dominicis, these proceedings;
C. de Dominicis, I. Kondor, Phys. Rev. B 27, 606 (1983)
/10/ H. Sompolinsky, Phys. Rev. Lett. 47, 935 (1981);
H. Sompolinsky, A. Zippelius, Phys. Rev. Lett. 47, 359 (1981); 50, 1297 (1983); J. Phys. C 15, L 1059 (1983); Phys. Rev. B 25, 6860 (1982);
H. Sompolinsky, these proceedings
/11/ D. J. Thouless, P. W. Anderson, R. G. Palmer, Phil. Mag. 35, 593 (1977)
/12/ L. R. Walker, R. E. Walstedt, J. Magn. Magn. Mater. 31-34, 1289 (1983);
R. E. Walstedt, L. R. Walker, J. Appl. Phys. 53, 7985 (1982)
/13/ H. Kett, W. Gebhardt, U. Krey, J. K. Furdyna, J. Magn. Magn. Mater. 25, 215 (1981);
H. Kett, PhD-thesis, Regensburg 1982
/14/ J. A. Gaj, R. R. Galazka, M. Nawrocki, Sol. State Comm. 25, 193 (1978)
/15/ R. R. Galazka, S. Nagata, P. H. Keesom, Phys. Rev. B 22, 3344 (1980)
/16/ T. Giebultowicz, H. Kepa, B. Buras, K. Klausen, R. R. Galazka, Solid State Comm. 40, 499 (1981)
/17/ M. Gabay, G. Toulouse, Phys. Rev. Lett. 47, 201 (1981);
G. Toulouse, M. Gabay, T. C. Lubensky, J. Vanninemus, J. Physique Lettres 43, L 109 (1982)
/18/ M. Gabay, T. Garel, C. de Dominicis, J. Phys. C 15, 7165 (1982)
/19/ If s is not $\gg 1$, then (1a) must be corrected, the dominant term at low fields being linear in H instead of quadratic, see chap. III.

/20/ R. V. Chamberlin, M. Hardiman, L. A. Turkevich, R. Orbach, Phys. Rev. B 25, 6720 (1982)
/21/ R. F. Schulz, E. F. Wassermann, J. Magn. Magn. Mater. 31-34, 1365 (1983)
/22/ P.Monod, H.Bouchiat, J.Physique Lettr. 43, L 45 (1982)
/23/ N.Bontemps, J.Rajchenbach, R.Orbach, preprint, 1983
/24/ M.B.Salamon, J.L.Tholence, J.Mag.Mag.Mater. 31-34, 1375 (1983); J.L.Tholence, M.B.Salamon, J.Mag.Mag.Mater. 31-34, 1340 (1983)
/25/ J.Hamida, C.Paulsen, S.J.Williamson, H.Maletta, to be published
/26/ W.Kinzel, K.Binder, Phys.Rev.Lett. 50, 1509 (1983)
/27/ R.Omari, J.J.Préjean, J.Souletie, preprint (1983); J.Souletie, these proceedings
/28/ J.Lauer, W.Keune, Phys.Rev.Lett. 48, 185 (1982); W.Marschmann, J.Lauer, W.Keune, J.Mag.Mag.Mater. 31-34, 1345 (1983); R.A.Brand, V.Manns, W.Keune, this conference
/29/ J.Stein, U.Krey, Z.Phys. B 34, 287 (1979); B 37, 13 (1980)
/30/ U.Krey, W.Maaß, J.Stein, Z.Phys. B 49, 199 (1982)
/31/ U.T.Höchli, L.A.Boatner, Phys.Rev. B 20, 266 (1979)
/32/ M.Fähnle, G.Herzer, T.Egami, H.Kronmüller, J.Appl.Phys. 53, 2326 (1982); J.Mag.Mag.Mater. 24, 175 (1981)
/33/ M.Fähnle, T.Egami, Solid State Comm. 44, 533 (1982); J.Appl.Phys. 53, 7693 (1982)
/34/ K.Binder, these proceedings
/35/ R.Barbara, A.P.Malozemoff, Y.Imry, Phys.Rev.Lett. 47, 1852 (1981); J.Appl.Phys. 53, 2205 (1982)
/36/ A.Berton, J.Chaussy, J.Odin, R.Rammal, R.Tournier, J.Phys. Lettr. 43,L153 (1982)
/37/ S.Nagata, R.H.Keesom, H.R.Harrison, Phys.Rev. B 19, 1633 (1979)
/38/ See e.g. W.Gebhardt, U.Krey, Phasenübergänge und kritische Phänomene, Vieweg, Wiesbaden, 1980, chapt.I.
/39/ N.D.Mackenzie, A.P.Young, preprint; see also P.Young, these proceedings
/40/ G.Toulouse, Comm.Phys. 2, 27 (1977)
/41/ P.W.Anderson, B.I.Halperin, C.M.Varma, Phil.Mag. 25, 1 (1975)
/42/ J.F.Fernandez, M.Medina, Phys.Rev. B 19, 3561 (1979)
/43/ D.Kumar, J.Stein, J.Magn.Mag.Mat. C 13, 3011 (1980)
/44/ Former doubts about the validity of the symmetry argument have been withdrawn in: A.P.Young, S.Jain, J.Phys. A 16, L 199 (1983)
/45/ K.H.Fischer, Sol. State Comm. 18, 1515 (1976)
/46/ H.J. Sommers, J.Physique Lett. in press; Z. Physik B 50, 97 (1983) J.Phys. A 16, 447 (1983)
/47/ $q(x_{max})$ is larger than the value obtained with the (replica symmetric) meanfield theory analogue q of Eqs. (15); e.g. in the Ising case $q(x_{max}) = \varepsilon + \varepsilon^2 + \dots$, see /10/, while $q = \varepsilon + 2\varepsilon^2/3 + \dots$, with $\varepsilon = (T_o - T)/T_o$.
/48/ C. de Dominicis, M.Gabay, H.Orland, J.de Physique Lett. 42, L 523 (1981); C. de Dominicis, M.Gabay, B.Duplantier, J.Phys. A 15, L 47 (1982)
/49/ D.L.Huber, Physica 87 A, 199 (1977)
/50/ R.G.Palmer, Adv. in Phys. 31, 669 (1982), and these proceedings
/51/ R.Kubo, J.Phys. Soc. Japan 12, 570 (1970)

SPIN GLASS BEHAVIOR IN FINITE NUMERICAL SAMPLES

by

R. E. Walstedt
Bell Laboratories
Murray Hill, NJ 07974/USA

ABSTRACT

The main points of a numerical simulation study of the spin glass transition in Ruderman-Kittel-Kasuya-Yosida (RKKY) systems are summarized. New results are also presented as follows. An investigation of the lifetime of spin freezing in a sample of 960 spins yields results which resemble qualitatively, if not quantitatively, the behavior of macroscopic systems. In the absence of anisotropy, a gradual spin freezing is found to set in at low temperatures when rotational decay of the Edwards-Anderson (EA) order parameter q is eliminated. However, this freezing exhibits no transition feature and is thought to be a finite sample effect. A study of 50 randomly selected ground states for a system of 500 spins is also presented. Evidence is given for a model of closely similar ground state pairs in which a small defect region occurs <u>inverted</u> in the two states concerned. Upper limit exchange barriers separating ground states are found to be substantially less than the mean thermal energy residing on the spins in the barrier region at reduced temperature $T^* = T_G^*$ in a number of cases. Thus, the possibility of barrier transitions, which underlie the observed decay of q, magnetic remanence, torque and EPR parameters, etc., in the spin glass state, is shown to be a natural feature of a disordered, exchange coupled spin system.

R. E. Walstedt

Bell Laboratories
Murray Hill, New Jersey 07974

I. INTRODUCTION

It is by now more than a decade since spin-glass ordering was first reported,[1] yet the precise nature of this phenomenon remains a fascinating and elusive puzzle. An enormous body of investigative work on the spin glass transition has been published.[2] Many mathematical models have been put forward. Unfortunately, those which have been solved in any detail up to this point[3-8] have the unrealistic feature of infinite range coupling. While their contribution to our understanding of spin glasses has been substantial, this feature distinguishes fundamentally their basic physics from that of a real, experimental spin glass system. With the coupling between all pairs of spins in the system on equal footing, such models possess a built-in infinite-range correlation, which in a real system must be developed arduously by a series of linkages between spatially neighboring spins. For real systems the range of correlated motion is difficult to measure and therefore generally unknown. Thus, although the spin glass transition is clearly a highly cooperative one, the question of whether it is actually a phase transition remains open.

Because of these continuing theoretical difficulties, there has been ample motivation to study the properties of spin glasses and spin glass models by numerical simulation methods.[2] In spite of the limitations of these methods, notably finite sample size, finite time scale, and the impracticality of quantum-statistical treatment, the observation of a finite-sample analogue of the spin freezing transition found (e.g.) in dilute Mn in Cu was recently reported.[9-10] Despite certain deviations of this transition from truly macroscopic behavior, the size-independence of the results and correlation with other properties, discussed below, give substantial evidence to establish its authenticity. The most remarkable property found is the precipitation of the freezing onset by a small dipolar anisotropy term in the Hamiltonian at a temperature apparently determined by the exchange. Without the anisotropy, no transition is found to occur.

This behavior is reminiscent of the successful prediction of the Néel temperature for a two-dimensional antiferromagnet by Stanley and Kaplan,[11] in apparent conflict with the proof by Mermin and Wagner[12] that this system, with pure Heisenberg exchange, would not support long-range antiferromagnetic order for $T > 0$. It is generally agreed that the (dipolar) anisotropy stabilizes the antiferromagnetic state.

Simulation of a spin-glass transition appears to open many interesting possibilities for the scrutinization of spin glass behavior inaccessible to current theory or, indeed, for testing theoretical predictions which are inaccessible to experiment. However, it is essential to examine the computer model carefully to establish what relationship its behavior bears to macroscopic spin glass behavior. Some aspects of this question have been considered in publications subsequent to our original report;[13-14] we shall review the main points here. In addition, two further studies relating to this question are reported in Sec. II.

The major questions regarding the model vs. macroscopic behavior can be briefly summarized as follows: (1) Is the freezing point T_G substantially independent of the number N of spins in the sample? (2) Is T_G only weakly dependent on anisotropy level as found experimentally?[15] (3) Does the freezing point obtained from simulation agree quantitatively with experiment? Considering questions (1) and (2) together, it was found in the original work on 500 spins[9-10] that the freezing temperature T_G^*[16] showed a distinct variation with (dipolar) anisotropy level, collapsing altogether for anisotropies less than an order of magnitude larger than the classical dipolar interaction. However, extension of the study to systems an order of magnitude larger[13] revealed two important features. First, in a larger system the variation of T_G^* with anisotropy level became flatter and the freezing effect extended to weaker anisotropies. Second, the curves of T_G^* vs. anisotropy for different sample sizes N had regions in common, suggesting no appreciable variation of T_G^* with N. The numerical model, therefore, appears to tend toward macroscopic behavior as its size increases.

Regarding the numerical prediction of a freezing temperature with this model, the situation is more complex. Using the Boltzmann-statistical temperature scale provided by conventional Monte Carlo methods leads[9] to an underestimate of T_G^* by a factor of ~3.[17] This discrepancy is not to be taken seriously, however, because of the clearly unphysical nature of Boltzmann statistics at low temperatures, leading for example to a nearly constant specific heat as $T \to 0$. As has been pointed out,[9] the freezing point found for this system can be rationalized if we abandon the temperature concept and regard the system energy (i.e., the thermodynamic internal energy) E as the independent variable for the spin freezing process. Then, using the RKKY constant determined by fitting the low-temperature specific heat[18] to effect a comparison, one finds the thermal energy $E(T_G) - E(0)$ of the model to lie within ~15% of that of CuMn (c = 0.88 at. %).[9,19] This result suggests that the transition point of the model is determined by the amount of thermal energy in the exchange bonds[20] rather than by the temperature. It would follow from this that quantum statistics play an essential role in determining T_G by determining the energy scale $E(T_G)$.

One further consequence of the unphysical Boltzmann temperature scale is a considerable broadening of the freezing transition for a finite sample. With, for example, the experimental specific heat curve the thermal energy decreases much more rapidly in the vicinity of T_G than the roughly linear classical law.[9] Curves for the Edwards-Anderson order parameter[21] $q(T)$ are therefore steeper by a factor ~2 with a realistic scale of E vs. T than with Boltzmann statistics.

Two other effects investigated with our simulation technique are as follows: (1) Peaking of the nonlinear susceptibility coefficient $d^2\chi/dH^2$ as a function of temperature near the freezing transition. The enhancement of $d^2\chi/dH^2$ over the free spin value can be interpreted in a simplified model[10] as the typical size of a coherently fluctuating cluster of spins. This enhancement is found to grow to nearly the size of the simulation sample as $T^* \to T_G^*$,[10] and in experiments on real systems grows to a number $>10^3$ as $T \to T_G$.[22] This behavior offers a simple physical mechanism for the freezing transition, namely, that spin clusters grow until their collective anisotropy energy becomes much larger than kT, at which point they become constrained to remain in a particular orientation. Because there are no physical boundaries in the system, this must happen everywhere at once. Mechanisms such as this were in fact suggested many years ago.[23] (2) Freezing of the transverse spin components in an applied magnetic field. This effect has been predicted by studies of the infinite-range Heisenberg model.[24] Specifically, one expects a sharp onset in the transverse freezing parameter q_T to appear at a temperature $T_G(H)$, even for fields much larger than that required to smear out the usual (longitudinal) transition completely. $T_G(H)$ is expected to vary initially as $T_G(H) \cong T_G(O) + aH^2$. The infinite-range models predict a phase transition at $T_G(H)$ because the long-range correlation is included ab initio. In simulation, we find a sharp transition in q_T[13] which gradually develops a small high-temperature tail as H^*[16] is increased to well above T_G^*. For $H^*/T_G^*(O) \sim 4$, $T_G^*(H^*)$ has only diminished by ~10%, in contrast with a much larger decrease predicted by the infinite-range model. This discrepancy is undoubtedly connected with the fact that for the infinite range model T_G is the order of a typical exchange field, whereas for our (RKKY) model T_G^* is an order of magnitude smaller.

Efforts have been made to identify the transverse freezing transition indirectly in terms of magnetization studies.[25] It is important to point out that this effect can also be probed microscopically, in principle, using the muon spin rotation and Mössbauer techniques. To carry out such studies one requires the distribution of q (or q_T) values among the individual spins of the sample. Such distributions have been obtained for the RKKY case.[14]

In the present paper we take up some additional aspects of our numerical simulation model. In Sect. IIA we consider the lifetime of q in the "frozen" state and the variation of that lifetime with temperature. This important question has not been addressed in previous work, where q has been calculated from runs of fixed length. In Sect. IIB the importance of purely rotational drift for the decay of q during a Monte Carlo run is discussed. This is a finite sample effect of potential importance for the understanding of the simulation results.

In Sect. III we present a study of ground state properties which extend considerably our earlier discussion of this topic.[18] In particular we consider the relation between ground states which differ only slightly and report upper limits on the energy barriers which separate them. Our results suggest that a slow migration among degenerate energy minima is possible at temperatures below T_G^*.

II. FURTHER ASPECTS OF THE SPIN FREEZING TRANSITION

In spite of successful observation of spin glass-like transition phenomena as reported in earlier publications, important questions remain open as to the precise properties of the freezing transition exhibited by this simple finite model. In this section we explore two questions not probed in previous work. First, we examine the behavior of the EA freezing parameter q as the length of the Monte Carlo run is varied, i.e., we study the Monte Carlo analogue of the time decay of q. Second, we look into the role played by random uniform rotations in the decay of q when anisotropy is absent.

II.A Time Decay of the EA Parameter

The EA parameter q(L) for a Monte Carlo run of L steps with a system of N spins is defined here as

$$q(L) = N^{-1} \sum_i \langle \vec{n}_i \rangle_L \cdot \langle \vec{n}_i \rangle_L, \tag{1}$$

where $\langle \vec{n}_i \rangle_L = L^{-1} \sum_{\ell=1}^{L} \vec{n}_{i\ell}$ is the vector average of spin i over the run, $\vec{n}_{i\ell}$ being the orientation of spin i after the ℓth iteration. In the case where the lifetime ℓ_0 of the spin orientation memory is short, then q(L) can be expressed in terms of the average time autocorrelation function $C(\ell)$ for spins,

$$q(L) \to L^{-1} \left[2 \sum_{\ell=0}^{\infty} C(\ell) - 1 \right], \quad \ell_0 \ll L, \tag{2}$$

where $C(\ell) = N^{-1}(L-\ell)^{-1} \sum_{i=1}^{N} \sum_{\ell'=1}^{L-\ell} \vec{n}_{i\ell'+\ell} \cdot \vec{n}_{i\ell'}$. In the absence of spin freezing, the sum $\sum_{\ell=0}^{L} C(\ell)$ will converge to a value independent of L, so that $q(L) \propto L^{-1}$. We anticipate that for finite systems q(L) will also decay with increasing L at all temperatures.[26]

However, only for temperatures $T^* > T_G^*$ will q(L) decay as fast as L^{-1}.

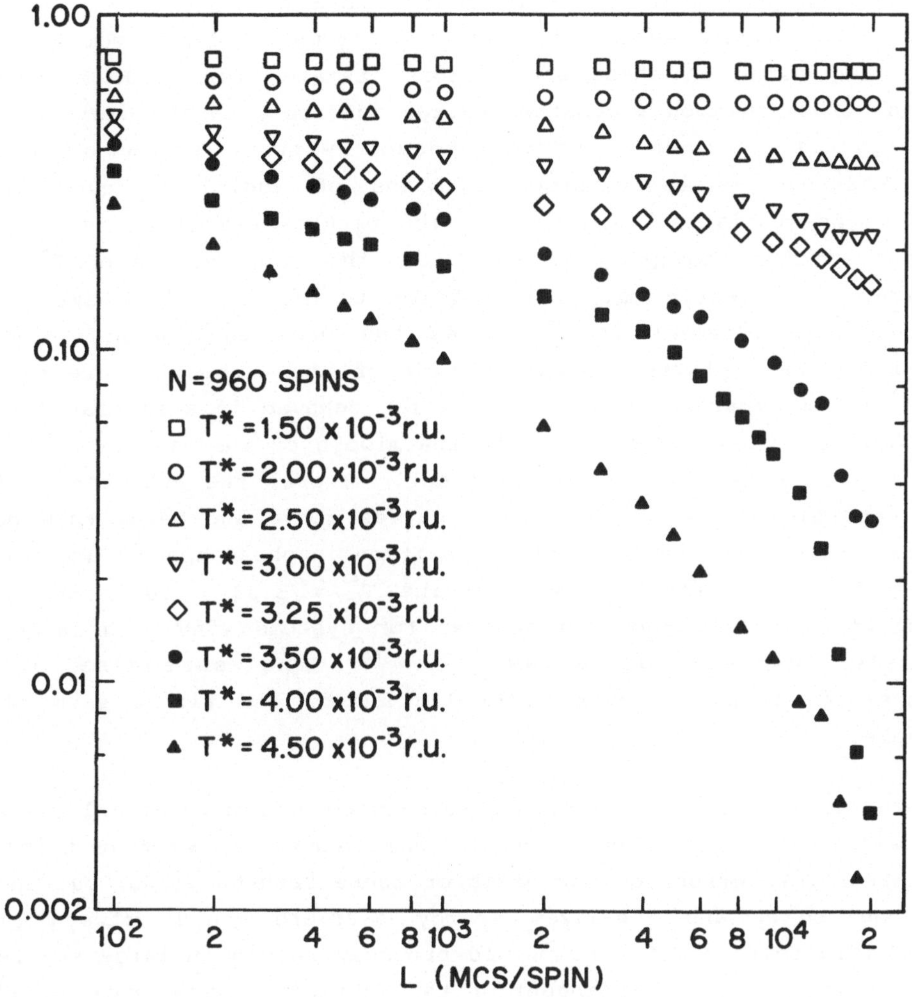

Figure 1

The EA parameter q is plotted as a function of the length of the (microcanonical) Monte Carlo run for a series of reduced temperatures, with open symbols for the $T^* < T_G^*$ data and closed symbols for $T^* > T_G^*$. For T^* values bordering the transition the data shown are averages over three rather similar runs.

The variation of q(L) with L has been studied in a series of microcanonical runs on a random system of 960 RKKY-coupled spins on an fcc lattice with periodic boundary conditions. The dipolar coefficient is taken to be 1% of the RKKY constant.[27] For these runs the system energy is brought gradually down from a random starting point in phase space, then iterated for 2.10^4 Monte-Carlo steps (MCS)/spin on a constant energy surface when the desired system energy is reached. Correspondence with a scale of reduced temperatures is established by means of additional runs with a Boltzmann-statistical algorithm. The results are shown in Fig. (1), where we plot log q(L) vs. log L for the eight values of T^* shown. There is a clearly marked transition to $q(L) \propto L^{-1}$ behavior, even though the interval in T^* values has been cut in half in that region. Data plotted on either side of the transition are averages over three separate runs, all of which behaved in a similar fashion. We have also carried out runs in the middle of the transition region, $T^* = 3.38 \cdot 10^{-3}$ r.u. (not plotted). There, the behavior is found to vacillate between that of the curves above and below this point, depending on the random starting position in phase space. We conclude that for this distribution of spins $T_G^* = 3.38 \cdot 10^{-3}$ r.u. ± 5%, similar to values reported earlier for systems with both larger and smaller numbers of spins.[10,13] The sharpness of the transition found is undoubtedly limited by the finite size of the sample.

The decay rate of q(L) with L is seen to be finite at all values of T^* within the spin-glass region. One cannot say whether this is a finite sample effect on the basis of these results alone, but in view of the small and localized energy barriers found between ground states in Sect. III.C, this would probably happen in large samples as well. It is quite reminiscent of the increasing decay rate found for remanent magnetization[28] as $T \to T_G$ in AuFe. What is striking here is the rapid increase in the decay time of q over a relatively small temperature decrease in the vicinity of T_G^*. This rapid slowdown is qualitatively similar to, although smaller than the increase in the fluctuation time constant at T_G measured by μSR.[29]

II.B The Uniform Rotation Decay Mode

In the absence of anisotropy, q has been found to decay on a relatively short time scale at all temperatures[10,17,30] for finite numerical samples. Because of rotational symmetry, however, we must consider the possibility that a major contribution to the decay of q arises from gratuitous rotation of the system as a whole caused

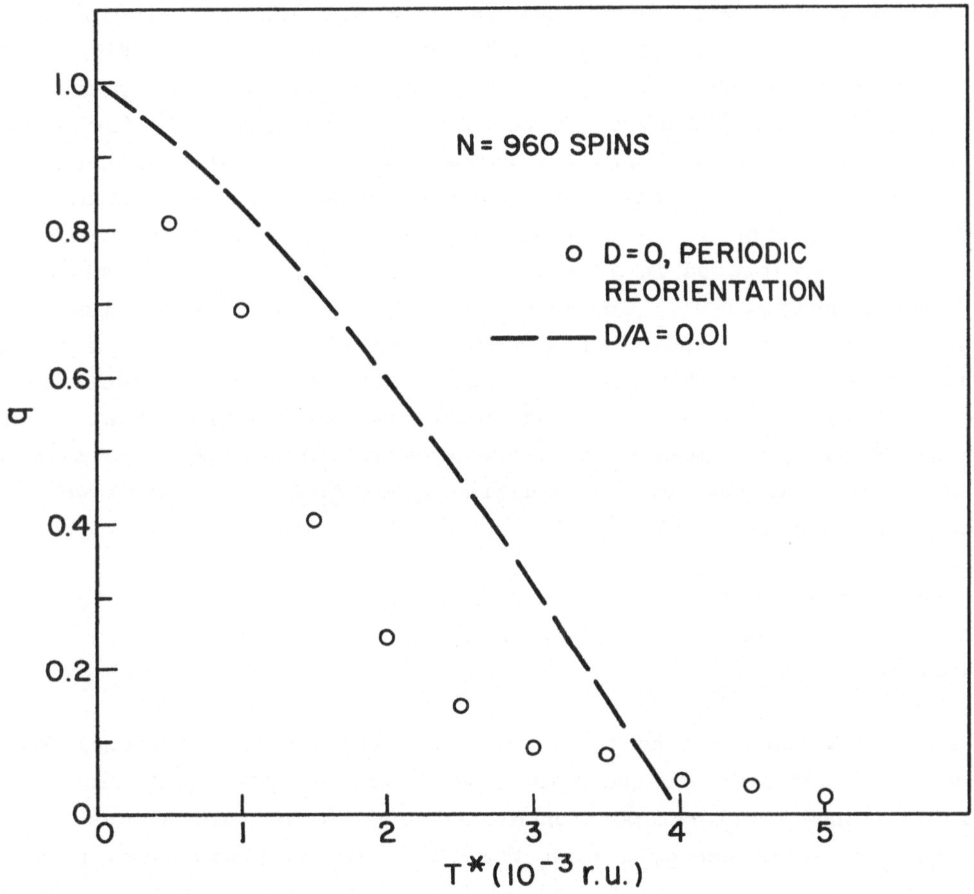

Figure 2

The variation of EA parameter q is plotted (open circles) vs. T^* for a system of 960 istropically coupled spins, in which rotational decay has been eliminated from 5000-step microcanonical data runs. Normal data for the spin-freezing transition, including dipolar anisotropy, is shown for comparison (dashed line).

by the Monte Carlo motions of the individual spins. Thus, q may appear to decay even below temperatures where the system is effectively frozen. It is important to investigate this point, because if found to be literally true, we would be compelled to reverse our conclusion[10] that anisotropic forces are required to stabilize the spin glass state.

We may examine the rotation hypothesis with the following technique. In order to extract from the Monte Carlo motion of the spins any effectively frozen component, we may use the methodology of Sect. III.A to rotate the system periodically so as to maximize its projection onto the initial state of the run, thus preventing rotational decay. The resulting variation of q vs. T^* for 5,000-step runs is shown in Fig. (2), where q vs. T^* obtained with dipolar anisotropy but no periodic rotations is shown for comparison. The periodic rotations produce a value of q which rises gradually as temperature is lowered in a fashion reminiscent of shattered susceptibility studies.[9] In fact, at $T^* \sim T_G^*$ where the spin configurations are known to resemble ground states,(9) a substantial non-zero value of q results simply from the mutual resemblance of ground states (see Fig. (7)). We therefore know a priori that q will exhibit a slow and gradual increase when rotational decay is eliminated. There is, however, no freezing transition of the sort we find with anisotropy present.

Thus, this "freezing" of the isotropic state appears to be a finite-sample effect not apparently related to a spin glass transition.

It is also interesting to examine the question of rotational decay from the standpoint of the normal mode frequency distribution. In Fig. 3 we show the lower portion of the spectrum of excitations around randomly selected energy minima for 20 independent configurations of 500 spins each, both with and without dipolar anisotropy. These results were obtained using methods described earlier;[18] they extend previously reported results to a larger sample size. Without anisotropy there is seen to be a substantial peak near zero frequency consisting of modes related to the rotational symmetry of the ground states.[18,31] The relative importance of these modes is inversely proportional to the sample size N. Thus, in macroscopic samples they become less important than finite wavelength "hydrodynamic" modes[32] for the purpose of deforming the spins from their T = 0 equilibrium orientations.

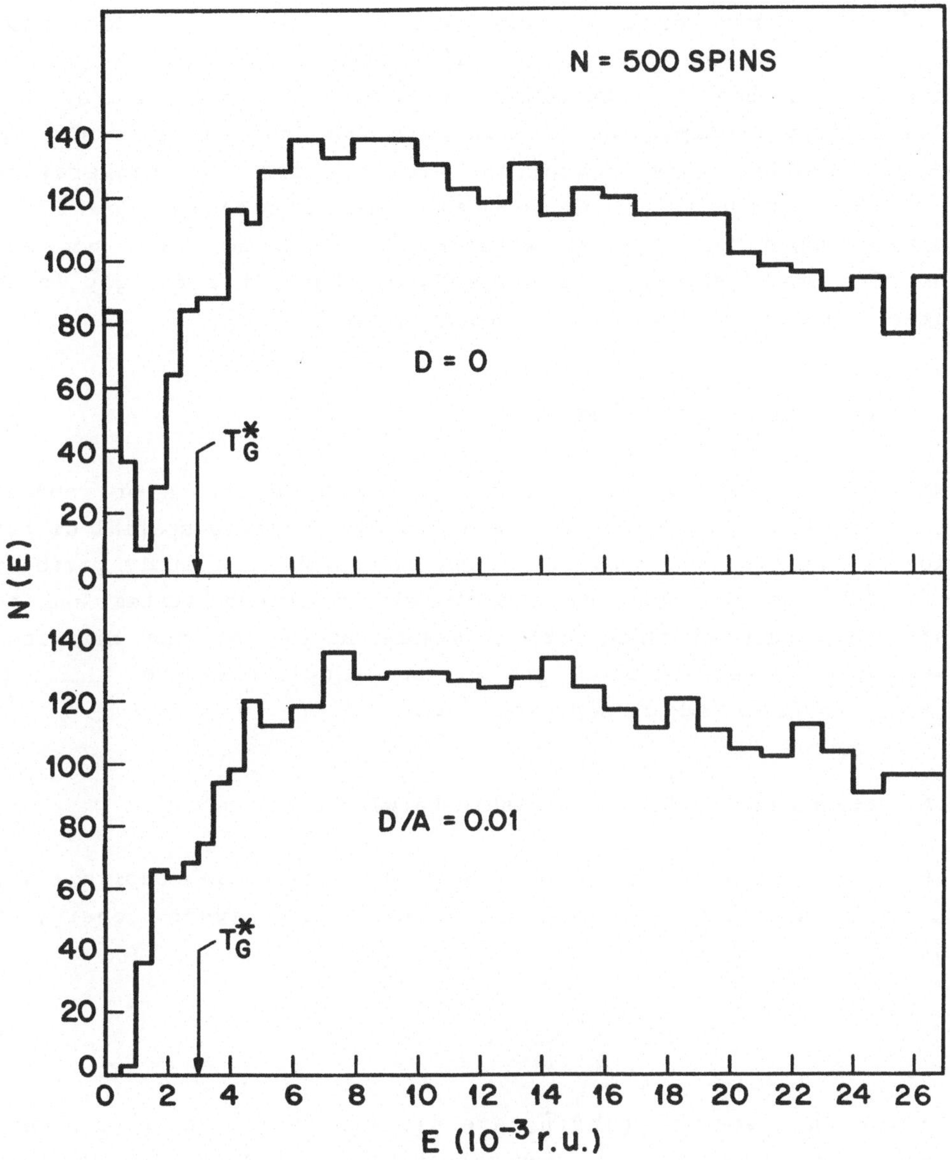

Figure 3

Distributions of zero-temperature oscillation mode frequencies from a series of 20 systems of 500 spins each (c = 0.9 at. %) are shown with and without dipolar anisotropy sufficient to precipitate the spin freezing transition. The fraction of modes having $E < T_G^*$ is seen to be small.

The effect of anisotropy on the excitation spectrum is to introduce a small gap at E = 0, i.e., to increase the energies of rotational and other long-wavelength modes to values which are a substantial fraction of T_G^*.[33] For example, one sees in Fig. 3 a noticeable increase in mode density at $E \sim T_G^*/2$ when anisotropy is present. This effect is thought to be closely related to the mechanism by which the anisotropy precipitates freezing, in that the gap prevents such low-lying modes from becoming over-excited at temperatures $T^* < T_G^*$. From Fig. 2 we see that rotational decay is important above T_G^*, in that without it a substantially non-zero value of q would appear. We conjecture, then, that the gap in the excitation spectrum (Fig. 3) disappears above $T^* \sim T_G^*$.

III. GROUND STATE PROPERTIES

We have carried out an extensive study of a series of randomly selected equilibrium configurations (EC's) for a single spatial distribution of 500 classical spins. In an extension of a study performed earlier,[18] we discuss (A) the number of unique states and the relationship between them, (B) a simple model for the difference between closely related EC's, and (C) some empirical upper limits for the energy barriers which separate them.

III.A Methodology: The Relationship Between EC's.

At the outset we consider only Heisenberg exchange coupling, with dipolar anisotropy to be added later. The system energy is thus[18]

$$E = -\frac{1}{2} \sum_{i \neq j} J_{ij} \vec{n}_i \cdot \vec{n}_j , \tag{1}$$

where the unit vector spins $\vec{n}_i$ are distributed randomly on a cubic section of fcc lattice at a concentration c = 0.9 at. %. The RKKY exchange[27] is taken in reduced units, defined by $A = 2\sqrt{2}\, a^3$, where a is the fcc lattice constant. This is better expressed in terms of the neighbor shell index $N_{ij} = 2(r_{ij}/a)^2$, giving

$$J_{ij} = \cos(6.945\, N_{ij}^{1/2})/N_{ij}^{3/2}, \tag{2}$$

where a value of k_Fa appropriate to Cu metal has been adopted. Periodic boundary conditions are employed throughout.

Combining Eq. (2) with the constraint $\vec{n}_i \cdot \vec{n}_i = 1$, it is easy to show that states representing energy extrema satisfy the condition

$$\lambda_i \vec{n}_i^o = \sum_j J_{ij} \vec{n}_j^o \ . \tag{3}$$

We have used an algorithm discussed earlier[18] to generate a series of 50 EC's which are energy minima and which satisfy Eq. (3), starting from random spin orientations. Since only 9 of the 50 obtained are duplicates, the number of distinguishable minima in this case appears to be at least several hundred. Thus, it did not appear feasible to obtain a complete set of EC's as was done in our previous study of 172 spins where there were only 7 EC's. We may also conclude that the number of distinguishable EC's rises very rapidly with the number N of spins in the sample, consistent with the exponential law derived by Bray and Moore.[34] The distribution of Heisenberg ground-state energies is confined to a range of $\sim \pm 0.02\%$ of the mean value, similar to what was found for N = 172 spins.

We have studied the variation of EC energy distributions with spin dimensionality D. Energy distributions for the XY and Ising model cases have been generated by a process of slow, progressive collapse of spin components in the Heisenberg EC's. By collapsing the α-component (α = x,y,z) of each spin to zero while maintaining quasi-equilibrium, one generates a set of XY EC's. The three sets of XY states so obtained can then undergo further collapse to obtain six sets of Ising EC's. As noted earlier,[18] one obtains unusually low-energy states in this fashion, although we have not attempted to prove they are the lowest. In any case, as shown in Fig. 4 the Ising states so obtained are well above the energy of the Heisenberg source states, but significantly lower than Ising states generated by simply flipping individual spins until they are all parallel to their local field. It is clear that there is a continuous distribution of Ising EC's filling the region in between as well.

Figure 4 also illustrates potential difficulties one might have in observing a spin-glass type of transition for an Ising system using Monte-Carlo methods. If the transition lies above the ground state by an amount of thermal energy comparable to what is found for the

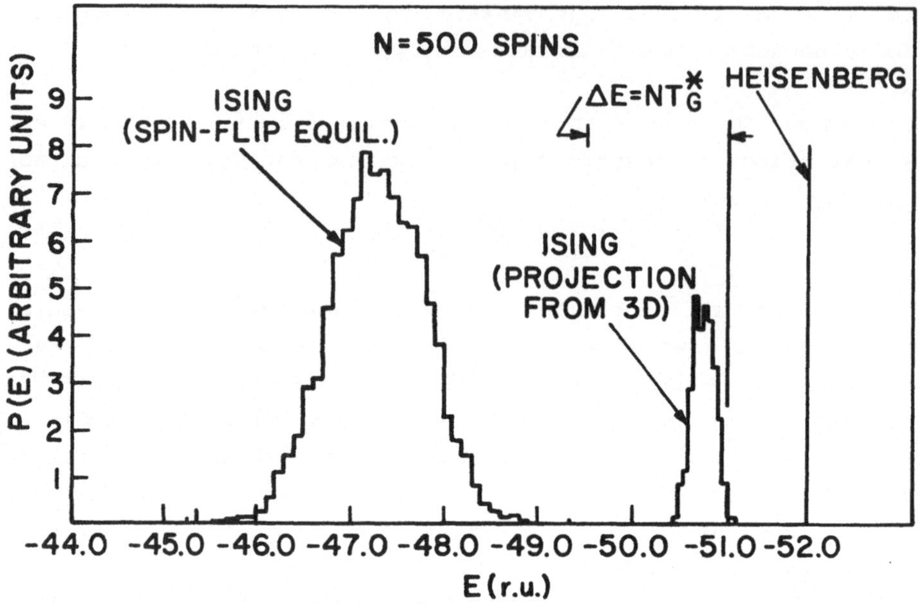

Figure 4

Distributions of EC energies are shown for vector (Heisenberg) spins, Ising states generated by progressive collapse of vector spin components, and Ising states derived by flipping spins progressively until each lies in the sense of its local field. Conditions are described in the text.

Heisenberg case,[10] then as shown in Fig. 4 the transition will lie in a region very dense with EC's. One might therefore expect to find long relaxation times in moving the system through these energies. The difficulties encountered in such simulation studies are well documented.[35,36]

In Fig. 5 are shown the Ising, XY, and Heisenberg energy distributions obtained as described above. One finds a rapidly converging progression of both widths and mean values of the energy distributions with increasing spin dimensionality. One might conjecture that for $D > 3$ the Heisenberg states would all collapse into a single ground state. This idea has been tested by adding a fourth spin dimension and further relaxing the Heisenberg EC's to equilibrium. As shown in Fig. 6, they all relaxed into one of two EC's with only a slight change in energy. We have not checked to see if there are additional EC's for $D = 4$. The near-uniqueness of the $D = 4$ ground state

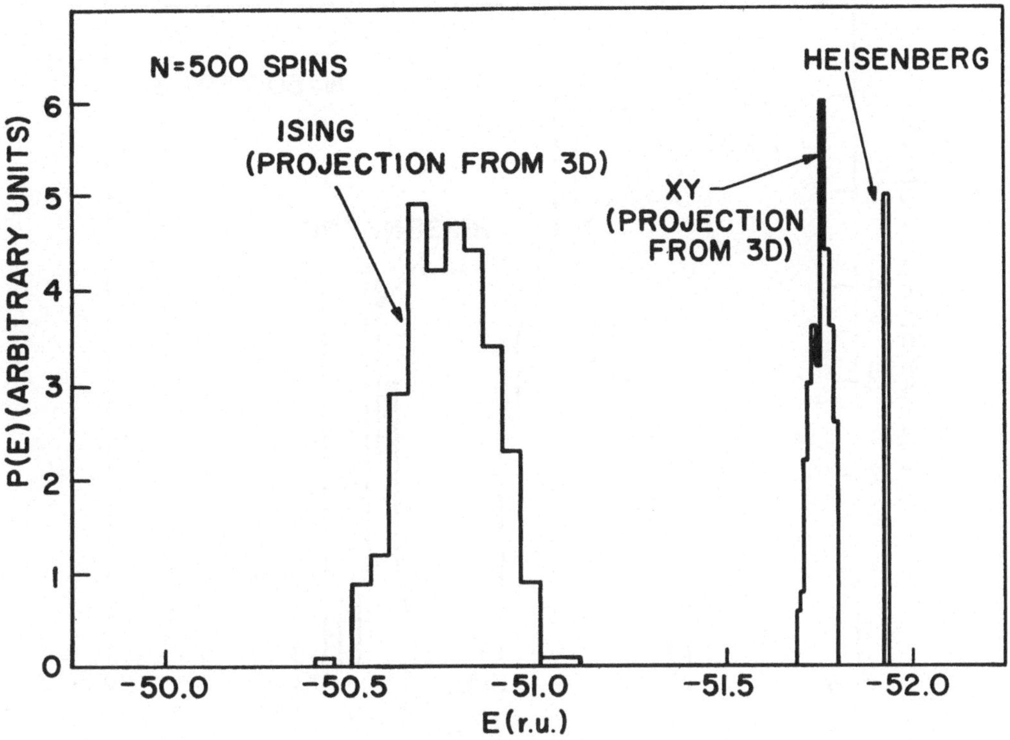

Figure 5

Expanded view of the right hand portion of Fig. 4 with XY states added. The width of the Heisenberg distribution is beginning to show.

suggests that the spin freezing transition for this case might be quite interesting.

The remainder of this sub-section is devoted to a study of relationships between and among the Heisenberg EC's found as described above. The general question we bear in mind here regards the phase space of near-degenerate states which the system occupies when it is below the freezing transition and the kind of motions, apart from harmonic oscillations studied earlier,[18] which the system is permitted to execute. It is useful to introduce the idea of the distance between EC's α and β in the phase space of unit vectors $\{\vec{n}_i\}$, namely

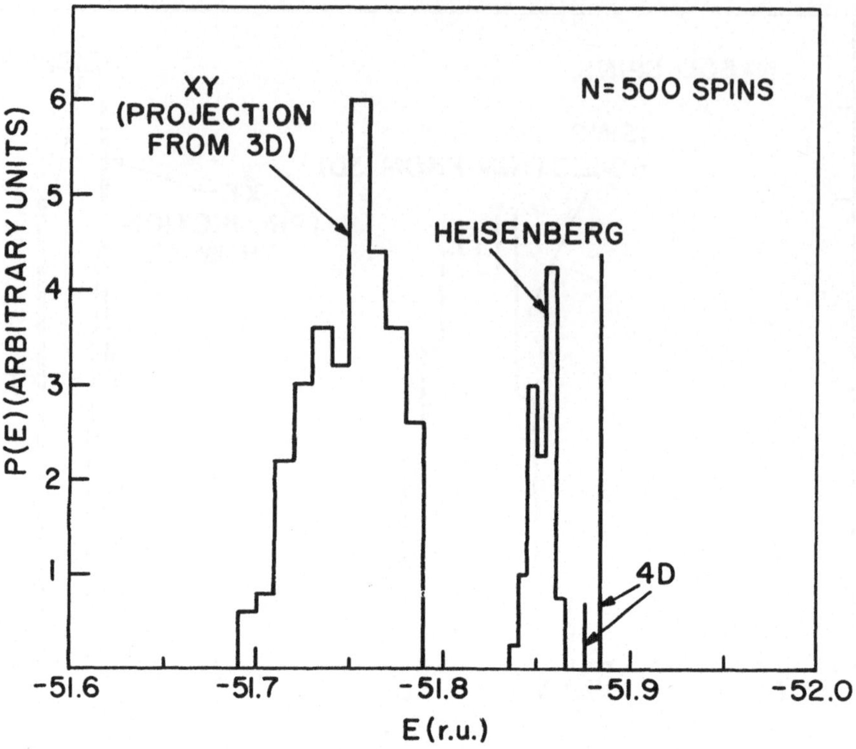

Figure 6

Expanded view of the right hand portion of Fig. 5, showing also the two states into which the Heisenberg distribution condenses when a fourth component is added to each spin.

$$d_{\alpha\beta}^2 = \sum_i (\vec{n}_{i\alpha}^o - \vec{n}_{i\beta}^o) \cdot (\vec{n}_{i\alpha}^o - \vec{n}_{i\beta}^0) \tag{4}$$

This may also be written $d_{\alpha\beta}^2 = 2N(1-P_{\alpha\beta})$, where $P_{\alpha\beta} = N^{-1}\sum_i \vec{n}_{i\alpha}^0 \cdot \vec{n}_{i\beta}^0$ is the mean dot product between corresponding spin vectors. We eliminate the ambiguity in $P_{\alpha\beta}$ caused by rotational symmetry by further defining

$$P_{\alpha\beta}^{max} = N^{-1} \left(\sum_i \vec{n}_{i\alpha} \cdot R\vec{n}_{i\beta}\right)_{max}, \tag{5}$$

where the rotation operator R is adjusted to maximize the mean projection $P_{\alpha\beta}$. From Eq. (4) we see that $P_{\alpha\beta}^{max}$ leads to the minimum separation $d_{\alpha\beta}^{min}$ between points α and β in phase space. In contrast with our previous discussion,[18] we shall retain here both a state $\{\vec{n}_{i\alpha}\}$ and its inverse $\{-\vec{n}_{i\alpha}\}$ as distinct EC's. The reason for this is that an EC and its inverse occupy rather remote points in phase space and are clearly separated by an energy barrier. The full manifold of states available in the spin glass phase must therefore include both.

The 500-spin EC's to be considered thus include 41 states and their inverses. Using a four-parameter representation for R in Eq. (5)

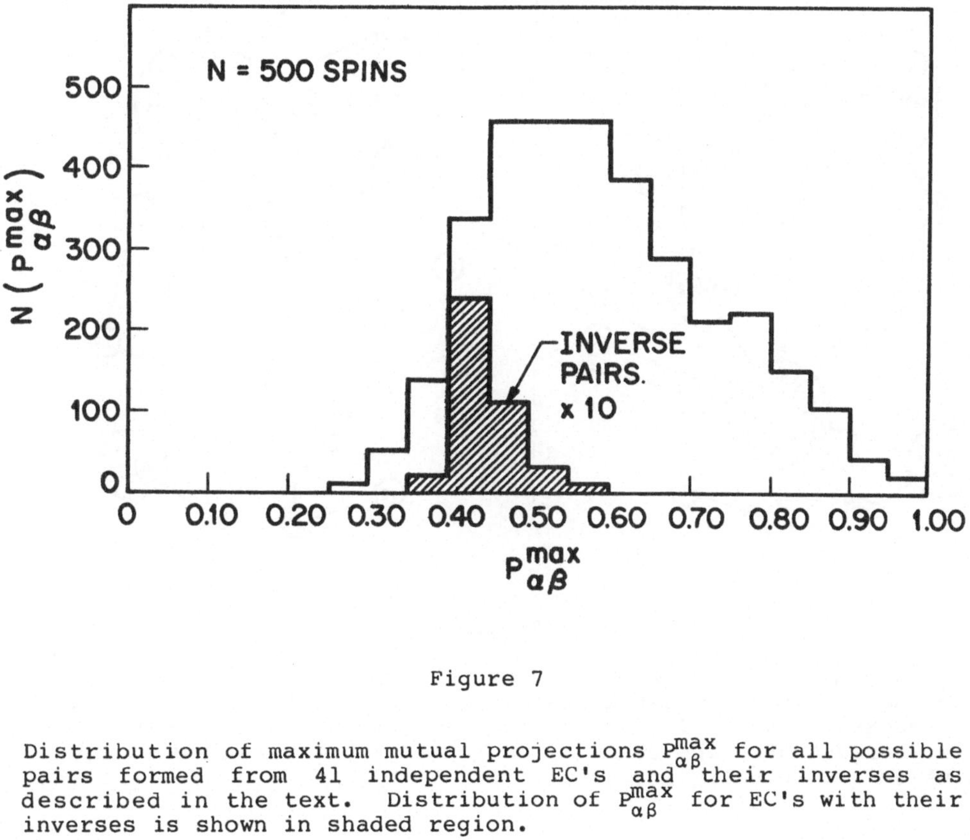

Figure 7

Distribution of maximum mutual projections $P_{\alpha\beta}^{max}$ for all possible pairs formed from 41 independent EC's and their inverses as described in the text. Distribution of $P_{\alpha\beta}^{max}$ for EC's with their inverses is shown in shaded region.

leads to a simple procedure to extract the $P_{\alpha\beta}^{max}$'s for all pairs of EC's. The 3321 values obtained are shown in a histogram in Fig. 7, where they are seen to range from $P_{\alpha\beta}^{max} < 0.25$ to $P_{\alpha\beta}^{max} > 0.95$. This is a decidedly wider range of values than found for N = 172 spins earlier,[18] but still much larger than random values $P_{\alpha\beta}^{(random)}$

$\sim N^{-1/2}$. Also shown in Fig. 7 is a subhistogram of $P_{\alpha\beta}$'s for EC's paired with their inverses. These are seen to be smaller on average than for pairs of EC's chosen at random. EC's and their inverses are therefore well removed from one another in phase space as noted above.

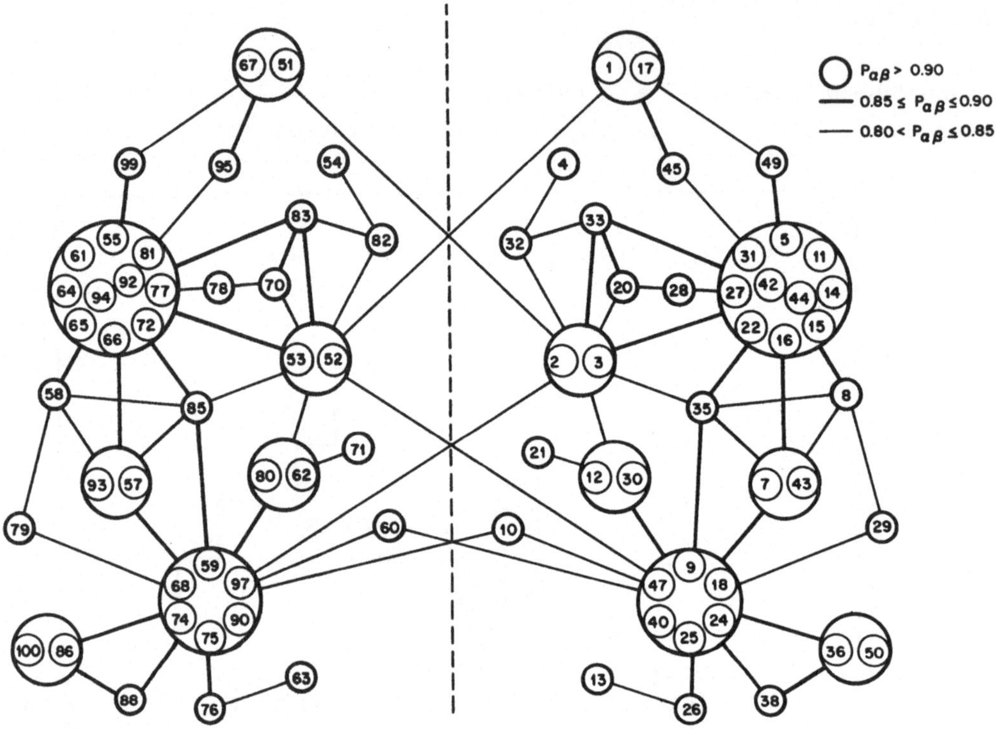

Figure 8

A diagram in which EC's are grouped according to the size of mutual projection bonds as indicated. Original EC's are numbered n = 1-50 with duplicates omitted. Inverses are numbered n+50. The dashed line separates mirror-image groups which are mutually inverted.

We can construct a diagram (Fig. 8) in which the EC's are grouped inside heavy circular boundaries by the occurrence of mutual projections $P_{\alpha\beta}^{max} > 0.90$. At this level of association, the groups (and singles) may be divided into two halves with the inverse of each entity in one half occurring also in the other. No connection between these halves occurs when linkages $0.85 < P_{\alpha\beta}^{max} < 0.90$ are also included in Fig. 8 (heavy lines). With projections $0.80 < P_{\alpha\beta}^{max} < 0.85$

inserted (light lines), connections appear between the two inverted halves and it becomes possible to reach any EC from any other by traversing bonds no smaller than $P_{\alpha\beta}^{max} = 0.80$, i.e., by a series of relatively small changes. Thus, there are no truly isolated EC's in phase space, a feature which would presumably be even more striking if one considered all EC's rather than our small subset.

III.B Inversion Model for Distinguishable EC's

The foregoing observations suggest a simple model for the occurrence of near-degenerate ground states in the Heisenberg spin glass. If an energy barrier separates any state and its inverse, it is quite likely that pairs of EC's will occur which differ by having only a small region inverted. Such pairs will have $P_{\alpha\beta}^{max} \lesssim 1$, with the difference residing in a "defect region" containing a small number of spins. Eight neighboring EC pairs with $P_{\alpha\beta}^{max} > 0.96$ (see Fig. 7) have been examined in some detail to test this idea. We define the number of spins N_{diff} upon which the difference between EC's α and β resides as the total of spins i for which the individual projection $P_{\alpha\beta}^{i} < P_o$ when the EC's are rotated to achieve the maximum projection $P_{\alpha\beta}^{max}$. The threshold P_o is a somewhat arbitrarily defined limit above which the spins are considered unperturbed. We find $\overline{N_{diff}} = 12.8$ with $P_o = 0.80$ and $\overline{N_{diff}} = 22.4$ with $P_o = 0.90$ for the 8 EC pairs considered. Individual values of N_{diff} are given in Table I. These numbers evidently represent typical minimum cluster sizes upon which the distinction between a pair of EC's can reside.

One may then ask whether the defect regions occur inverted in these EC pairs. It is difficult to devise a conclusive test of this question because of the involvement of the surrounding medium. Some evidence may be obtained as follows. For each pair α,β of EC's the spins identified as residing in the defect region of state α were inverted, then rotated using the algorithm described earlier to find their maximum projection both onto the original state α and the other state β. The results, shown in Table I, support the inversion model idea to the extent that in all but one case the inverted defect region has a greater projection onto the corresponding spins of state β than onto those of the original state. If we attempt to go further and re-equilibrate the spins into state β after inverting the defect region, ambiguous results are obtained. For example, if the inverted region

TABLE I

For eight close-neighbor EC pairs ($P_{\alpha\beta}^{max} > 0.96$) are given the values of N_{diff} (see text) for two different thresholds P_o. Also listed are the optimal projections of inverted defect region spins (using $P_o = 0.8$) from state α onto states α (P_{defect}^{α}) and β (P_{defect}^{β}).

EC Pair	$P_{\alpha\beta}^{max}$	N_{diff} (P_o = 0.8)	N_{diff} (P_o = 0.9)	P_{defect}^{α}	P_{defect}^{β}
1	0.969	14	22	0.825	0.914
2	0.969	19	30	0.704	0.693
3	0.979	10	14	0.933	0.964
4	0.972	11	22	0.798	0.953
5	0.980	10	20	0.835	0.944
6	0.980	11	21	0.802	0.976
7	0.982	9	10	0.917	0.984
8	0.960	18	30	0.567	0.906

is rotated to maximize projection onto the corresponding region in state α(β), then re-equilibration to state α(β) occurs with 100% efficiency. The apparent reason for this is that the initial state with an inverted and rotated defect region has an energy much higher than the energy barrier separating EC's α and β, as discussed in the following subsection.

III.C Exchange Barriers Separating Energy Minima

Because of the many time-dependent phenomena which occur in the spin-glass state, it is important to consider the energy barriers which separate pairs of EC's in relation to the thermal energy available to cause a transition between them. What is required is the minimum energy path; that, of course, is extremely difficult to determine. On the other hand, it is straightforward to establish a locally minimized energy path, which then stands as an upper limit to the minimum energy path. If this limit is sufficiently low, then it will be of relevance to the discussion of barrier transitions.

Locally minimized energy paths between EC's in phase space have been constructed as follows. Beginning with a pair of EC's $\{\vec{n}_i^\alpha\}$ and $\{\vec{n}_i^\beta\}$, a series of 99 intermediate states $\{\vec{n}_i^\ell\}$ is formed by linear superposition; thus,

$$\vec{n}_i^\ell = A_\ell \left[\vec{n}_i^\alpha (100-\ell) + \vec{n}_i^\beta \ell \right] , \qquad (6)$$

where $0 < \ell < 100$ and A_ℓ is a normalization factor. These states are then iterated with an energy-lowering algorithm to minimize the energy of this path through phase space. Between iterations, a smoothing algorithm is used to ensure that a continuous path through phase space is maintained, i.e., that $1 - P_{\ell,\ell+1} \ll 1$. In the results described below, the defect $1 - P_{\ell,\ell+1}$ is always of order 10^{-3} or smaller. A typical example of the resulting upper limit barrier shapes is shown in Fig. 9 for a pair of states with $P_{\alpha\beta}^{max} = 0.969$ (EC pair #1 from Table I). The barrier converges rapidly to an essentially constant value with this procedure after ~25 iterations.

A histogram of all the energy barriers involving the eight pairs of neighboring states in Table I is given in Fig. 10. We show the barrier distribution for the case of added dipolar anisotropy energy $D/A = 0.01$[27] as well as for purely isotropic exchange. The anisotropic values were obtained by introducing the dipolar terms into the equilibrium condition[18] and re-equilibrating with the isotropic EC's as a starting point. For D = 0 in Fig. 10 we see that the bulk of the barriers are less than a typical T_G^*, thus there is clearly no inhibition to migration between these states for temperatures T^* well into the spin glass region. Even for barriers $\Delta E > T_G^*$ the thermal energy of the N_{diff} spins which occupy the defect region ($\Delta E_{thermal} \sim N_{diff} T^*$) is sufficient to overcome the barrier. These pairs of states therefore constitute a set of two-level systems which can account for the time decay of remanent magnetization, of q as studied in Sect. IIA (Fig. 1) and other time-dependent phenomena found in spin glasses.

The addition of dipolar anisotropy in Fig. 10 is seen to elevate the barriers by noticeable amount, raising the bulk of them to values above T_G^*. However, there is not a big enough change to modify our conclusion that many barrier transitions remain within the range of thermal energies present at temperatures below T_G^*. This point is further corroborated by studies of barrier heights between

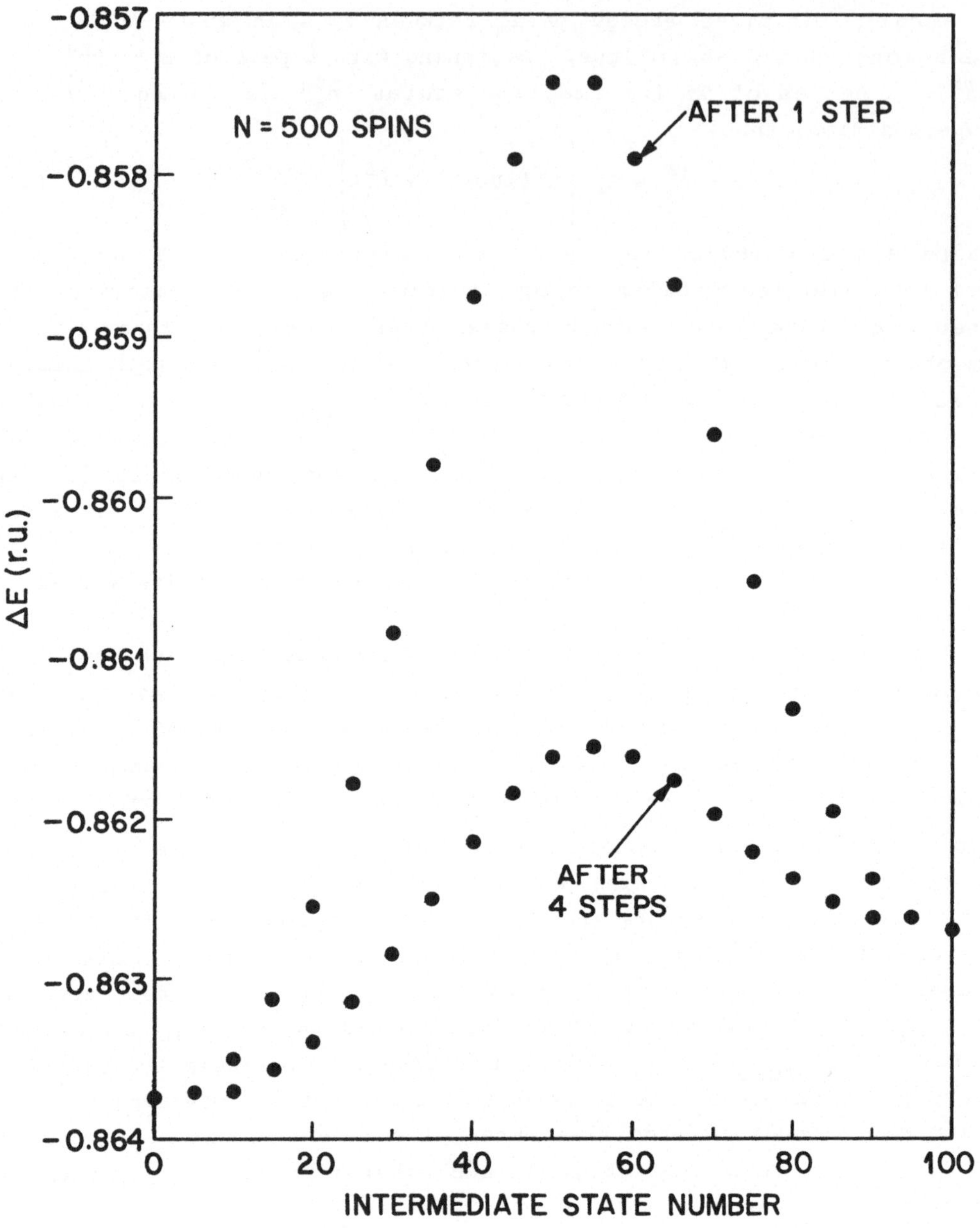

Figure 9

The convergence of an energy barrier profile between two EC's (see text) is illustrated by the curves found after one and four iterations, respectively. Only every fifth intermediate state energy is shown.

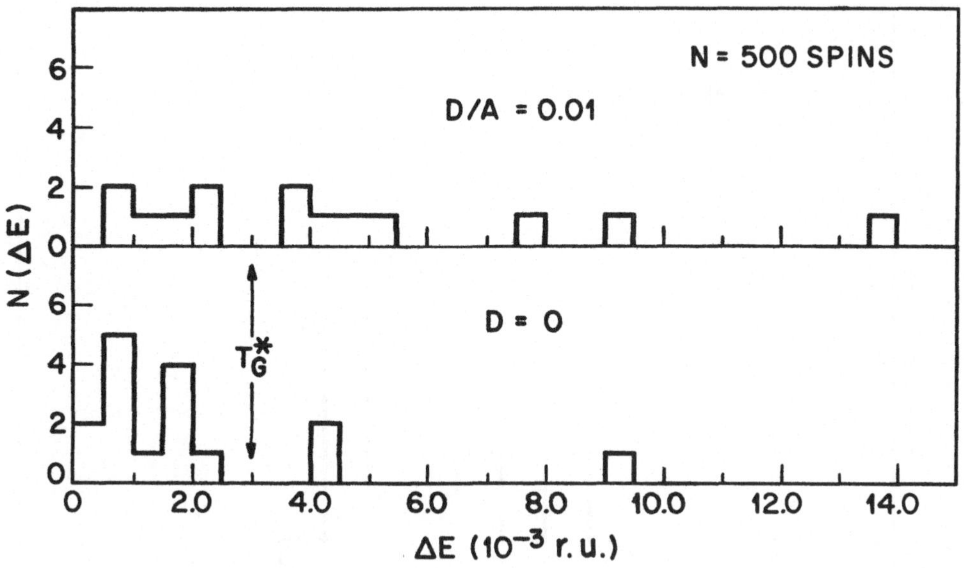

Figure 10

Distribution of upper limit energy barriers for eight pairs of neighboring EC's both with and without a small admixture of dipolar anisotropy. The spin freezing temperature T_G^* is shown for comparison.

each of the 41 EC's and their inverses, shown in Fig. 11. The upper-limit paths found in these cases involve a major fraction of the spins in the sample, yet a typical barrier is more than an order of magnitude less than the thermal energy NT_G^*. It must be emphasized that a direct transition between an EC and its inverse is rather improbable, i.e., that the system can achieve this result more easily by executing a series of smaller steps in phase space with presumably lower barriers.

Our major conclusion here is that this system of classical exchange-coupled spins, with or without anisotropy, is not energetically inhibited from making a wide variety of barrier transitions at temperatures within the spin glass phase. Experimentally, one finds a distribution of decay rates for the spin correlation function which extends over many orders of magnitude.[37] We see the "microscopic" equivalent of this result in Fig. 1. In either case, one needs the possibility of barrier transitions to understand the

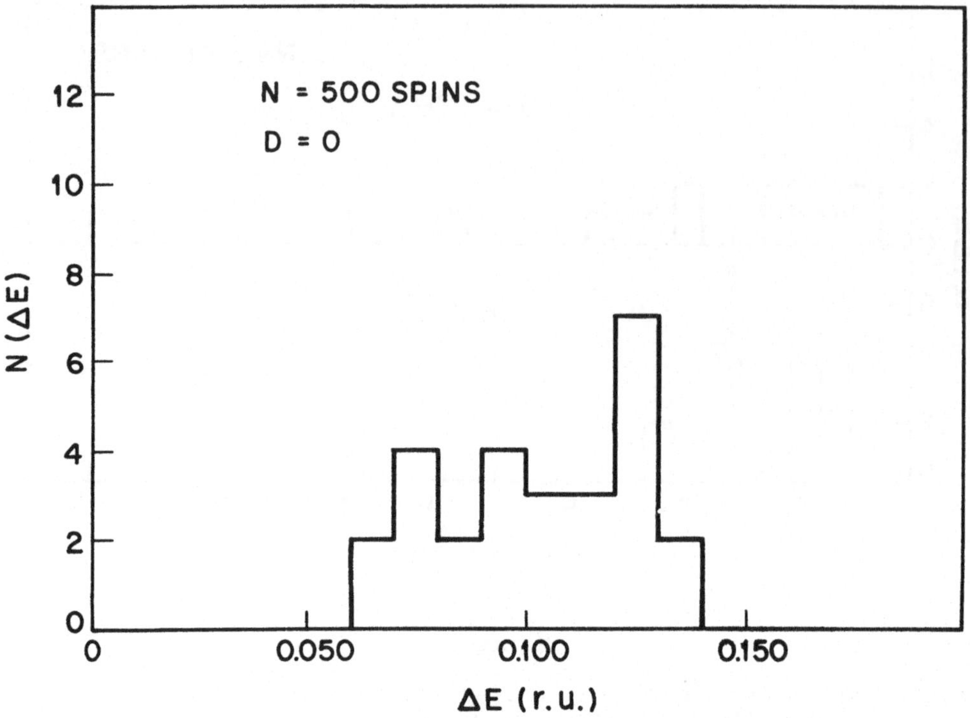

Figure 11

Distribution of upper limit energy barriers between the 41 independent EC's and their inverses. Substantially, all of the spins are involved in such a barrier transition. For comparison, the thermal energy in the system at $T^* = T_G^*$ is approximately $NT_G^* \sim 1.5$ r.u.

slow time decay processes involved. The results of this section show that the required occurrence of low barriers separating energy minima is a natural feature of the model of randomly positioned, Heisenberg-coupled spins.

Acknowledgment

The author would like to thank L. R. Walker, who participated in much of the work discussed in this paper, for many illuminating discussions.

REFERENCES

1. V. Cannella and J. A. Mydosh: Phys. Rev. B6, 4220 (1972).
2. K. Binder and D. Stauffer, to appear in Monte Carlo Methods in Statistical Physics II, Springer, Berlin-Heidelberg-New York; K. Binder and D. Stauffer: in Monte Carlo Methods in Statistical Physics (K. Binder, ed.) p. 301, Springer, Berlin-Heidelberg-New York 1979; K. Binder: in Fundamental Problems in Statistical Mechanics V, p. 21, ed. by E. G. D. Cohen, North-Holland, Amsterdam 1980; K. H. Fischer: Phys. Status Solidi (b).
3. D. Sherrington and S. Kirkpatrick: Phys. Rev. Lett. 32, 1972 (1975); Phys. Rev. B17, 4384 (1978).
4. D. J. Thouless, P. W. Anderson and R. G. Palmer: Phil. Mag. 35, 593 (1977).
5. G. Parisi: Phys. Lett. A 73, 203 (1979); Phys. Rev. Lett. 43, 1754 (1979); J. Phys. A 13, 1101 (1980).
6. G. Parisi, G. Toulouse: J. Phys. Lett. 41, L361 (1980).
7. D. J. Elderfield and D. Sherrington: J. Phys, A 15, L437 (1982); J. Phys. A 15, L513 (1982); J. Phys. C 15, L783 (1982); J. Phys. C, to be published (1983).
8. J. L. van Hemmen: Phys. Rev. Lett. 49, 409 (1982).
9. R. E. Walstedt: Physica 109 & 110B, 1924 (1982).
10. R. E. Walstedt and L. R. Walker: Phys. Rev. Lett. 47, 1624 (1981).
11. H. E. Stanley and T. A. Kaplan: Phys. Rev. Lett. 17, 913 (1966).
12. N. D. Mermin and A. Wagner: Phys. Rev. Lett. 17, 1133 (1966).
13. R. E. Walstedt and L. R. Walker: J. Appl. Phys. 53, 7985 (1982).
14. L. R. Walker and R. E. Walstedt: J. Mag. Mag. Mat. 31-34, 1289 (1983).
15. The anisotropy of 0.3 at % Mn in Cu is increased by a factor ~5 by adding 0.1 at % Au (J. J. Préjean, M. J. Joliclerc and P. Monod: J. Phys. 41, 1127 (1980)). The resulting change in T_G is an increase by ~5% (F. Milliken and S. J. Williamson: private communication).
16. The asterisk is used to denote quantities expressed in the reduced units of our model.
17. T_G^* is estimated by scaling the experimental transition temperature by a factor $2\sqrt{2}\, V_o S(S+1)/k_B a^3$ (W. Y. Ching and D. L. Huber: J. Phys. F8, L63 (1978)) where the RKKY exchange term is written $-J_{ij}\vec{S}_i \cdot \vec{S}_j$, with $J_{ij} = V_o \cos(2k_F r_{ij})/r_{ij}^3$.

18. L. R. Walker and R. E. Walstedt: Phys. Rev. B 22, 3816 (1980).
19. D. L. Martin: Phys. Rev. B20, 368 (1979).
20. This argument assumes the equivalence of classical and quantum thermal energies and ignores changes in the zero-point energy, which is large. The former assumption is reasonable for large spin quantum numbers. On the latter point, changes in the zero-point energy may be small if relatively few modes are excited for $T^* < T_G^*$. That this is the case may be seen in Fig. 3.
21. S. F. Edwards and P. W. Anderson: J. Phys. F5, 965 (1975).
22. J. Souletie: Heidelberg Colloquium on Spin Glasses 1983.
23. D. A. Smith: J. Phys. F 4, L26 (1974); 5, 2168 (1975); F. A. Rozario and D. A. Smith: J. Phys. F 7, 439 (1977).
24. G. Toulouse and M. Gabay: J. Phys. Lett. (Paris) 42, L163 (1981); G. Toulouse, M. Gabay, T. C. Lubensky and J. Vannimenus: J. Phys. Lett. (Paris) 43, L109 (1982).
25. R. V. Chamberlin, M. Hardiman, L. A. Turkevich, and R. Orbach: Phys. Rev. B 25, 6720 (1982).
26. H. Sompolinsky and A. Zippelius: Phys. Rev. B 25, 6860 (1982).
27. The RKKY and dipolar interaction are defined here as $-A\vec{n}_i \cdot \vec{n}_j \cos(2k_F r_{ij})/r_{ij}^3$ and $-D[\vec{n}_i \cdot \vec{n}_j/r_{ij}^3 - 3(\vec{n}_i \cdot \vec{r}_{ij})(\vec{n}_j \cdot \vec{r}_{ij})/n_{ij}^5]$, respectively. Dipolar interaction terms are limited to nearest neighbor pairs only. The results in this paper were obtained using D/A = 0.01.
28. J. Souletie and R. Tournier: J. Low Temp. Phys. 1, 95 (1969).
29. R. H. Heffner, M. Leon, M. E. Schillaci, D. E. MacLaughlin and S. A. Dodds: J. Appl. Phys. 53, 2174 (1982); R. H. Heffner, M. Leon and D. E. MacLaughlin: Proceedings of the Yamada Conference on Muon Spin Rotation and Associated Problems, Shimoda, Japan 1983.
30. K. Binder, Z. Phys. B. 26, 339 (1977).
31. In previous work (Ref. 18) the rotational modes were omitted from the spectra shown because of their limited importance for the macroscopic case.
32. B. I. Halperin and W. M. Saslow: Phys. Rev. B 16, 2154 (1977).
33. S. A. Roberts: J. Phys. C 15, 4755 (1981).
34. A. J. Bray and M. A. Moore: J. Phys. C 14, 2629 (1981).
35. I. Morgenstern and K. Binder: Phys. Rev. Lett. 43, 1615 (1980); Phys. Rev. B 22, 288 (1980).
36. A. P. Young: Phys. Rev. Lett. 50, 917 (1983).
37. F. Mezei: J. App. Phys. 53, 7654 (1982).

EQUILIBRIUM THEORY OF SPIN GLASSES:
MEAN-FIELD THEORY AND BEYOND

J.L. van Hemmen
Sonderforschungsbereich 123
Universität Heidelberg
Im Neuenheimer Feld 294
6900 Heidelberg 1
Federal Republic of Germany

Abstract: In this paper a new technique is analyzed to solve certain mean-field models, with particular emphasis on the spin-glass case. We also present an *extended* Gibbs formalism which is based on the observation that every equilibrium state can be decomposed uniquely into its ergodic components, and apply it to spin-glasses.

Contents

1. Introduction

Conventional statistical mechanics looks paradoxical. Suppose a macroscopic but finite system is in contact with a heat bath at inverse temperature $\beta=1/k_BT$ ($k_B=1$). According to the Gibbs Ansatz its thermal equilibrium state is given by the density matrix $\rho_N=C^{st}\exp\{-\beta H_N\}$, where H_N is the system's Hamiltonian. The density matrix ρ_N contains all the symmetries of H_N, so that many expectation values have to vanish. However, if a phase transition occurs at T_c (or T_f), we know that below T_c symmetries are broken and many expectation values should not vanish. For example, in zero external field the Edwards-Anderson [1] spin-glass order parameter

$$q_{EA} = \langle\langle S(i)\rangle_\beta^2\rangle_J , \tag{1.1}$$

where the inner average is taken with respect to ρ_N and the outer average over the coupling constants $\{J_{ij}\}$, vanishes identically - if taken literally. On the other hand, if we interpret [2] a spin glass as a disordered, magnetic system with a well-defined freezing temperature T_f such that for $T<T_f$ the magnetic moments are frozen in random orientations without a conventional long-range order, we expect that $\langle S(i)\rangle_\beta\neq 0$ below T_f and, hence,

$$q'_{EA} = \frac{1}{N}\sum_{i=1}^{N} \langle S(i)\rangle_\beta^2 \tag{1.2}$$

is non-zero. The question then is: are the thermal averages in (1.1) and (1.2) the same? They are not. In this paper we will study an extended Gibbs formalism [3] which also applies to random systems and offers a consistent description of the spin-glass transition and the spin-glass phase. In this section I will indicate which types of systems can be dealt with and what characteristics are to be described. Sections 2 and 3 are devoted to an exactly soluble spin-glass model [4], which includes both randomness and frustration but whose solution can be obtained without replicas. A new technique to analyze the model is presented. It also offers a nice illustration of some of the general principles of the extended Gibbs formalism, which will be explained in section 4. The results are discussed in section 5. In the appendix I present the proofs of some theorems which are of direct relevance to large deviations in random systems.

1.1 The nature of the spin-glass transition.

There is increasing evidence [5-8] that metallic and insulating spin glasses are in different universality classes, perhaps even with different values of the lower critical dimensionality d_L. If true, this implies that metallic RKKY systems such as AuFe and CuMn exhibit an equilibrium phase transition at a positive freezing temperature T_f. So, what is equilibrium? According to Feynman [9], a system in contact with a heat bath is in thermal equilibrium if all the "fast" things have happened and all the "slow" things not. The experimentalist's "trick" to obtain a spin glass in thermal equilibrium is careful field cooling [2,10]. This we assume throughout the following. At T_f, which is reproducible, we have a transition with well-defined critical exponents [7]. If field cooled, the system allows an experimental time scale τ_{obs} of at least several days [5].

1.2 The range of the interaction

In insulating spin glasses such as $(Eu_xSr_{1-x})S$ the main interaction between the magnetic moments is short range; typically, only between nearest and next-nearest neighbours [11]. Asymptotically, the Ruderman-Kittel-Kasuya-Yosida (RKKY) interaction in metallic spin glasses behaves like $\cos(2k_FR)/R^3$, which is of long range. Nevertheless, the folklore is that this is really short range. Let us examine this statement a bit more closely.

Suppose the coupling constants J_{ij} are given by $J_{ij}=\varepsilon_{ij}/R_{ij}^{\alpha d}$ where $R_{ij}=|i-j|$ is the distance between the spin at i and the one at j, d is the dimension, and the ε_{ij} are <u>in</u>dependent random variables <u>with mean zero</u>. Then the free energy exists and is a non-random number [12] if $\sum_j R_{ij}^{-2\alpha d}<\infty$, i.e., $\alpha>\frac{1}{2}$. If, however, the mean does not vanish, $R_{ij}^{-\alpha d}$ must be summable itself, i.e, α has to exceed 1. So independent random variables ε_{ij} whose mean is zero effectively <u>de</u>crease the interaction in such a way that it falls off with an exponent which is twice as large as αd. Accordingly, folklore apparently expects the RKKY interaction to behave like R^{-6}, which is short range. However, the assumptions used to arrive at this conclusion do not apply. We have

$$J_{ij}(\mathrm{RKKY}) \sim c_i c_j \frac{\cos(2k_F R_{ij})}{R_{ij}^3}, \quad \text{as } R_{ij}\to\infty , \tag{1.3}$$

where the c_i are random variables which are either one or zero according to whether the site is occupied or not, and we end up with a random site, not with a random bond problem. Since $\varepsilon_{ij}=c_ic_j\cos(2k_FR_{ij})$, the ε_{ij} are neither independent nor do they have mean zero. The same type of argument applies to Dzyaloshinsky-Moriya interactions [13]. And in both cases there is no a priori reason why the randomness effectively decreases the range of the potential .

1.3 Thou shalt not average

If a macroscopic random system is taken twice as big, is its thermodynamics twice as random? No, as we all know, it's not. Its thermodynamics is the same as before and does not depend on the specific random configuration. We know that thermodynamic observables like T_f itself, the magnetization m(h), the susceptibility χ, and the magnetic part of the specific heat are reproducible, i.e., two alloys whose microscopic structure differs but whose macroscopic constitution and preparation are identical (same concentration) give the same experimental outcomes. Therefore one can simply take a specific sample and need not average over an ensemble of them. As

Anderson [14] aptly observed: "No real atom is an average atom, nor is an experiment ever done on an ensemble of samples." Any viable spin-glass theory has to take this into account. It is important to take the thermodynamic limit because thermodynamic observables become reproducible <u>only if</u> $N \to \infty$ [15,12]. Since a macroscopic system is nearly infinite on a microscopic scale we will study a priori infinite systems ($N=\infty$) in section 4. As we shall see, such an approach has definite advantages.

2. An exactly soluble mean-field model

2.1 *The model*

To model a metallic spin glass such as CuMn we have to take into account (a) the long-range, indirect, RKKY interaction, which oscillates strongly as $R \to \infty$, and (b) a short-range, direct, interaction which is taken ferromagnetic [16]. We, therefore, start with the Hamiltonian

$$H_N = -\frac{J_o}{N} \sum_{(i,j)} S(i)S(j) - \sum_{(i,j)} J_{ij} S(i)S(j) - h\sum_i S(i) \tag{2.1}$$

This describes N Ising spins interacting with an external magnetic field h, and with each other in pairs (i,j). The generalization to n-vector models is immediate but will not be given here. A direct ferromagnetic coupling has been incorporated via J_o. Though not short-range, we take this type of interaction because it allows an analytic solution. The J_{ij} contain the randomness,

$$J_{ij} = \frac{J}{N} \{\xi_i \eta_j + \xi_j \eta_i\} , \tag{2.2}$$

where the ξ's and η's are independent, identically distributed random variables with mean zero and, say, variance one. Furthermore, we assume their distribution to be even. The ξ's and η's are to be chosen in such a way that the <u>distribution</u> of the J_{ij} resembles the one in a real metal [17]: (nearly) symmetric and highly peaked at $J_{ij}=0$, since the long tail of the RKKY interaction samples many small J-values. Instead of a long-range coupling with oscillating sign we have taken an infinite-range coupling with randomly fluctuating sign.

Given N spins, the J_{ij} contain 2N and not $\frac{1}{2}N(N-1)$ independent random variables. So the interaction (2.2) represents a random-site rather than a random-bond problem. This is also true for the original physical problem. Nevertheless about half of the spins belong to a fully frustrated configuration, as will be shown in section 2.3. We now turn to the calculation of the free energy.

2.2 *The free energy (solution of the model)*

The free energy $f(\beta)$ is given by

$$-\beta f(\beta) = \lim_{N\to\infty} N^{-1} \ln \mathrm{Tr} \exp\{-\beta H_N\}. \tag{2.3}$$

The trace is a finite sum over all spin configurations, and the ξ's and η's have fixed values, randomly chosen according to their distribution. We show that $f(\beta)$ exists with probability one and does not depend on the specific sample of ξ's and η's we have taken.

In calculating the free energy we must take the thermodynamic limit $N\to\infty$. Only then does the model become exactly soluble. It is a mean-field model where, in the limit $N\to\infty$, the "means" (or order parameters) are given by

$$m_N = \frac{1}{N}\sum_{i=1} S(i),\quad q_{1N} = \frac{1}{N}\sum_{i=1}^{N} \xi_i S(i),\quad q_{2N} = \frac{1}{N}\sum_{j=1}^{N} \eta_j S(j) . \tag{2.4}$$

Using (2.4) we rewrite the Hamiltonian,

$$-\beta H_N = N[\tfrac{1}{2} K_o m_N^2 + K q_{1N} q_{2N} + H m_N] \equiv NQ(\underset{\sim}{m}), \tag{2.5}$$

with $\beta J_o = K_o$, $\beta J=K$, and $\beta h=H$. Instead of the N spins we will use the three components of the vector $\underset{\sim}{m}=(m_N, q_{1N}, q_{2N})$ as summation variables in the trace and, hence, need something like a Jacobian. To see how this may be accomplished we must make a small detour [18].

Suppose we have a sequence of independent, identically distributed stochastic variables $\sigma(i)$ with mean zero and finite variance (not to be confused with the ξ's and η's; from now on the latter are fixed numbers, which do not play any role yet). As $N\to\infty$,

$$S_N = \frac{1}{N}\sum_{i=1}^{N} \sigma(i) \to 0 \tag{2.6}$$

with probability one [19]. The event $\{S_N \geq \varepsilon\}$ with $\varepsilon>0$ is called a large deviation since it becomes highly improbable as $N\to\infty$. We wish to estimate its probability Prob $\{S_N \geq \varepsilon\}$. To this end we introduce two functions,

$$c(t) = \ln \mathbb{E}\{\exp(t\sigma)\} \tag{2.7}$$

where $\mathbb{E}\{\ldots\}$ denotes the expectation with respect to σ, and

$$c^*(m) = \sup_{-\infty<t<+\infty} \{mt - c(t)\}. \tag{2.8}$$

The function $c(t)$ is convex (differentiate twice) and so is its Legendre transform $c^*(m)$ [20]. We have [21]

$$\lim_{N\to\infty} N^{-1} \ln \text{Prob}\{S_N \geq \varepsilon\} = -c^*(\varepsilon) , \tag{2.9}$$

where $c^*(m) \geq 0$, with equality if and only if $m=0$. For the events $\{S_N \leq \varepsilon\}$ with $\varepsilon<0$ the same formula holds. See the appendix. In fact, one can show that events with $S_N \approx \varepsilon$ have the same probability so that, as $N \to \infty$,

$$\text{Prob}\,\{m \leq S_N < m+dm\} \sim \exp\{-Nc^*(m)\}dm\ . \tag{2.10}$$

The limit $N \to \infty$ is essential in obtaining (2.10).

As a first, and simple, application of the above formalism we calculate the free energy of the model (2.1) with J=h=0. The Ising spins S(i) are considered as independent, identically distributed random variables which are ±1 with equal probability $\frac{1}{2}$. Given N, we divide the trace by 2^N so as to stress this new point of view. We then get a normalized trace, $\mathbb{E}_S\{X\} = 2^{-N}\text{Tr}\{X\}$, which is the mathematical expectation with respect to the probability distribution of the spins. In sections 2 and 3 we will use this convention. One only has to add $\ell n2$ to obtain the physical entropy. For the present purposes we can identify S(i) and σ(i) and, hence, m_N and S_N. The partition function may be written $\mathbb{E}_S\{\exp NQ(S_N)\}$ with $Q(m)=\frac{1}{2}(\beta J_o)m^2$, and the c-function is readily obtained, $c(t)=\ell n[\cosh(t)]$, and so is its Legendre transform $c^*(m)$,

$$c^*(m) = \frac{1}{2}[(1+m)\ell n(1+m)+(1-m)\ell n(1-m)] \tag{2.11}$$

if $|m|\leq 1$, and $c^*(m)=+\infty$ elsewhere. There is a straightforward analog for n-vector models. By (2.10) we may write, as $N\to\infty$,

$$\mathbb{E}_S\{\exp NQ(S_N)\} \sim \int_{-\infty}^{+\infty} dm\,\exp N\{Q(m)-c^*(m)\} \tag{2.12}$$

and thus, apart from a trivial $\ell n2$,

$$-\beta f(\beta) = \lim_{N\to\infty} \frac{1}{N}\,\ell n\,\mathbb{E}_S\{\exp NQ(S_N)\} = \max_m\,\{Q(m)-c^*(m)\} \tag{2.13}$$

We now return to the original problem.

Define $\underset{\sim}{W}_N = (Nm_N, Nq_{1N}, Nq_{2N})$. The three-vector $\underset{\sim}{W}_N$ contains the Ising spins as stochastic variables while the ξ's and η's are fixed numbers. As before we define a c-function,

$$c(\underset{\sim}{t}) = \lim_{N\to\infty} \frac{1}{N}\,\ell n\,\mathbb{E}_S\{\exp(\underset{\sim}{t}\cdot\underset{\sim}{W}_N)\} \tag{2.14}$$

where $\underset{\sim}{t}=(t_1,t_2,t_3)$ is also a three-vector. For Ising spins we obtain

$$c(\underset{\sim}{t}) = \lim_{N\to\infty} \frac{1}{N}\,\ell n \prod_{i=1}^{N} \cosh(t_1+t_2\xi_i+t_3\eta_i)$$
$$= \lim_{N\to\infty} \frac{1}{N} \sum_{i=1}^{N} \ell n[\cosh(t_1+t_2\xi_i+t_3\eta_i)]$$

$$= \langle \ell n[\cosh(t_1+t_2\xi+t_3\eta)]\rangle \quad , \tag{2.15}$$

with probability one [19]. The angular brackets denote an average with respect to one ξ and η. The function $c(\underset{\sim}{t})$ is convex and so is its Legendre transform $c^*(\underset{\sim}{m})$ [20]. Using equation (2.5) and a slight generalization of (2.12) we then find, as $N\to\infty$,

$$\mathbb{E}_S\{\exp(-\beta H_N)\} \sim \int_{\mathbb{R}^3} d\underset{\sim}{m} \exp N\{Q(\underset{\sim}{m})-c^*(\underset{\sim}{m})\} \tag{2.16}$$

and thus, apart from a trivial $\ell n2$,

$$-\beta f(\beta) = \max_{\underset{\sim}{m}}\{Q(\underset{\sim}{m})-c^*(\underset{\sim}{m})\}. \tag{2.17}$$

Full details are given in the appendix.

The maximum in (2.17) is realized for a certain $\underset{\sim}{m}=(m,q_1,q_2)$ and the negative of $c^*(\underset{\sim}{m})$ is the mean entropy. Using the convexity of $c^*(\underset{\sim}{m})$ one shows [22] that $q_1=q_2\equiv q$ maximizes the free energy functional in (2.17). The remaining order parameters m and q satisfy the equations

$$m = \langle \tanh\{K_0 m+H+Kq(\xi+\eta)\}\rangle \quad , \tag{2.18a}$$

$$q = \langle \tanh\{K_0 m+H+Kq(\xi+\eta)\}(\xi+\eta)/2\rangle \quad . \tag{2.18b}$$

Putting $H=0$ we quickly recognize three phases as special solutions of (2.18). The trivial solution $m=q=0$ represents a paramagnet (P). If $q=0$ and $m\neq 0$, we have a ferromagnet (F), and when $m=0$ and $q\neq 0$, a spin-glass phase (SG) appears. Finally $m\neq 0$ and $q\neq 0$ characterize a mixed phase (II); it need not always occur. See Figs. 1 and 2.

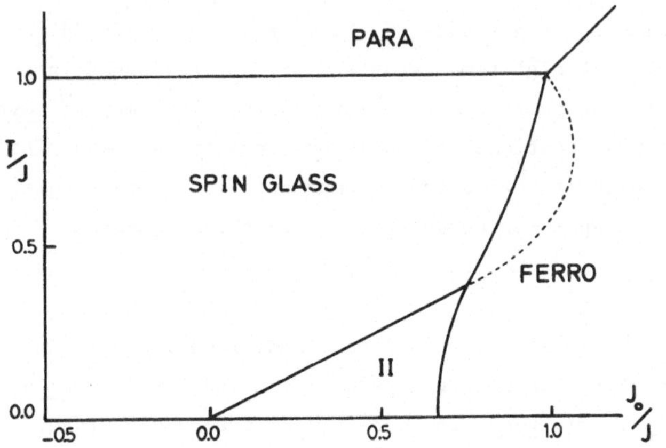

Fig. 1.

Phase diagram for ξ and η $=\pm 1$ with equal probability. II is the mixed phase. There is no external field. The critical line SG-II and its continuation, the broken line, represent the curve where the spin-glass fixed point bifurcates.

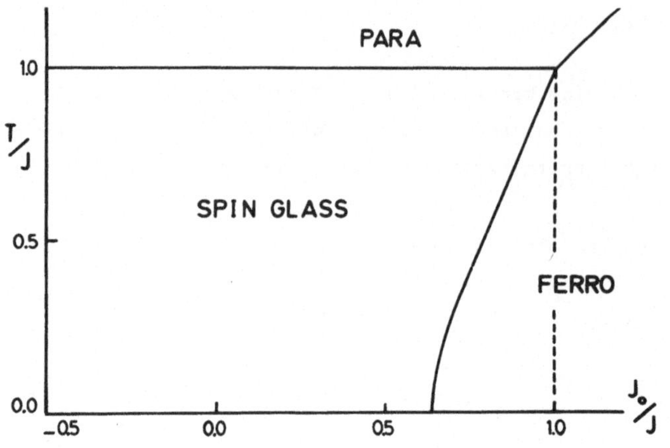

Fig.2. As Fig.1, for ξ and η Gaussian. There is no mixed phase.

In Fig.2 we show the phase diagram for Gaussian random variables ξ and η. There is no external field. The line which goes downward from (1,1) to $(\frac{2}{\pi},0)$ separates the spin-glass phase from the ferromagnetic phase and represents a first-order transition. All other transitions correspond to bifurcations of solutions of (2.18) and are second-order. The only way to reach a point below the first-order line experimentally is by lowering the temperature at fixed $\alpha=J_o/J$. When we cross the line and go further downward, no bifurcation is involved. It, therefore, is not to be expected that the system will jump spontaneously to the ferromagnetic phase [23]. It simply remains in the (metastable) spin-glass phase down to T=0. A similar argument applies to Fig.1.

Two final remarks are in order. First, the expression (2.17) may be rewritten in a slightly more convenient form [22],

$$-\beta f(\beta) = \max_{m,q}\{<\ln[\cosh\{K_o m+H+kq(\xi+\eta)\}]> - \frac{1}{2} K_o m^2 - Kq^2\} . \qquad (2.19)$$

Second, the m and q which maximize the right side of (2.19) satisfy the fixed point equations (2.18) - as they should. At high temperatures there is only one solution {m,q} and one phase. If, however, the temperature is low enough, there may be several {m,q} that maximize the free energy functional but only one need to be taken. The system, so to speak, is free to pick its own ergodic component. The corresponding free energy depends neither on the specific component, nor on the randomness.

2.3 *Frustration*

Let us for the moment fix N. The N lattice points can be divided into two disjoint subsets according to the sign of $\xi_i\eta_i$. We call the points with $\text{sgn}(\xi_i\eta_i)=+1$ blue and the remaining ones, where $\text{sgn}(\xi_i\eta_i)=-1$, red. (The case $\xi_i\eta_i=0$ can be discarded.) The probability for a point to be either blue or red is $\frac{1}{2}$. The model (2.1) is classical, so a Mattis transformation [24] makes sense. Apply the Mattis transformation $S(i) \to \mathrm{sgn}(\xi_i)S(i)$ and consider the interaction between two blue points i and j,

$$\mathrm{sgn}(\xi_i)\mathrm{sgn}(\xi_j)J_{ij} \Leftrightarrow \{|\xi_i\eta_j|+|\xi_j\eta_i|\}>0, \tag{2.20}$$

which is ferromagnetic. There is no frustration [25]. On the other hand, the coupling constant between two red points is transformed into

$$\mathrm{sgn}(\xi_i)\mathrm{sgn}(\xi_j)J_{ij} \Leftrightarrow -\{|\xi_i\eta_j|+|\xi_j\eta_i|\}<0 , \tag{2.21}$$

which is antiferromagnetic. All the simple closed loops (i.e., all triangles) are fully frustrated. As a Mattis transformation is frustration preserving about half of the spins (the red ones) are fully frustrated. A bond between a blue and a red spin, however, with

$$\mathrm{sgn}(\xi_i)\mathrm{sgn}(\xi_j)J_{ij} \Leftrightarrow \{|\xi_i\eta_j|-|\xi_j\eta_i|\} , \tag{2.22}$$

where i is red and j is blue , may have either sign with equal probability.

Colouring the spins may have interesting consequences if we want to explain the existence of a mixed phase [4] or if the J_{ij} are taken to be finite-range, say with i and j nearest-neighbours (see section 5.5).

3. Performance of the model

There are ample data to compare the predictions of the model with experiment. In this section we study the linear susceptibility χ, the spin-glass magnetization m as a function of h and T, and the magnetic specific heat c(T) as $T \to 0$. The following definition will be useful:

$$d_n(\beta) = \langle \cosh^{-2}\{Kq(\xi+\eta)\}(\xi+\eta)^n \rangle , \tag{3.1}$$

where n is a nonnegative integer.

3.1 The linear susceptibility

To calculate the zero-field susceptibility in the paramagnetic (P) and spin-glass (SG) phase we return to the fixed point equations (2.18), with $K_0=\beta J_0$, $K=\beta J$, $H=\beta h$ and $h\neq 0$, and differentiate equation (2.18a) with respect to h,

$$\frac{\partial m}{\partial h} = \langle \cosh^{-2}\{\dots\}\beta[J_0 \frac{\partial m}{\partial h} + 1 + J(\xi+\eta) \frac{\partial q}{\partial h}]\rangle \tag{3.2}$$

where $\{\dots\}=\{K_0 m+H+Kq(\xi+\eta)\}$. If we let h go to zero, use (3.1), and take advantage of the fact that m=0 in the paramagnetic and spin-glass phase for h=0 (and thus $d_1(\beta)=0$), we find

$$\chi_o(T) = Kd_o(\beta)[J-J_o(Kd_o(\beta))]^{-1} \ . \tag{3.3}$$

Let us now discuss the behaviour of the zero-field susceptibility $\chi_o(T)$ for the three temperature domains $T>T_f$, $T\approx T_f$, and $0\leq T<T_f$ separately.

(a) $T>T_f$: The Curie-Weiss regime. In the paramagnetic region q=0 and we have, as one easily verifies,

$$\chi(h,T) = [T \cosh^2\{\beta(J_o m+h)\}-J_o]^{-1} \tag{3.4}$$

with

$$m = \tanh\{\beta(J_o m+h)\} \tag{3.5}$$

and $(h\to 0)$

$$\chi_o(T) = [T-J_o]^{-1} \ . \tag{3.6}$$

We obtain a pure Curie-Weiss behaviour. This type of behaviour has been verified experimentally by Morgownik and Mydosh [26]. Increasing, for instance, in CuMn the Mn-concentration c we also increase the ferromagnetic (short-range [16]) interaction between the spins and, therefore, J_o. In fact, here it is natural to assume J_o to be proportional to c - in agreement with Ref.26.

(b) $T\approx T_f$: "The cusp." Lowering the temperature we have a second-order phase transition at $T_f=J$ and, hence, a discontinuity [4] in χ_o', the T-derivative of $\chi_o(T)$,

$$\Delta\chi_o'(T_f) = \chi_o'(T_f^+)-\chi_o'(T_f^-) = \frac{12}{<(\xi+\eta)^4>}\chi_o'(T_f^+) < 0 \ . \tag{3.7}$$

We note that $\chi_o(T_f)=[J-J_o]^{-1}$.

(c) $0\leq T<T_f$: The plateau. In general $\chi_o(T)$ as given by (3.3) cannot be evaluated explicitly since a nice expression for $d_o(\beta)$ is not available. We now show that $\chi_o(T)$ is independent of T for $0\leq T\leq T_f$ if ξ and η are Gaussian. See Fig.3.

It suffices to show that $Kd_o(\beta)=1$ for $0\leq T<T_f$. Then, by (3.3), $\chi_o(T)=[J-J_o]^{-1}$. For the proof we first note that ξ,η and, thus, $(\xi+\eta)/\sqrt{2}$ are Gaussians with mean zero and variance one. In the fixed point equation for q, viz.Eq.(2.18b),

$$q = <\tanh\{Kq(\xi+\eta)\}(\xi+\eta)/2> \quad , \tag{3.8}$$

we now perform a partial integration with respect to $x = (\xi+\eta)/\sqrt{2}$,

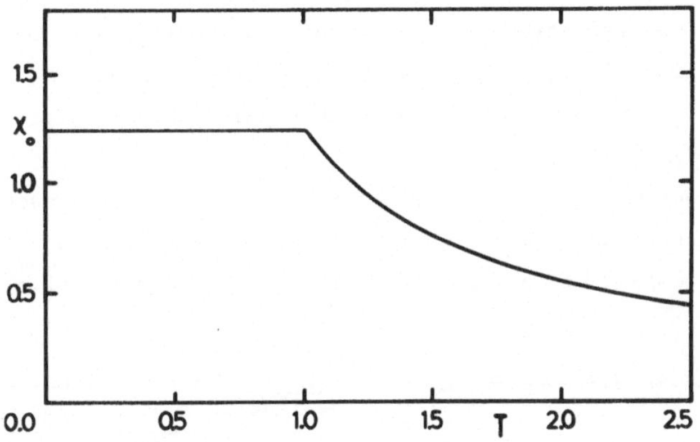

Fig.3. The zero-field susceptibility χ_o as a function of the temperature T. The random variables ξ and η are Gaussian and $J_o/J=0.2$. We have a Curie-Weiss behaviour for $T>T_f=J$.

$$
\begin{aligned}
q &= \int_{-\infty}^{+\infty} \frac{dx}{\sqrt{2\pi}}\, e^{-x^2/2}\, \frac{x}{\sqrt{2}} \tanh\{Kq\sqrt{2}\, x\} \\
&= Kq \int_{-\infty}^{+\infty} \frac{dx}{\sqrt{2\pi}}\, e^{-x^2/2} \cosh^{-2}\{Kq\sqrt{2}x\} = Kqd_o(\beta) \quad ,
\end{aligned} \tag{3.9}
$$

and find $q[1-Kd_o(\beta)]=0$. In the spin-glass phase q is nonzero, hence $Kd_o(\beta)=1$. I think it is satisfying that the rather subtle interplay of equilibrium formalism and randomness is responsible for this explanation of the experimentally well-known plateau in the susceptibility [27].

3.2 The magnetization

Fixing a temperature T, below T_f, one can study the spin-glass magnetization m(h) as a function of the external field h. In Fig.4 we show m(h) as a solution to the fixed point equations (2.18); q vanishes for $h>h_t$.

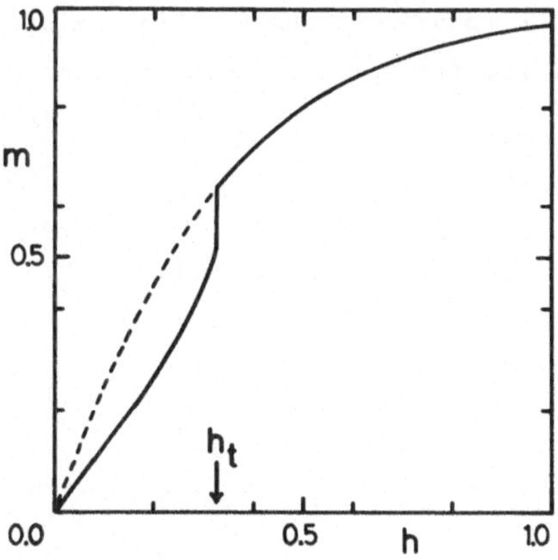

Fig.4. The spin-glass magnetization m as a function of the external field h, with $J_o/J=0.2$ and $T/J=0.6$ where the units are such that J=1. The random variables ξ and η are Gaussian. At h_t there is a field-induced transition to a state of higher magnetization. The broken line indicates the metastable continuation of the high-magnetization state;cf. Fig.2 of Ref.28.

The magnetization has a distinct S-shape character. At h_t there is a field induced transition to a state of higher magnetization (P). The magnetization is convex on the left of h_t and concave on the right, i.e., h_t acts as a point of inflection. The broken line indicates the metastable continuation of the high-magnetization state; cf. Fig.2 of Ref.28.

Fixing the external field h one may also vary the temperature. In Fig.5 we show the spin-glass magnetization m as a function of the temperature T. The plateau and field dependence of the "knee" temperature are in reasonable agreement with Ref.6.

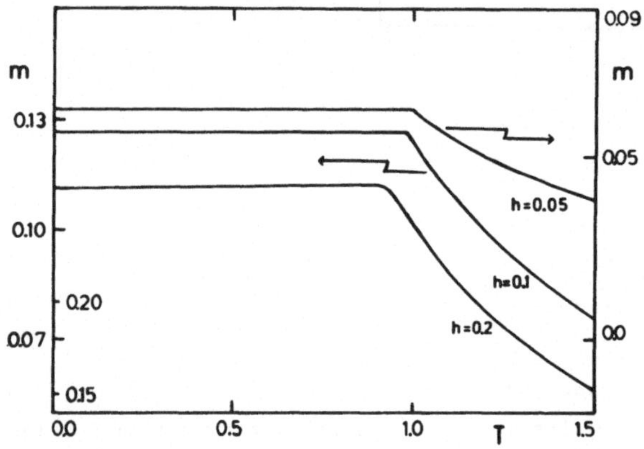

Fig.5. The spin-glass magnetization m as a function of the temperature T in the Gaussian case with $J_o/J=0.2$ and for fixed h. Note the difference in scale: on the right h=0.05, on the left h=0.1, and in the lower left-hand corner h=0.20. The field dependence of the "knee" temperature is in fair agreement with Ref.6.

3.3 The magnetic specific heat

Whatever the inverse temperature β, the energy per spin is given by

$$u(\beta) = -\frac{1}{2} J_o m^2 - Jq^2 - hm \tag{3.10}$$

where m and q satisfy (2.18). We take h=0. In the spin-glass phase, which exists for $\beta>\beta_c$ with $\beta_c J=1$, the order parameter m vanishes and the specific heat is given by

$$c(\beta) = -\frac{1}{T^2} \frac{\partial}{\partial\beta} u(\beta) = \frac{J}{T^2} 2q \frac{\partial q}{\partial\beta} \tag{3.11}$$

Differentiating (2.18b) or (3.8) with respect to β we find, using (3.1),

$$\frac{\partial q}{\partial\beta} = \frac{1}{2} d_2(\beta) Jq [1 - \frac{1}{2} \beta J d_2(\beta)]^{-1} \tag{3.12}$$

and thus

$$c(\beta) = k_B (\beta Jq)^2 d_2(\beta) [1 - \frac{1}{2} \beta J d_2(\beta)]^{-1} \quad . \tag{3.13}$$

Here k_B is Boltzmann's constant; we usually take it as one. For a discrete probability distribution of the coupling constants, $d_2(\beta)$ goes to zero exponentially fast as $\beta \to \infty$ or, equivalently, $T \to 0$.

In metallic spin glasses at low temperatures the long range of the RKKY interaction samples many small J_{ij}-values [17] so that in the neighbourhood of J=0 their distribution is better approximated by a strongly peaked continuous curve. Let us therefore suppose that $\xi+\eta$ has a probability density $\rho(x)$ which is continuous at x=0. Then we obtain, via the substitution $\beta Jq(0)x=y$,

$$d_2(\beta) = \int_{-\infty}^{+\infty} dx\rho(x)x^2 \cosh^{-2}\{(\beta Jq)x\} \sim \frac{\pi^2}{6}\frac{\rho(0)}{[\beta Jq(0)]^3} \tag{3.14}$$

as $\beta J \to \infty$. Here $q(0)=\frac{1}{2}<|\xi+\eta|>$ is the value of q at T=0; for example, in the Gaussian case $q(0)=1/\sqrt{\pi}$. Hence the specific heat at low temperature is linearly dependent on T,

$$c(\beta) \sim \frac{\pi^2}{6} k_B^2 \, T\rho(0)/Jq(0) \ . \tag{3.15}$$

Apart from the factor $Jq(0)$, (3.15) is just Eq.(1) of Anderson et al. [29]. Note that in this argument the continuity of $\rho(x)$ at x=0 is essential. The linear dependence upon T signals the existence of many low-lying excitations in the neighbourhood of the groundstate(s).

Calorimetric experiments which show the linear dependence upon T have been done mainly on CuMn [30] and AuFe [31]. The short-range interaction between the magnetic moments, which varies with the concentration, is taken into account by J_o. Since by normalization $\rho(0)$ and $q(0)$ do not depend on the concentration, the linear temperature specific-heat coefficient is essentially concentration independent [30,31]. This independence has also been predicted by Anderson et al. [29], who used a scaling argument.

The high-temperature behaviour of $c(\beta)$, as deduced from (3.13), is characterized by a discontinuity determined by $c(\beta_c^-)=0$ and $c(\beta_c^+)=6k_B/<(\xi+\eta)^4>$ (twice as large as claimed in Ref.4). This is most easily seen by returning to $u(\beta) = -Jq^2$ and noting that, as $T \uparrow T_f=J/k_B$,

$$q^2 \approx 6(\beta J-1)/<(\xi+\eta)^4> \ . \tag{3.16}$$

In our opinion the discontinuity is not a serious drawback since the inclusion of a short-range interaction presumably gives the high-temperature behaviour required by experiment [30]. However, at low temperatures most short-range interactions are not expected to contribute significantly to (3.15) [32].

4. The extended Gibbs formalism

The Ising ferromagnet in zero external field has a discrete spin-flip symmetry. Hence

$$\langle S(i)\rangle_\beta = \mathrm{Tr}\ S(i)\ \exp(-\beta H_N)/\mathrm{Tr}\ \exp(-\beta H_N) = 0 \tag{4.0}$$

whatever β and whatever the dimension. According to the Gibbs prescription $\langle S(i)\rangle_\beta$ is the magnetic moment we measure when the system is in thermal equilibrium at inverse temperature β. In spite of that, we know the symmetry is broken below T_c ($d \geq 2$) and experiment shows a nonzero magnetization. How can we explain this paradox? A way out was already indicated by Ising in his classic 1925 paper [33]. Trying to explain why the magnetization m of the one-dimensional chain vanishes in zero external field h, he observed the following:

"Two arguments could explain this behaviour. If one plots for h=0 the probabilities as a function of the magnetic moment m, then this curve either has *two, presumably very steep maxima, situated symmetrically with respect to m=0,* or a single maximum at m=0. Though thermal averaging produces m=0, we are not forced to conclude that the moments we want to observe vanish if the probabilities have two maxima. In fact, it would be much more plausible that the magnetization of the system spends quite a long time, *much longer than the observation time,* at either of the maxima."

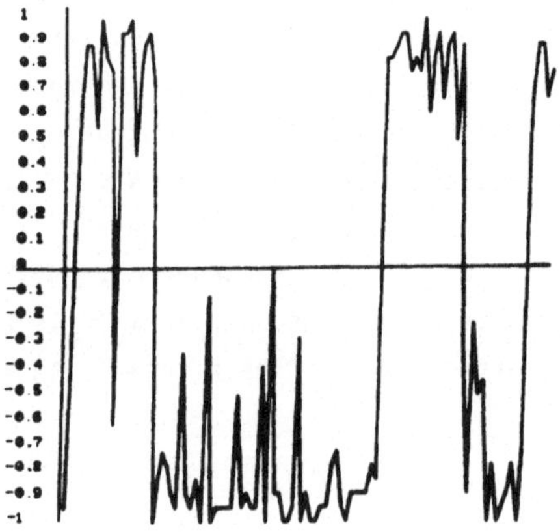

Fig.6. Ferromagnetic Ising model: magnetization every 1,000 steps for a Monte Carlo simulation with 500,000 steps 6x6 lattice with periodic boundary conditions and T slightly below T_c; after Kindermann and Snell [56]. Both ergodic components, (+) and (-), are equally allowed. Though the system is rather small it clearly exhibits the tendency to be in *one* ergodic component: most of the time we have almost all spins up or all spins down. [Reprinted from Contemporary Mathematics, "Markov Random Fields and Their Applications", Volume 1 (1980) p. 59, by permission of the American Mathematical Society].

The idea Ising put forward was right (see Ref.34 and Fig.6), though he correctly observed that it did not apply to the one-dimensional chain. In this section I want to explain how it emerges from the basic postulates of equilibrium statistical mechanics.

Throughout this section we assume that the Hamiltonian may be written

$$H_N = -\sum_{(i,j)} J_{ij} S(i)S(j) - h\sum_i S(i) \tag{4.1}$$

where the J_{ij} are coupling constants which describe the strenght of the interaction between the spins. In real life interactions are not infinite-range. For instance, the conduction electrons in a metallic spin-glass have a finite mean free path λ and thus $J_{ij}(\mathrm{RKKY}) \to J_{ij}(\mathrm{RKKY})\exp\{-R_{ij}/\lambda\}$, i.e., the RKKY interaction is exponentially damped [35]. Hence

$$||J|| = \sup_i\{\sum_j |J_{ij}|\} < \infty \tag{4.2}$$

Though somewhat too restrictive [12], (4.2) will be assumed to simplify the discussion. In section 4.1 we analyze the conventional Gibbs Ansatz. The extended Gibbs formalism is treated in section 4.2 . Section 4.3 is devoted to the notions of phase transition and order parameter as they occur in the newly developed context. We allow both discrete and continuous symmetries, and randomness. A simple example is given in section 4.4. Below we study classical systems only, with special emphasis on Ising spins. Though rather analogous, quantum systems will be considered elsewhere [3].

4.1 The Gibbs Ansatz (Canonical Formalism)

Suppose a finite spin system with Hamiltonian (4.1) is in contact with a heat bath at inverse temperature β. If the system is in thermal equilibrium, the expectation values are determined by the probability density, or density matrix,

$$\rho_N = C\exp(-\beta H_N) \ . \tag{4.3}$$

Moreover, the expectation values correspond to the values one determines experimentally. This is the Gibbs Ansatz [36,37]. The Hamiltonian H_N contains the full symmetry group of the system (spin-flip or rotation invariance, etc.) so that expectation values à la (4.0) always vanish, i.e., the symmetry is never broken. Nevertheless it is well-known that in experiments on macroscopic systems the symmetry is broken below T_c. A finite system with density matrix (4.3) never allows for broken symmetry. We, therefore, take the practical point of view that an infinite system is a reasonable approximation of a macroscopic system and assume the system to be infinite from the very beginning. Moreover, taking the system infinite is unavoidable if one wants to study equilibrium phase transitions [38]. One may wonder, though, what to do with $\exp(-\beta H_N)$ as $N \to \infty$.

4.2 The extended Gibbs formalism (N=∞).

Let μ_β or simply μ be an equilibrium state of the infinite system. It is a probability measure on the system's phase space Ω, the set of all possible spin configurations. If A and B are two events in Ω with $\mu(B)>0$ (say A is spin up in i and B is spin down in j), then the conditional probability

$$\mu(A|B) = \mu(A\cap B)/\mu(B) \tag{4.4}$$

is the probability that A happens given B.

What do we expect from an equilibrium state? At least, that each spin is in thermal equilibrium with its surroundings. It turns out that this simple criterion suffices. More precisely, pick an arbitrary point i and suppose that the range of the interaction is r, i.e., $J_{ij}=0$ if $|i-j|>r$. Let Λ' be an arbitrary domain containing i and the sphere $\Lambda(i)$ around i with radius r. Finally+, let S_Λ be a set of spin values S_i, $i\in\Lambda$, and denote by $\mu(S_\Lambda)$ the probability of getting S_Λ. Then μ is said to be an equilibrium or Gibbs state if, whatever i, μ satisfies the Dobrushin-Lanford-Ruelle (DLR) condition [39]

$$\mu(S_i\cap S_{\Lambda'-\{i\}}) = \mu(S_i|S_{\Lambda'-\{i\}})\mu(S_{\Lambda'-\{i\}}) \tag{4.5}$$

where $\mu(S_i|S_{\Lambda'-\{i\}})$ is given by

$$\mu(S_i|S_{\Lambda'-\{i\}})= C\exp\beta\{\sum_{j\in\Lambda'} J_{ij}S_iS_j+hS_i\} , \tag{4.6}$$

the constant C in (4.6) being chosen in such a way that $\mu(S_i=+1|S_{\Lambda'-\{i\}}) + \mu(S(i)=-1|S_{\Lambda'-\{i\}})=1$. To derive (4.5) and (4.6) we imagine Λ' to be contained in a big volume Λ, apply (4.3) and (4.4), and let $\Lambda\to\infty$. The spins S_j with j in $\Lambda(i)$ produce an external field $\sum_j J_{ij}S_j$ at the site i and hence, together with h, the Boltzmann factor (4.6) - as advertized [40].

Alternatively one could argue that an equilibrium state μ should be obtained as a limit of finite-volume Gibbs states, $\langle A\rangle=\lim_{N\to\infty} \mathrm{Tr}(\rho_N A)$ with ρ_N given by (4.3), so that $\langle A\rangle=\int d\mu(\omega)A(\omega)$. Then μ satisfies the DLR condition. However, in the spin-glass case there is nothing that guarantees the existence of this limit. The DLR conditions themselves are a much better alternative since they do not involve a limit any more.

Any solution to the DLR equations (4.5)-(4.6) is called a Gibbs state (Gibbs measure). Gibbs states have some interesting properties. First, for a finite system there is a unique solution to the DLR equations: the Gibbs state (4.3) we started with. Second, any convex combination of Gibbs states gives another solution to the DLR equations, as is directly seen by noticing that the right-hand side of (4.6) does not de-

+We take here subscripts i and Λ instead of S(i) and S(Λ) to simplify the notation.

pend on μ. Conversely, every Gibbs measure μ may be decomposed uniquely into ergodic components $\mu^{(\alpha)}$ (D,LR 1969):

$$\mu = \sum_\alpha p_\alpha \mu^{(\alpha)}, \qquad p_\alpha \geq 0 \quad \text{and} \quad \sum_\alpha p_\alpha = 1 \ . \tag{4.7}$$

This is the ergodic decomposition of μ, which is basic to all that follows. The $\mu^{(\alpha)}$ live on disjoint regions Ω_α of the phase space Ω. Finally, ergodic components are space-clustering, i.e., if $\mu^{(\alpha)}$ is ergodic, then

$$\lim_{|x|\to\infty} |\mu^{(\alpha)}(A\tau_x B) - \mu^{(\alpha)}(A)\mu^{(\alpha)}(\tau_x B)| = 0 \ , \tag{4.8}$$

where τ_x is a space translation by x.

It is often convenient to think of the components $\mu^{(\alpha)}$ or, identifying them, Ω_α as being separated by infinitely high free energy barriers. Of course, for a macroscopic but finite system, infinitely high is also a matter of time scales. For practical purposes a free energy barrier may become "infinitely" high if it cannot be passed within τ_{obs}, the time available for the experiment. Here we need not go into this problem since it has already been discussed, for instance, by Münster [41], and ... Ising [33].

4.3 Phase transitions and order parameters

If the temperature T is high enough, there is a unique solution μ_β to the DLR equations (Dobrushin's uniqueness theorem). By the uniqueness μ_β is ergodic and the limit of finite-volume Gibbs states. We say the system exhibits a (conventional) phase transition at T_c if below T_c more than one solution to the DLR equations exists. So the following picture emerges. For $T>T_c$ there is a unique, ergodic, equilibrium state. Imagine that μ_β is obtained as a limit of finite-volume Gibbs states with free boundary conditions. Below T_c the Gibbs state μ_β is never ergodic, but it can be decomposed uniquely, à la (4.7), into ergodic components, which represent the pure thermodynamic phases. That is, *below* T_c *the ergodicity is broken.*

N.B.: We say a phase transition is "unconventional" if several solutions to the DLR equations exist only at T_c (energy-entropy transition [42]) or if the equilibrium state μ_β is unique for all values of β but the free energy is not analytic at β_c (Kosterlitz-Thouless transition in the two-dimensional XY-model [43]). We assume that neither possibility applies in the spin-glass case.

The free energy per spin $f(\beta)$ is given by (2.3). For random potentials which satisfy (4.2) or a slightly more general condition, $f(\beta)$ depends neither on the boundary conditions nor on the specific random configuration [15,12]. As a first consequence of this result we note that if a system has a "phase transition" in the form of a nonanalyticity in $f(\beta)$ at β_c, then all samples have the same T_c.

We can also assign a free energy $f(\beta;\alpha)$ to each of the ergodic components $\mu^{(\alpha)}$ of a Gibbs state μ_β,

$$f(\beta;\alpha) = \lim_{N\to\infty} \frac{1}{N} F_N(\beta;\alpha) = \lim_{N\to\infty} \frac{1}{N} [\mu^{(\alpha)}(H_N)-\beta^{-1}S_N(\alpha)] \quad , \tag{4.9}$$

where $\mu^{(\alpha)}(H_N)$ is the expectation of H_N with respect to $\mu^{(\alpha)}$ and $S_N(\alpha)$ is the entropy of $\mu^{(\alpha)}$ restricted to a finite domain with N spins. [Roughly, if A is an arbitrary local observable, then $\mu^{(\alpha)}(A)=\mathrm{Tr}(\rho_N^\alpha A)$ with ρ_N^α not necessarily identical with (4.3), and $S_N(\alpha)=-\mathrm{Tr}\rho_N^\alpha \ell n\rho_N^\alpha$.] It can be shown [3] that in spite of the randomness *all the components have the same free energy, which agrees with $f(\beta)$ as defined by (2.3)*, i.e., $f(\beta;\alpha)=f(\beta)$, whatever α.

Let Ω^N be the phase space of a finite system. It is tempting to suppose that the component structure already exists (at least, in some sense) for finite N so that $\Omega^N \approx \bigcup_\alpha \Omega_\alpha^N$ with the $\mu^{(\alpha)}$ living on the (approximately) disjoint Ω_α^N. Then $f(\beta;\alpha)= \lim_{N\to\infty} N^{-1}\ell n\mathrm{Tr}_\alpha(\exp-\beta H_N)$ where $\mathrm{Tr}_\alpha(\ldots)$ denotes the restriction of the trace to Ω_α^N. If the number of ergodic components is finite (say, two), this idea might be sensible but for classical systems with a continuous symmetry and for quantum systems the above decomposition is problematic. A better way of understanding (4.9) and the fact that $f(\beta;\alpha)=f(\beta)$ is, for classical systems, provided by the following argument. The ergodic state $\mu^{(\alpha)}$ may be obtained as the thermodynamic limit of C $\exp(-\beta H_N)$ where H_N is equipped with suitable boundary conditions (nontrivial; cf. section 4.4). The free energy, $N^{-1}\ell n\mathrm{Tr}\exp(-\beta H_N)$, does not depend on boundary conditions as $N\to\infty$. Hence $f(\beta;\alpha)=f(\beta)$.

According to (4.7), below T_c a Gibbs state μ_β is a convex combination of ergodic components $\mu_\beta^{(\alpha)}$, each with its own weight (probability) p_α, and the system certainly picks one component α. We just do not know which one. The probabilities p_α are consistent with our information (ignorance) about the system: it is in thermal equilibrium.

Since an ergodic component has at best a probability p_α and, hence, one does not know which one the system is in, one might argue that thermodynamic observables such as T_f itself, the magnetization, the susceptibility, and the specific heat, cannot be predicted with certainty. That is, they are not reproducible. This is not the case. All components have the same free energy $f(\beta)$ or more precisely, if $h\neq 0$, $f(\beta,h)$. The singularities of this function signal a phase transition (so all components have the same phase transition at the same time) and the derivatives with respect to β and h give the magnetization, etc.. The function $-\beta f(\beta,h)$ is convex in β and h and, hence, almost everywhere differentiable with respect to these variables. Accordingly all components have the same magnetization, etc., at these values of β and h. Summarizing: Thermodynamic observables are reproducible since, in general, they depend neither on the specific random configuration nor on the ergodic component.

Suppose now that we have determined all the ergodic components $\mu_\beta^{(\alpha)}$. How do we discriminate between them? The answer is simple: by finding an order parameter. More specifically, an order parameter is an observable A, or a group of observables A_i, $1\leq i\leq n$, such that the numbers $\mu_\beta^{(\alpha)}(A)$ or $\mu_\beta^{(\alpha)}(A_i)$, $1\leq i\leq n$, determine α and, hence,

$\mu_\beta^{(\alpha)}$ uniquely. At the moment we do not specify n. However, finding suitable order parameters may be highly nontrivial.

4.4 An example

The two-dimensional Ising model with translationally invariant and ferromagnetic nearest-neighbour interactions is defined by (4.2) with $J_{ij}=J>0$ if i and j are nearest neighbours, and $J_{ij}=0$ otherwise. We assume h=0. Onsager showed in his classical paper [44] that the model has a phase transition at a positive T_c. More precisely, he showed that $f(\beta)$ is nonanalytic at β_c. Below T_c the model has two ergodic equilibrium states, $\mu_\beta^{(+)}$ with positive magnetization and $\mu_\beta^{(-)}$ with negative magnetization. If $T>T_c$, then $\mu_\beta^{(+)}=\mu_\beta^{(-)}$ and the spontaneous magnetization vanishes. The states $\mu_\beta^{(\pm)}$ can be obtained as thermodynamic limits of finite-volume Gibbs states with (±) boundary conditions. In three dimensions there are many more ergodic components which can be obtained by mixed (±) boundary conditions with an interface.

The natural symmetries of the model are translational invariance and spin-flip symmetry $S(i)\to -S(i)$ for all i. By flipping the spins we map $\mu_\beta^{(+)}$ onto $\mu_\beta^{(-)}$ and conversely. If μ_β is the Gibbs state obtained via the thermodynamic limit with free or periodic boundary conditions, then μ_β is spin-flip invariant. Hence

$$\mu_\beta = \frac{1}{2}\,\mu_\beta^{(+)} + \frac{1}{2}\,\mu_\beta^{(-)} \quad . \tag{4.10}$$

Below T_c the ergodicity of μ_β is broken. Once we know the system is in thermodynamic equilibrium we find with probability $\frac{1}{2}$ the state $\mu_\beta^{(+)}$ and with probability $\frac{1}{2}$ the state $\mu_\beta^{(-)}$. But we do not find both at the same time.

An order parameter is easily found. We take S(0), the spin at 0. If $\mu_\beta^{(\alpha)}(S(0))>0$, we have $\alpha=+$, and if $\mu_\beta^{(\alpha)}(S(0))<0$, we are left with $\alpha=-$. The sign of $\mu_\beta^{(\alpha)}(S(0))$ already determines α completely. In the case of the three-dimensional isotropic XY or Heisenberg model we take $\underset{\sim}{S}(0)$, the two or three components of the spin at 0 (n=2 or 3). For full details, see Ref.3.

5. Discussion

It is time to harvest some corollaries. In section 5.1 we compare the new formalism with conventional statistical mechanics. The maximal number of components is estimated in section 5.2, the role of order parameters is analyzed in section 5.3, and the fluctuation-dissipation "theorem" is discussed briefly in section 5.4. In section 5.5 we return to the mean-field model of sections 2 and 3 and indicate its short-range version which for a particular probability distribution also gives rise to some exact results. A conclusion can be found in the final subsection. Since the appearance of the preprint of Ref.3 (summer 82) several applications [45-48,52] of the extended Gibbs formalism have appeared, and in passing I will comment on some aspects of this work.

5.1 Comparison with conventional statistical mechanics

The reader might wonder whether the extended Gibbs formalism as outlined in section 4 is compatible with Boltzmann's ideas on ergodicity [37]. Indeed, it's not. Boltzmann imagined that the system would visit all hamlets during its journey on the energy hypersurface. And we have just seen that this is not to be expected if $N \to \infty$ and the system is *in thermal equilibrium*. At low enough temperature the energy hypersurface decomposes into several disjoint regions Ω_α and the system stays in only one of them. For example, though the Ising model does not have a dynamics by itself, one may give it a Glauber dynamics [49]. If two components are separated by a free energy barrier of height proportional to $N^{1/2}$, say (true for the Ising model of section 4.4), then the time it takes to climb the free energy barrier between them is proportional to $\exp(N^{1/2})$, which is "infinite" for a human observer of a macroscopic system with $N=10^{23}$. One may call this absolute confinement [50]. There is no chance of the system visiting all the components within the experimenter's observation time τ_{obs}. In the next paper Palmer will extend parts of our formalism to the case where the free energy barriers have a finite height.

It is to be noted, however, that the extended Gibbs formalism displays its full strength when applied to models with continuous symmetries, where a decomposition of the phase space for $N < \infty$ is not that evident, and to random systems. The point is that the free energy depends neither on the specific random configuration nor on the ergodic component, hence guaranteeing that thermodynamic observables are reproducible.

5.2 The number of components

How many components can we expect? Quite a few, but I will give a simple argument to show that there cannot be $\exp(aN)$ disjoint ergodic components as $N \to \infty$. To simplify the discussion we take N finite but very large and estimate the free energy $f(\beta)$ of the canonical Gibbs state μ_β, i.e., (4.3) with free boundary conditions. If the above suggestion were true, the phase space Ω^N could be composed into $M=\exp(aN)$ disjoint ergodic components Ω_α^N, and

$$\mu_\beta = \sum_{\alpha=1}^{M} p_\alpha \mu_\beta^{(\alpha)} \quad , \quad \mu_\beta^{(\lambda)}(\Omega_\nu) = \delta_{\lambda,\nu} \quad . \tag{5.1}$$

The entropy of μ_β is given by

$$S_N(\mu_\beta) = -\mathrm{Tr}\rho_N \ln\rho_N = \sum_{\alpha=1}^{M} p_\alpha S_N(\mu_\beta^{(\alpha)}) - \sum_{\alpha=1}^{M} p_\alpha \ln p_\alpha \tag{5.2}$$

where $S_N(\mu_\beta^{(\alpha)})$ is the entropy of the αth component. To estimate the last term in (5.2), which Palmer calls the complexity [50],

$$I_N = -\sum_{\alpha=1}^{M} p_\alpha \ln p_\alpha \quad , \tag{5.3}$$

we note that $p_\alpha \approx \exp(-aN)$ with $a>0$ (by assumption), and find $I_N \approx Na$. Thus the entropy per spin may be written

$$s(\mu_\beta) = \sum_{\alpha=1}^{M} p_\alpha s(\mu_\beta^{(\alpha)}) + a \ . \tag{5.4}$$

The energy per spin is easily found,

$$u(\mu_\beta) = \frac{1}{N} \mu_\beta(H_N) = \sum_{\alpha=1}^{M} p_\alpha u(\mu_\beta^{(\alpha)}) \ , \tag{5.5}$$

and hence, since all the ergodic components have the same free energy and $f(\beta)=u(\mu_\beta)-Ts(\mu_\beta)$,

$$f(\beta) = \sum_{\alpha=1}^{M} p_\alpha f(\beta) - aT = f(\beta) - aT \Rightarrow a=0, \tag{5.6}$$

which contradicts our assumption $a>0$. In fact, we have shown that in the context of equilibrium statistical mechanics $N^{-1} I_N = o(N)$. If the disjointness of the components were only approximately true, the same conclusion would still hold [51].

5.3 Order parameters

Though we have estimated the maximal number of components, we have not characterized them yet. A first distinction is made through order parameters (section 4.3). Instead of characterizing ergodic components uniquely (labeling) one could require [52] that order parameters should signal a phase transition and not that much more. That is, order parameters should vanish for $T>T_c$ only. Let us write m_i^α for $\mu_\beta^{(\alpha)}(S(i))$, the expectation of $S(i)$ in the ergodic component α. As we have seen, $q_{EA} = \langle\langle S(i)\rangle_\beta^2\rangle_J$ is not quite sensible, but a slightly modified version does make sense, at least formally,

$$q'_{EA}(\alpha) = \lim_{N\to\infty} \frac{1}{N} \sum_{i=1}^{N} (m_i^\alpha)^2 \ , \tag{5.7}$$

which is nothing but (1.2). However, now the average is taken with respect to <u>one</u> component, labeled by α. In the spin-glass phase one expects $m_i^\alpha \neq 0$ provided $T<T_f$. This need not be true for all i, but should certainly hold for the spatial average (5.7). A second candidate was put forward by De Dominicis and Young [46,48],

$$q''_{EA} = \sum_\alpha p_\alpha q'_{EA}(\alpha) \ . \tag{5.8}$$

Several other, equivalent, candidates have been considered by van Enter and Griffiths [52], who also proposed

$$q'''_{EA} = \lim_{N\to\infty} \max_\alpha \{ \frac{1}{N} \sum_{i=1}^{N} (m_i^\alpha)^2 \}. \tag{5.9}$$

The existence of the limit in (5.9) is guaranteed. It is identical with a duplicating proposal of Blandin et al. [55]. The order parameters (5.8) and (5.9) do not depend on α and, hence, do not discriminate between the different ergodic components.

We now turn to the Parisi scheme [45]. Define the overlap between the magnetization in two different components by $(N\to\infty)$

$$q_{\alpha\beta} = \frac{1}{N}\sum_{i=1}^{N} m_i^{\alpha} m_i^{\beta} \tag{5.10}$$

$-1\leq q_{\alpha\beta}\leq +1$. Given $\{p_\alpha\}$, the probability "density" $P(q)$ of $q_{\alpha\beta}$ is given by

$$P(q) = \sum_{\alpha,\beta} p_\alpha p_\beta \delta(q-q_{\alpha\beta}). \tag{5.11}$$

Then another, monotonic, function is introduced,

$$x(q) = \int_{-1}^{q} dqP(q) , \tag{5.12}$$

whose inverse $q(x)$, $0\leq x\leq 1$, is the order parameter we are looking for. Note that $x(q)$ increases by at least $2p_\alpha p_\beta$ at each $q_{\alpha\beta}=q_{\beta\alpha}$.

For the Ising ferromagnet with h=0 and spontaneous magnetization m we have the following picture, where the function $q(x)$ has been indicated by heavy vertical lines. It is odd with respect to $x=1/2$, and $q(1)=m^2$. The function $q(x)$ is constant if and only if there is a unique ergodic state $(T>T_c)$. This is also true in the spin-glass case. And here too $q(x)$ has to be <u>odd</u> by spin-flip symmetry. Parisi [45] asserts

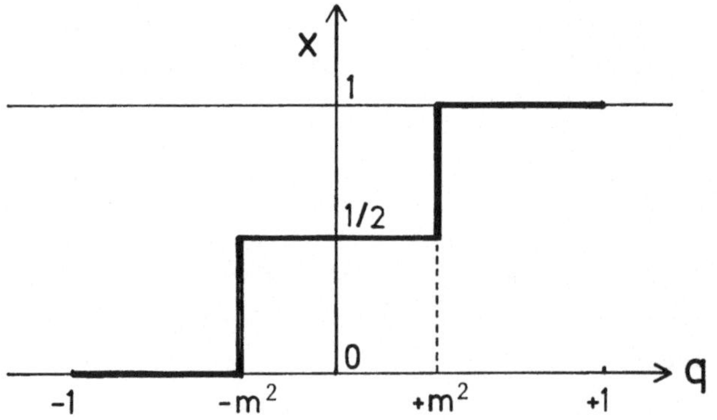

Fig.7. The order parameter function $q(x)$ for the two-dimensional Ising ferromagnet. The horizontal lines refer to $x(q)$, $-1\leq q\leq 1$, and the heavy vertical lines to $q(x)$, $0\leq x\leq 1$.

$q_{EA}=\int_0^1 dxq(x)$. Hence $q_{EA}=0$. Moreover, by Cauchy-Schwarz,

$$q_{\alpha\beta} < \max_{\alpha,\beta} \{q_{\alpha\alpha}, q_{\beta\beta}\} , \qquad \alpha \neq \beta \quad , \tag{5.13}$$

so that

$$q_N(1) = \max_{\alpha} \{q_{\alpha\alpha}\} \quad . \tag{5.14}$$

Sending $N \to \infty$ we find

$$q(1) = q'''_{EA} \quad . \tag{5.15}$$

5.4 The fluctuation-dissipation theorem

This theorem is supposed to be a "pièce de résistance" of equilibrium statistical mechanics, but what does it mean? The most frequently quoted version is that, as $N \to \infty$, the susceptibility is given by

$$\chi(T,h) = \frac{1}{N} \sum_{i=1}^{N} \left[\sum_{j=1}^{N} \{\mu_\beta(S(i)S(j)) - \mu_\beta(S(i))\mu_\beta(S(j))\}\right] \quad , \tag{5.16}$$

where μ_β is the canonical Gibbs state with free boundary conditions. Let us take a two-dimensional ferromagnetic Ising model with h=0. Then $\mu_\beta(S(i))=0$ by spin-flip symmetry and $\mu_\beta(S(i)S(j)) \to m^2 \neq 0$ if $T<T_c$, so that the right side of (5.16) diverges as $N \to \infty$. As it stands the relation simply does not make sense. One could argue that one has to take the limit $h \downarrow 0$ after $N \to \infty$ so as to pick the positive component $\mu_\beta^{(+)}$, but this is somewhat dissatisfying. The way out is that, fixing h, we are in a specific component α so that we have to take $\mu_\beta^{(\alpha)}$ instead of μ_β in (5.16). Then

$$\mu_\beta^{(\alpha)}(S(i)S(j)) - \mu_\beta^{(\alpha)}(S(i))\mu_\beta^{(\alpha)}(S(j)) \tag{5.17}$$

is expected to be summable in j, for i fixed; cf. relation (4.8). Indeed, for the two-dimensional Ising model at h=0 and $T \neq T_c$, the truncated correlation function is known to decrease exponentially fast as $|j| \to \infty$. The same holds true for a random Ising ferromagnet, the correlation length being independent of the specific random configuration [53]. But for systems with a broken continuous symmetry this type of fall-off cannot be expected [54].

5.5 Beyond mean-field theory once again

The mean-field model of section 2 nicely exemplifies many of the ideas put forward in section 4. As $N \to \infty$ the system picks a specific component whose free energy depends neither on the component itself nor on the specific random configuration.

What, however, happens if we take the J_{ij} finite-range, say, nearest-neighbour,

$$J_{ij} = J\,\{\xi_i \eta_j + \xi_j \eta_i\} \quad ? \tag{5.18}$$

Suppose first that the ξ's and η's are ±1 with equal probability. Then

$$\{\xi_i\eta_j + \xi_j\eta_i\} = \xi_i\eta_j\{1 + (\xi_i\eta_i)(\xi_j\eta_j)\} \tag{5.19}$$

vanishes if i and j have a different colour. The blue spins give rise to a phase transition if the blue colour percolates. It does so if p_c, the site-percolation threshold, is less than $\frac{1}{2}$. There is no infinite blue cluster in two dimensions (p_c=0.59) but there is in three dimensions (p_c=0.307) and, hence, a phase transition. For Ising spins the lower critical dimensionality has been raised by 1.

Suppose next that we take a continuous distribution of ξ and η. Then, with probability one, the J_{ij} never vanish. Moreover, they can have either sign; see section 2.3, in particular (2.22). In this way quite a bit of frustration is introduced. Spins grouped into clusters are connected by partially frustrated bonds, and are all members of an infinite percolating network. It is an open question whether and, if so, how the above (±) picture persists.

5.6 Summary

There is increasing evidence that metallic (RKKY) and insulating spin glasses are in different universality classes, even perhaps with different values of the lower critical dimensionality d_L. It is suggested that d_L(RKKY)=3. If so, not only mean-field theory but also the extended Gibbs formalism offers a surprisingly consistent description of a metallic spin glass in thermal equilibrium.

Every equilibrium state can be decomposed uniquely into ergodic components. Below T_f one has to take a thermal average with respect to one of them and need not average over the whole phase space, i.e., all of them. Since the free energy is identical for all components and does not depend on the specific random configuration, thermodynamic observables are reproducible.

Understanding the equilibrium behaviour of a spin glass means characterizing the ergodic components. This is highly nontrivial and a challenge for future research.

Acknowledgments

First of all I would like to thank my collaborators, A.C.D. van Enter and J. Canisius, with whom I did most of the work described in this paper. Moreover, I gratefully acknowledge stimulating discussions with J.T. Chalker, O.J. Indekeu , I. Morgenstern, R.G. Palmer, and Ph. de Smedt.

Appendix on large deviations

In this appendix I give the proofs of some arguments on large deviations which were used in section 2 and Ref.4. The main ideas are outlined in subsection A.1, the existence of certain limits is proven in subsection A.2, and the explicit answers are provided in subsection A.3. The case of independent, identically distributed stochastic variables has been considered by Lanford [21] and I will heavily draw on some of his ideas.

A.1. *The main ideas.* Let us assume Ising spins and take a discrete probability distribution for ξ and η. This is convenient but not essential. In calculating the free energy (2.3) we have to evaluate the trace, which is a sum over all spin configurations. As we already noted in section 2.2, it is advantageous to make a "coordinate transformation", and sum over the variables m_N, q_{1N}, and q_{2N} instead. That is, writing $-\beta H_N = NQ(\underset{\sim}{m}_N)$ with $\underset{\sim}{m}_N = (m_N, q_{1N}, q_{2N})$,

$$-\beta f(\beta) = \lim_{N\to\infty} N^{-1} \ln 2^{-N} \sum_{\{S(i)=\pm 1\}} \exp\{-\beta H_N\}$$

$$= \lim_{N\to\infty} N^{-1} \ln \sum_{\underset{\sim}{m}_N} 2^{-N} W(\underset{\sim}{m}_N) \exp\{NQ(\underset{\sim}{m})\} \qquad (A.1)$$

where $W(\underset{\sim}{m}_N)$ is the number of spin states for a <u>given</u> $\underset{\sim}{m}_N$, and the sum is over all admissible $\underset{\sim}{m}_N$.

Now the ξ's and η's are fixed and the Ising spins are to be considered as *independent* , stochastic, variables. Then $2^{-N} W(\underset{\sim}{m})$ is just the probability Prob$\{\underset{\sim}{m}\}$ that $N^{-1}\sum_i S(i)$, $N^{-1}\sum_i \xi_i S(i)$, and $N^{-1}\sum_j \eta_j S(j)$ assume a particular value $\underset{\sim}{m}$. Hence

$$-\beta f(\beta) = \lim_{N\to\infty} N^{-1} \ln \sum_{\underset{\sim}{m}} \exp N\{Q(\underset{\sim}{m}) - (-\frac{1}{N} \ln \text{Prob}\{\underset{\sim}{m}\})\}. \qquad (A.2)$$

Since we have to divide the logarithm in (A.2) by N we get a well-defined, nonzero, contribution only if the sum contains terms of the form exp N{...} and {...} converges to a limit as $N\to\infty$. If so, $-\beta f(\beta) = \max_{\underset{\sim}{m}}\{\ldots\}$. More precisely, let J be a <u>convex</u> open set in $\mathbb{R}^3$; for our purposes it suffices to take, say, an infinitesimally small volume surrounding $([m,m+dm),[q_1,q_1+dq_1),[q_2+dq_2))$. We show that

$$\lim_{N\to\infty} -\frac{1}{N} \ln \text{Prob}\{\underset{\sim}{m}_N \in J\} \qquad (A.3)$$

exists and does not depend on the specific random configuration of ξ's and η's we started with. Once we know the limit (A.3) exists we can determine it explicitly. This will be done in subsection A.3 where we relate the limit (A.3) to the Legendre transform of a c-function; cf. Eqs. (2.7)-(2.17).

If J does not contain the mean spin value $\mathbb{E}_S\{S\}=0$, the event $\{\underset{\sim}{m}_N \in J\}$ is called a *large deviation* since it entails a deviation away from the mean (here zero) which is proportional to N. It becomes more and more improbable as $N\to\infty$. Small deviations are associated with the central limit theorem where one scales by $\sqrt{N}$ instead of N. On the other hand, if J contains zero, Prob$\{\underset{\sim}{m}_N \in J\}$ converges to one. This may be seen by using the following version of the strong law of large numbers [19], which is a mainstay for what follows:

Let $X_1, X_2, \ldots$ be independent random variables with mean zero and finite variance, i.e., $\mathbb{E}\{X_k\}=0$ and $\mathbb{E}\{X_k^2\} < \infty$. If

$$\sum_{k=1}^{\infty} k^{-2} \mathbb{E}\{X_k^2\} < \infty , \qquad (A.4)$$

then, as $N\to\infty$,

$$N^{-1}\sum_{i=1}^{N} X_i \to 0 \qquad \textit{with probability one.}$$

Condition (A.4) is satisfied if, for instance, $\mathbb{E}\{X_k^2\}$ is uniformly bounded. Here $X_i=\xi_i S(i)$ or $\eta_i S(i)$ and we have to check that $\sum_{k=1}^{\infty} k^{-2}\xi_k^2<\infty$ and $\sum_{k=1}^{\infty} k^{-2}\eta_k^2<\infty$. But that's easy. Take the expectation value with respect to ξ or η and use the monotone convergence theorem:

$$<\sum_{k=1}^{\infty} k^{-2}\xi_k^2> \;=\; \sum_{k=1}^{\infty} k^{-2}<\infty \;.$$

Hence the sum inside the angular brackets is finite with ξ-probability one.

A.2 *The existence of the limit (A3).* We show that , for J fixed, $-\ln \mathrm{Prob}\{\underset{\sim}{m}_N \in J\}\geq 0$ is subadditive in N or, equivalently, that

$$\mathrm{Prob}\{\underset{\sim}{m}_{N+M}\in J\}\geq \mathrm{Prob}\{\underset{\sim}{m}_N\in J\}\cdot\mathrm{Prob}\{\tau_N \underset{\sim}{m}_M\in J\} \tag{A.6}$$

where $\tau_N(\xi_i)=\xi_{i+N}$ and $\tau_N(\eta_j)=\eta_{j+N}$. Since we will return to the notion of subadditivity shortly, we first prove (A.6). By the independence of the spin variables (!) the right-hand side may be written

$$\mathrm{Prob}\{\underset{\sim}{m}_N\in J \;,\; \tau_N \underset{\sim}{m}_M\in J\} \;. \tag{A.7}$$

Suppose now that a certain spin configuration satisfies the condition inside the curly brackets in (A.7). Then

$$\underset{\sim}{m}_{N+M} = \left(\frac{N}{N+M}\right)\underset{\sim}{m}_N + \left(\frac{M}{N+M}\right)\tau_N \underset{\sim}{m}_M \in J \tag{A.8}$$

by the convexity of J. This and (A.7) imply (A.6).

Let us assume for the moment that our problem were one-dimensional, without any ξ and η. By (A.6) and translational invariance there exists g satisfying

$$0\leq g(N+M)\equiv -\ln\,\mathrm{Prob}\{m_{N+M}\in J\}\leq g(N)+g(M) \;. \tag{A.9}$$

One says the function g is subadditive. There is a simple argument to prove that $\lim N^{-1}g(N)$ exists and is finite as $N\to\infty$. Fix $b>0$. We can write $N=nb+c$ with $0\leq c<b$. By the subadditivity of g,

$$g(N)\leq ng(b)+g(c) \Rightarrow N^{-1}g(N)\leq (n/N)g(b)+N^{-1}g(c).$$

Since $(n/N)\to b^{-1}$ as $N\to\infty$,

$$0 \leq \lim_{N\to\infty} \sup\{N^{-1}g(N)\} \leq b^{-1}g(b) \ , \tag{A.10}$$

and taking the $\lim\inf$ with respect to b we are done. We now return to the general case.

By (A.6) and the subadditive ergodic theorem, which has been discussed extensively by van Enter and myself in Ref.12 (see also Ref.[57]), the limit (A.3) exists and does not depend on the specific random configuration of ξ's and η's.

A.3. *The role of the Legendre transform.* For the time being we suppose that we again have a one-dimensional problem: estimate the probability of the events

$$q_{1N} = N^{-1}\sum_{i=1}^{N} \xi_i S(i) \in J \tag{A.11}$$

as $N\to\infty$. For J we take an open interval, to be specified later on. Let μ be the original, normalized, spin distribution (in the Ising case concentrated at ± 1 with equal weight) with expectation value $\mathbb{E}_S\{\ldots\}$, and denote by $\mu^N = \bigotimes_1^N \mu$ the probability distribution of N independent spins. We have, whatever $t \geq 0$,

$$\begin{aligned} \mathrm{Prob}\{N^{-1}\sum_{i=1}^{N}\xi_i S(i) \geq \varepsilon\} &= \int d\mu^N(\underset{\sim}{S})\, \mathbb{1}_{\{\sum_i \xi_i S(i) \geq N\varepsilon\}} \\ &\leq \int d\mu^N(S)\, \mathbb{1}_{\{\sum_i \xi_i S(i) \geq N\varepsilon\}} \exp\, t(\sum_i \xi_i S(i) - N\varepsilon) \\ &\leq \int d\mu^N(\underset{\sim}{S}) \exp\, t(\sum_i \xi_i S(i) - N\varepsilon) \\ &= e^{-Nt\varepsilon} \prod_{i=1}^{N} \int d\mu(S(i))\, e^{t\xi_i S(i)} \ , \end{aligned} \tag{A.12}$$

where $\mathbb{1}_{\{\ldots\}}$ is 1 for the event $\{\ldots\}$ and 0 elsewhere. For $\varepsilon \geq \mathbb{E}_S\{S\} = 0$ and $t<0$ the last term in (A.12) exceeds 1; use Jensen's inequality to prove

$$\int d\mu(S) \exp(t\xi_i S) \geq \exp(t\xi_i \int d\mu(S) S) = 1 \ , \tag{A.13}$$

and the assertion follows. For $\varepsilon \leq 0$ an analogous argument applies. Hence (A.12) holds for all t and

$$\begin{aligned} &\lim_{N\to\infty} N^{-1} \ln \mathrm{Prob}\{N^{-1}\sum_{i=1}^{N}\xi_i S(i) \geq \varepsilon\} \\ &\quad \leq -t\varepsilon + \lim_{N\to\infty} N^{-1}\sum_{i=1}^{N} \ln\int d\mu(S)\, e^{t\xi_i S} \\ &\quad = -t\varepsilon + \langle \ln \mathbb{E}_S\{e^{t\xi S}\}\rangle \\ &\quad \equiv -t\varepsilon + c(t) \ ; \end{aligned} \tag{A.14}$$

cf. (2.14)-(2.15). Taking the infimum in (A.14) with respect to t we find that (A.3) is majorized by

$$\inf_t\{-t\varepsilon+c(t)\} = -\sup_t\{t\varepsilon-c(t)\} = -c^*(\varepsilon) \tag{A.15}$$

where c^* is the Legendre transform of c, as defined in (2.8). I now prove that (a) instead of an *in*equality we have equality in (A.14)-(A.15) and (b) only the events $q_{1N}\approx\varepsilon$ are important.

Let $Z(t\xi_i)=\mathbb{E}_S\{\exp(t\xi_i S)\}$ and define a new probability measure μ_{ξ_i} at the site i by

$$d\mu_{\xi_i}(S(i)) = \frac{e^{t\xi_i S(i)}}{Z(t\xi_i)}\, d\mu(S(i)) . \tag{A.16}$$

Note that μ_{ξ_i} depends on t. As yet we have t at our disposal. Differentiating c(t) twice we find

$$c'(t) = \langle\xi\mathbb{E}_S\{Se^{t\xi S}\}/Z(t\xi)\rangle \equiv \langle\xi\mu_\xi(S)\rangle \tag{A.17}$$

and

$$c''(t) = \langle\xi^2[\mu_\xi(S^2)-\mu_\xi^2(S)]\rangle > 0. \tag{A.18}$$

That is, c(t) is <u>strictly</u> convex (c"(t)>0) and its derivative c'(t) is monotonically increasing. When t varies from $-\infty$ to $+\infty$ the function c'(t) ranges through a certain interval, say I. For a given ε in I there is a <u>unique</u> t such that

$$c'(t) = \langle\xi\mu_\xi(S)\rangle = \varepsilon \quad . \tag{A.19}$$

One easily shows that for Ising spins $I=(-\langle|\xi|\rangle,+\langle|\xi|\rangle)$. Moreover, by Cauchy-Schwarz $\langle|\xi|\rangle\leq\langle\xi^2\rangle^{1/2}=1$ so that $I\subseteq(-1,+1)$. Since c'(t) is differentiable in t, the inverse function theorem implies that $t(\varepsilon)$ is differentiable in ε.

We choose J to be an open interval (a,b) containing ε. Our aim is to estimate

$$\mathrm{Prob}\{a<q_{1N}=N^{-1}\sum_{i=1}^N \xi_i S(i)<b\} = \bigotimes_1^N\mu\{a<q_{1N}<b\} \tag{A.20}$$

with $a<\varepsilon<b$. To this end we first note that the probability

$$\lim_{N\to\infty} \bigotimes_1^N \mu_{\xi_i}\{q_{1N}\in J\} = 1 \tag{A.21}$$

because by (A.5) and (A.19)

$$N^{-1}\sum_{i=1}^{N}\xi_i\mu_{\xi_i}(S)\rightarrow\langle\xi\mu_\xi(S)\rangle=\varepsilon\in J\ . \tag{A.22}$$

Here we exploited our freedom to choose t in (A.19) suitably. From (A.16) we have that:

$$\bigotimes_1^N\mu = e^{-t\sum_{i=1}^N\xi_i S(i)}\prod_{i=1}^N Z(t\xi_i)\bigotimes_1^N\mu_{\xi_i}\ , \tag{A.23}$$

and combining (A.20)-(A.22) we find that $\bigotimes_1^N\mu\{a<q_{1N}<b\}$ lies between ($a<b\Rightarrow e^{-ta}>e^{-tb}$)

$$e^{-tbN}\prod_{i=1}^N Z(t\xi_i)\bigotimes_1^N\mu_{\xi_i}\{a<q_{1N}<b\}\ \text{and}\ e^{-taN}\prod_{i=1}^N Z(t\xi_i)\bigotimes_1^N\mu_{\xi_i}\{a<q_{1N}<b\} \tag{A.24}$$

Taking logarithms, dividing by N, and using (A.21) together with the definition of c(t) as it occurs in (A.14) we get that the limit (A.3) is between $-tb+c(t)$ and $-ta+c(t)$. We now contract the interval (a,b) to ε and obtain ($\varepsilon>0$)

$$\lim_{N\to\infty} -\frac{1}{N}\ln \text{Prob}\{\varepsilon\leq q_N<\varepsilon+d\varepsilon\} = t\varepsilon-c(t) = c^*(\varepsilon)\ . \tag{A.25}$$

To see why the last equality in (A.25) is true we return to the definition of $c^*(\varepsilon)$,

$$c^*(\varepsilon) = \sup_t\{\varepsilon t-c(t)\}\Rightarrow c'(t)=\varepsilon\ . \tag{A.26}$$

Since there is a unique $t=t(\varepsilon)$ satisfying $c'(t)=\varepsilon$ and we have chosen this one in (A.22) and (A.25), we are left with $\varepsilon t-c(t)=c^*(\varepsilon)$. And this is what had to be shown.

Due to (A.13) the function c(t) is nonnegative. Hence $c^*(0)=0$. In fact, the function $c^*(\varepsilon)$ assumes its unique minimum at $\varepsilon=0$. To see this we take advantage of the fact that $t=t(\varepsilon)$ is a differentiable function of ε and differentiate $c^*(\varepsilon)$ twice,

$$\frac{d}{d\varepsilon}c^*(\varepsilon) = \frac{d}{d\varepsilon}\{t\varepsilon-c(t)\} = t+[\varepsilon-c'(t)]\frac{dt}{d\varepsilon} = t\ , \tag{A.27}$$

and, by (A.26) and (A.18),

$$\frac{d^2}{d\varepsilon^2}c^*(\varepsilon) = \frac{dt}{d\varepsilon} = [\frac{d\varepsilon}{dt}]^{-1} = [\frac{d}{dt}c'(t)]^{-1} > 0\ . \tag{A.28}$$

That is, $c^*(\varepsilon)$ is strictly convex. It has a minimum when $t(\varepsilon)=0$. Using (A.26) we then get $\varepsilon=c'(0)=0$, and this solution is unique. Since $c^*(0)=0$ and c^* is strictly convex, we have that $c^*(\varepsilon)\geq 0$ and the inequality is strict if $\varepsilon\neq 0$ - as advertized in section 2.

We finally return to our original problem which is three-dimensional. The function $c(\underset{\sim}{t})$ now depends on the three-vector $\underset{\sim}{t}$ and in the definition of c^* we have to replace εt by the inner product $\underset{\sim}{\varepsilon}\cdot\underset{\sim}{t}$ (or $\underset{\sim}{m}\cdot\underset{\sim}{t}$ in the notation of section 2). Except for these trivial changes all the previous arguments can be repeated nearly word for word. Here too, $c(\underset{\sim}{t})$ is strictly convex [58].

References

1. S.F. Edwards and P.W. Anderson, J.Phys.F 5 (1975) 965
2. J.A. Mydosh, in: Springer Lecture Notes in Physics 149 (1981) 87-106
3. A.C.D. van Enter and J.L. van Hemmen, Phys.Rev.A 29 (1984)
4. J.L. van Hemmen, Phys.Rev.Lett. 49 (1982) 409; J.L. van Hemmen, A.C.D. van Enter, and J. Canisius, Z.Phys.B 50 (1983) 311
5. A.P. Malozemoff and Y. Imry, Phys.Rev.B 24 (1981) 489
6. P. Monod and H. Bouchiat, J.Phys. (Paris) 43 (1982) L 45
7. R. Omari, J.J. Préjean, and J. Souletie, to be published; J. Souletie, these proceedings
8. N. Bontemps and J. Rajchenbach, to be published; H. Maletta, G. Aeppli, and S.M. Shapiro, J.Magn.Magn.Mater. 31-34 (1983) 1367
9. R.P. Feynman, *Statistical Mechanics* (Benjamin, Reading, 1972)p.1
10. L. Lundgren, P. Svedlindh, and O. Beckmann, Phys.Rev.B 26 (1982) 3990
11. H. Maletta, in: Excitations in disordered systems, edited by M.F. Thorpe (Plenum, New York, 1982) pp.431-462
12. A.C.D. van Enter and J.L. van Hemmen, J.Stat.Phys. 32 (1983) 141 and work in preparation; K.M. Khanin and Ya.G. Sinai, J.Stat.Phys. 20 (1979) 573
13. P.M. Levy and A. Fert, Phys.Rev.B 23 (1981) 4667
14. P.W. Anderson, Rev.Mod.Phys. 50 (1978) 199
15. J.L. van Hemmen and R.G. Palmer, J.Phys.A: Math.Gen. 15 (1982) 3881
16. J.A. Mydosh, these proceedings
17. K. Binder and K. Schröder, Phys.Rev.B 14 (1976) 2142; in particular Fig.1
18. See also J.P. Provost and G. Vallée, Phys.Rev.Lett. 50 (1983) 598
19. By the strong law of large numbers applied to the ξ's and the η's. See L. Breiman, Probability (Addison-Wesley, Reading, 1968) Sec. 3.6
20. A.W. Roberts and D.E. Varberg, *Convex functions* (Academic Press, New York, 1973) pp. 30 and 110. This excellent book contains a wealth of information.
21. H. Cramér, Act.Sci.Ind., vol.736 (Herman, Paris, 1938)pp.5-23; H. Chernoff, Ann. Math.Stat. 23 (1952) 493. Textbook versions have been given by O.E. Lanford, Springer Lecture Notes in Physics 20 (1973) 35-49, and R.J. Serfling, *Approximation theorems of mathematical statistics* (Wiley, New York, 1980)pp. 326-328.
22. Ref.4 (Z.Phys.), Sec. VI
23. R.B. Griffiths, C.-Y. Weng, and J.S. Langer, Phys.Rev. 149 (1966) 301
24. D.C. Mattis, Phys.Lett. 56A (1976) 421
25. G. Toulouse, Commun.Phys. 2 (1977) 115
26. A.F.J. Morgownik and J.A. Mydosh, Phys.Rev.B 24 (1981) 5277
27. S. Nagata, P.H. Keesom, and H.R. Harrison, Phys.Rev.B 19 (1979) 1633
28. R.W. Knitter and J.S. Kouvel, J.Magn.Magn.Mater. 21 (1980) L316
29. P.W. Anderson, B.I. Halperin, C.M. Varma, Phil.Mag. 25 (1972)1
30. L.E. Wenger, and P.H. Keesom, Phys.Rev.B 13 (1979) 4053
31. D.L. Martin, Phys.Rev.B 21 (1980) 1906
32. K. Binder, In: *Fundamental problems in statistical mechanics*, vol.V, edited by E.G.D. Cohen (North-Holland, Amsterdam, 1980) pp.21-51
33. E. Ising, Z.Phys. 31 (1925) 253, in particular pp.256-7
34. The idea reappears in the more recent literature; e.g. R.B. Griffiths, Phys.Rev. 152 (1966) 240, Fig. 1. The formalism of section 4.2 nicely explains the main results of this paper. See also R.G. Palmer, Adv.Phys. 31 (1982) 669, Fig. 2
35. D.C. Mattis, *The theory of magnetism I* (Springer, New York, 1981) §6.6. Note that, if λ is large, the effective range of the interaction greatly exceeds that of $1/R^6$; cf. section 1.2
36. K. Huang, *Statistical Mechanics* (Wiley, New York, 1963) chapters 8 and 9
37. G.E. Uhlenbeck and G.W. Ford, *Lectures in Statistical Mechanics* (American Mathematical Society, Providence, R.I., 1963) chapter I
38. C.N. Yang and T.D. Lee, Phys.Rev. 87 (1952) 404; Ref.36, §§15.1 and 15.2
39. R.B. Israel, *Convexity in the theory of lattice gases* (Princeton University Press, Princeton, N.J., 1979)
40. In passing we note that a finite range of the interaction is not strictly necessary to derive the DLR equations.
41. A. Münster, *Statistische Thermodynamik* (Springer, Berlin, 1956) §5.12

42. R.L. Dobrushin and S.B. Shlosman, Sel.Math.Sov. 1 (1981) 317-338
43. J.M. Kosterlitz and D.J. Thouless, J.Phys.C 6 (1973) 1181;
J. Fröhlich and T. Spencer, Commun.Math.Phys. 81 (1981) 527
44. L. Onsager, Phys.Rev. 65 (1944) 117; cf. Ref.36, chapter 17
45. G. Parisi, Phys.Rev.Lett 50 (1983) 1946
46. C. de Dominicis and A.P. Young, J.Phys.A: Math.Gen., 16 (1983) 2063
47. A. Houghton, S. Jain, and A.P.Young, J.Phys.C 16 (1983) L375
48. A.P. Young,these proceedings
49. *Monte Carlo methods in statistical physics*, edited by K. Binder (Springer Verlag, Berlin-Heidelberg-New York, 1979)
50. R.G. Palmer, Adv.Phys. 31 (1982) 669, and these proceedings
51. B. Simon and A. Sokal, J.Stat.Phys. 25 (1981) 679, sections 1 and 2
52. See, for instance, A.C.D. van Enter and R.B. Griffiths, Commun.Math.Phys. 90 (1983) 319
53. A.C.D. van Enter and J.L. van Hemmen, Ref.12, section 3
54. L. Landau, J.F. Perez, and W.F. Wreszinski, J.Stat.Phys. 26 (1981) 755;
Ph.A. Martin, Nuovo Cimento 68B (1982) 302
55. A. Blandin, M. Gabay, and T. Garel, J.Phys.C 13 (1980) 403.
56. R. Kindermann and J.L. Snell, Markov random fields and their applications (American Mathematical Society, Providence, R.I. 1980). This is their Fig.18 on p.59. I thank the authors for their permission to reproduce it here.
57. All we need here is Kingman's subadditive ergodic theorem. See J.F.C. Kingman, in: Springer Lecture Notes in Mathematics 539 (1976) 168-223; also Y. Derriennic, C.R. Acad.Sci.Paris 281A (1975) 985-988
58. See Roberts and Varberg [20], section 51.

BROKEN ERGODICITY IN SPIN GLASSES

R.G. Palmer

Dept. of Physics, Duke University, Durham NC 27706, USA.

Abstract

The nature of the frozen spin glass phase is discussed, paying particular attention to the methods of statistical mechanics needed to treat it. A phenomenological approach is advocated. Suggestions are made for the characterization of the possible frozen components and their time, temperature, field, and system-size (N) dependence. The paper is based on reference 1 but some discussion and Monte Carlo results are new.

I. Free Energy Surface and Timescales

A common conceptual picture for a spin glass is a many-valleyed free energy surface, as depicted in Fig. 1. The details may be disputed, but there is now widespread

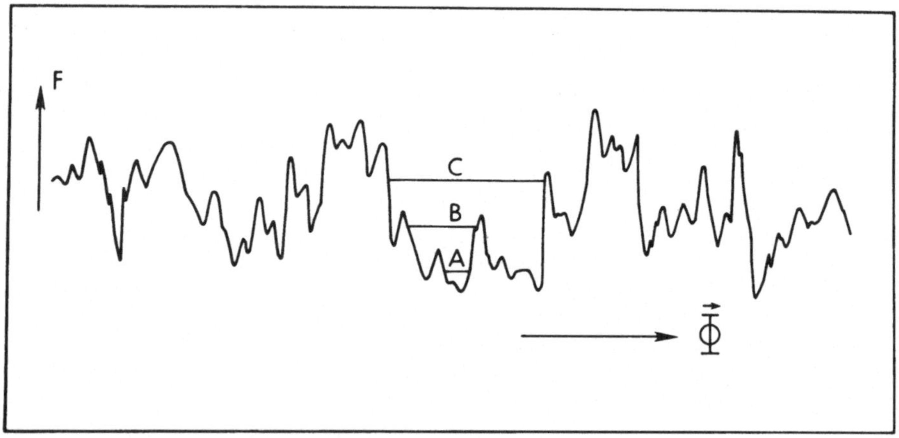

Fig. 1: A schematic free energy surface.

agreement on some such picture. The horizontal axis represents in one dimension the whole free energy space spanned by a set $\vec{\Phi}$ of appropriate configurational coordinates Φ_i. $F(\vec{\Phi})$ is in reality a very high dimensional surface since many Φ_i's are expected to be relevant. The idea behind Fig. 1 is that the physical system breaks ergodicity and becomes stuck, or frozen, in a particular valley. This depends on temperature and observation time. If we wait long enough only the lowest valley is stable, but within a practical waiting time there can be many possible frozen states that are technically only metastable. With increased time or temperature the conceptual picture may be changed in one of two ways; either one may regard the lower free energy barriers as passable, so that the system is stuck only in a larger valley composed of subvalleys (A → B → C in Fig. 1), or (better) one may redefine the free energy space using fewer coordinates Φ_i. The crucial question, of course, is what to use for these Φ_i's -- we must be able to characterize the frozen states. In a broken symmetry system this amounts to knowing the order parameter.

The free energy surface $F(\vec{\Phi})$ is well defined once $\vec{\Phi}$ is known:

$$F(\vec{\Phi}) = - kT \ln \mathrm{Tr}[x|\vec{\Phi}(x) = \vec{\Phi}] \exp(-\beta H(x)) \qquad (1)$$

where x is a point in configuration space Γ and the trace is restricted to those points x that satisfy $\vec{\Phi}(x) = \vec{\Phi}$. One can imagine a succession of $\vec{\Phi}$'s, appropriate for increasing times, starting at $\vec{\Phi}(x) = x$ (so $F(\Phi) = H(\Phi)$, $\Phi \in \Gamma$, and free energy space is configuration space), and eliminating more and more fast degrees of freedom until only a few thermodynamic variables remain. This selection of the relevant degrees of freedom is the ultimate aim of much -- perhaps most -- condensed matter physics. What distinguishes the spin glass is the large number of Φ_i's needed, and their apparently hierarchical structure.

The maximum free energy barrier heights separating valleys are almost certainly finite[2] in short range spin glasses for dimension 2 and 3, but diverge with N in the SK model.[3] The situation is less clear in experimental systems, but there are without doubt some very long relaxational timescales of the order of days at least. With shorter observation times some barriers are too high to cross with any reasonable probability.

In general, the allowed frozen states depend crucially on the observation time τ_{obs}. Conventionally one looks for thermal equilibrium by taking $\tau_{obs} \to \infty$ (or, equivalently, using an ergodic phase average), but this is not appropriate for a system with broken ergodicity. It is actually rarely appropriate without modification, as some examples will show:

a. Coffee and cream: Fig. 2a shows a time-line for the process of adding cream (or milk) to an uncovered cup of hot coffee. There are relaxational timescales for mixing, cooling to room temperature, and evaporation, separating relatively wide domains of equilibrium. In each equilibrium domain "all the fast things have happened and all the slow things have not",[4] with the meaning of "fast" and "slow" dependent on the domain considered. Given a system whose microscopic description (Hamiltonian) includes the whole room, it is clearly not appropriate to take $\tau_{obs} \to \infty$ if one wishes to study hot coffee-cream mixtures. Instead one must keep τ_{obs} in the appropriate domain, as in Fig. 2a. It is however easy (and less

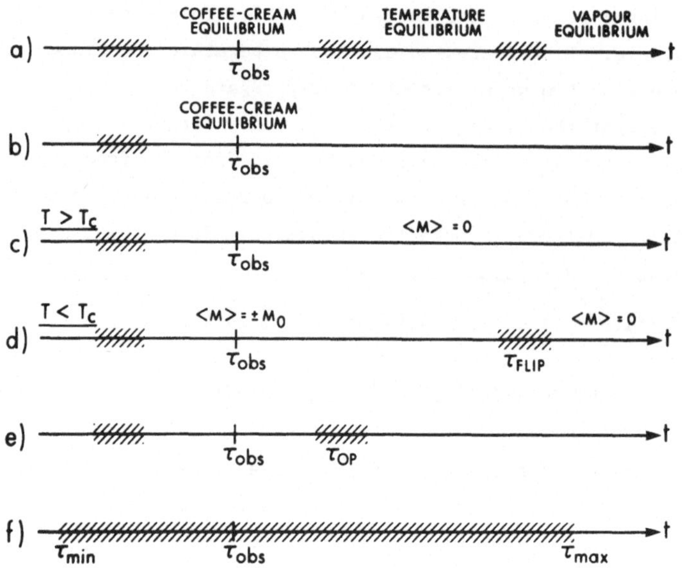

Fig. 2: Time-lines (logarithmic scale) for several systems. Shaded regions are characteristic system timescales.

perverse) to redefine the Hamiltonian to ignore the degrees of freedom for evaporation and to introduce a heat bath to inhibit cooling. Fig. 2a then becomes 2b and the usual methods of statistical mechanics ($\tau_{obs} \to \infty$) may be used with impunity. The example is only trivial because we understand how to separate the fast and slow degrees of freedom, or how to define an appropriately idealized Hamiltonian. We have not yet reached that understanding for the spin glass.

b. Ising ferromagnet with nearest-neighbour interactions and free boundary conditions: Figs. 2c and 2d illustrate this simple system with broken symmetry. A Glauber dynamic is assumed, to give time a meaning. For $T>T_c$ (Fig. 2c) the time-averaged magnetization $\langle M \rangle$ is zero (in order N) beyond some short relaxation time, whatever the initial state. For $T<T_c$ (Fig. 2d) there is a wide time domain in which there are two possible frozen states with $\langle M \rangle = \pm M_0(T)$. In a finite system there is a long relaxation time τ_{flip} for crossing the barrier between these states, and for $\tau_{obs} \gg \tau_{flip}$ one finds $\langle M \rangle = 0$, the true $\tau_{obs} \to \infty$ result. However, the time τ_{flip} diverges with N; the free energy barrier diverges at least as fast as $N^{1-1/d}$ in d dimensions,[1] and is thus impassable in the thermodynamic limit $N \to \infty$. Fig. 3 shows the relevant regimes in the t-N plane, plotted with axes chosen to show the $t = \infty$, $N = \infty$ point. Note that $\lim_{N\to\infty} \lim_{t\to\infty}$ and $\lim_{t\to\infty} \lim_{N\to\infty}$ lead to quite different results. The first is the conventional (ergodic) order of limits in statistical mechanics, whereas the second is the order appropriate for the study of ferromagnetism. Since the broken ergodicity is understood here, it is relatively easy to modify

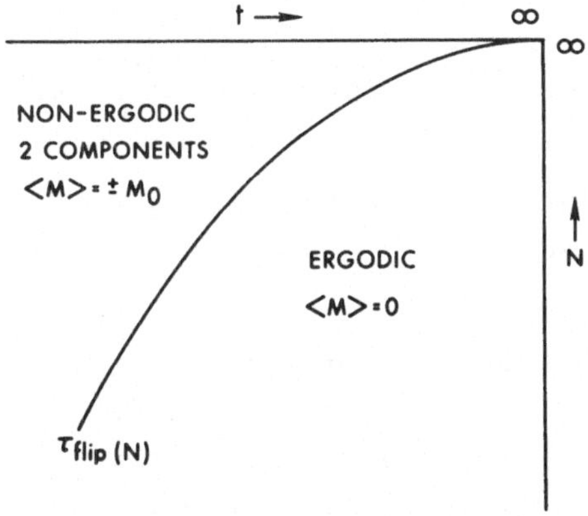

Fig. 3: Ergodic and non-ergodic regimes for an Ising ferromagnet below T_c.

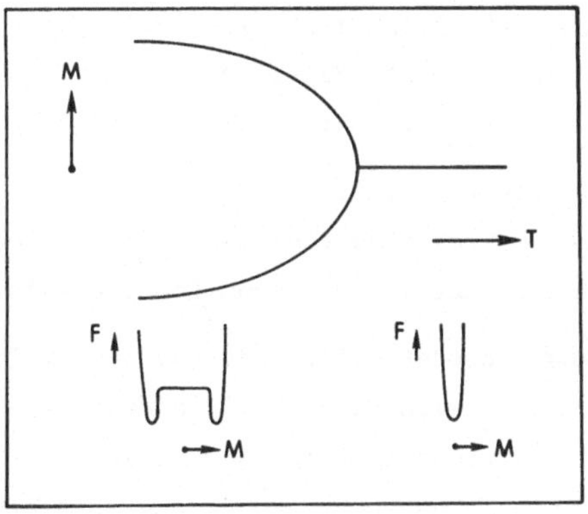

Fig. 4: Component structure and schematic free energy surfaces for an Ising ferromagnet.

the system to eliminate the eventual equilibrium beyond τ_{flip}, as in Fig. 2a → 2b. Application of an infinitesimal field ($h \gtrsim 1/N$) is the best-known method.

Henceforth I shall call the different frozen states components. The terms valleys, phases, and solutions (of the TAP equations) have also been used at this conference. Fig. 4 provides a map of the component structure versus temperature at fixed τ_{obs} (as in Figs. 2c and 2d), with one component $\langle M \rangle = 0$ above T_c and two ($\langle M \rangle = \pm M_o$) below T_c. Also sketched is F(M) in the two regimes, the magnetization M being the appropriate configurational coordinate, or order parameter, in Fig. 1 and Eqn. 1. These sketches employ a subextensive scale for F. On a conventional order N scale the free energy barrier is of negligible height and F(M) appears flat (convex) between $-M_o$ and $+M_o$. Mean field theory (and the infinite range ferromagnet[5]) gives an order N barrier.

c. Ortho/para hydrogen: Fig. 2e applies to molecular hydrogen, discussed in detail in ref. 1. The ortho/para conversion time τ_{OP} is of the order of years in the absence of a catalyst. The $\tau_{obs} \to \infty$ result corresponds to ortho-para equilibrium, which disagrees totally with practical experiments. Even purely thermal quantities, such as the specific heat, are miscalculated if full equilibrium is used; the problems are not limited to quantities like order parameters which take different values in different components. Again, it is easy to rectify the problem, by imposing ΔJ = even, because we understand the nature of the frozen states.

Turning back to the spin glass, we expect a wide range of relevant relaxational timescales, say from τ_{min} to τ_{max}. The very long non-exponential relaxations of remanences and correlation functions point strongly to such a continuum. Mezei and Murani[6] show relaxation of $\langle S_i(0)S_i(t)\rangle$ over more than ten time decades. If τ_{obs} lies within a timescale continuum, as in Fig. 2f, the system's properties are strongly τ_{obs} dependent. In contrast, Figs. 2a - 2e have gaps between relevant timescales, within which a change of τ_{obs} has little effect.

To treat a system described by Fig. 2f at the τ_{obs} shown, one must find a way of freezing or removing those degrees of freedom slower than τ_{obs}, as in Fig. 2a → 2b. No complete way is yet known for the spin glass.

Most theoretical effort has been spent on the SK model,(7) in which both τ_{min} and τ_{max} diverge with N, probably as $\exp(aN^{1/4})$.(3,8) There is also an even longer timescale τ_{eq}, probably diverging as $\exp(aN^{1/2})$, beyond which true ergodicity holds.(3) Note: in each usage herein of the form exp(aX), or the equivalent c^X, a or c represents a constant independent of X but not necessarily equal to other a's or c's. Fig. 5 shows the various regimes on a t-N diagram. Besides the uninteresting ergodic phase (q = 0), there are two relatively simple spin glass phases labelled I and II, separated by a region of complex time-dependent behavior for $\tau_{min} < \tau_{obs} < \tau_{max}$ as in Fig. 2f. The essence of the Sompolinsky theory(8) is a connection from II (x = 0) to I (x = 1) via a continuum of intermediate timescales, all diverging as $\exp(a(x)N^b)$ (b = 1/4?) as $N \to \infty$. The phase I is easily understood as a fully frozen state corresponding to a single solution(9) of the TAP equations.(10) Phase II involves an average over many such solutions, or over many free energy valleys. Figure 5 shows that the joint limit $t \to \infty$ and $N \to \infty$ may be taken in several ways with different results. The limit required for phase II ($\ln t \sim N^b$, $1/4 < b < 1/2$) is unusual. Further questions concerning order of limits arise as soon as temperature and magnetic field are included.

Fig. 5: Regimes with different degrees of ergodicity in the SK model spin glass. q is the Edwards-Anderson order parameter.

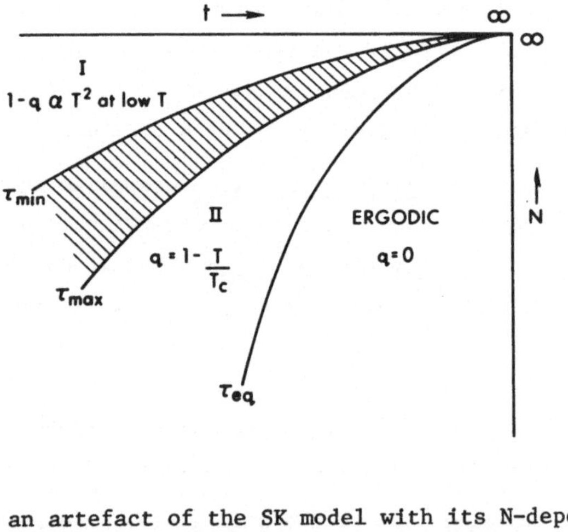

It is not clear how much of Fig. 5 is an artefact of the SK model with its N-dependent interaction strengths. A real spin glass may have an N-independent upper limit τ_{max}, as in short range models.(2) Even so, $\tau_{obs} < \tau_{max}$ may be physically appropriate. In principle $\tau_{obs} < \tau_{min}$ should be used for a "real" SK system (since τ_{min} diverges), but this is not likely to apply elsewhere. There is currently no evidence for an equilibrium time τ_{eq} distinct from τ_{max}. It seems probable that neither phase I nor phase II applies to a realistic spin glass, which is fundamentally stuck in a continuum of timescales, as in Fig. 2f.

II. Breaking Ergodicity

The examples of the previous section show that:

i. Physical systems often break ergodicity, especially when τ_{obs} is held finite; their behavior differs qualitatively and quantitatively from that predicted by unmodified equilibrium statistical mechanics.

ii. In analyzing theoretical models, we must break ergodicity. Definition and selection of a component must be put into the analysis; they do not fall from it.

Point (i) is not really surprising. A system only visits a tiny fraction (vanishing as $N \to \infty$) of its accessible phase space within a reasonable observation time (less than the age of the universe, say). It is *a priori* more surprising that some systems *are* effectively ergodic -- the relatively few phase points visited portray an accurate picture of the whole. This can be understood most easily in systems with translational invariance or many spatially independent parts.

Point (ii) requires emphasis. There is still sometimes heard the attitude that our ultimate goal is the computation of a partition function and that this constitutes a closed procedure for obtaining macroscopic behavior from microscopic description. In fact this applies only to a few effectively ergodic systems (e.g. gases, paramagnets) that tend to be rather dull. Moreover, there is no known general way of telling in advance whether a given system is effectively ergodic.

This does not mean that the usual canonical prescription is valueless. The existence of broken ergodicity may be reflected in some quantities calculated without any inserted ergodicity breaking, even though they and/or other quantities may be calculated incorrectly. Thus, for example, Morgenstern and Binder[11] were able to examine spin-spin correlation functions to look for a form of spin glass ordering, even though their full canonical average implies $\langle S_i \rangle = 0$.

There are many ways we can break ergodicity. Some, such as mean field theory, are second nature to us, and barely regarded as a modification of the canonical prescription. But this they are, and not solely an approximation to a canonical ideal. In applying mean field theory we use knowledge (or guesses) of the nature of the broken ergodicity (e.g., "ferromagnetic order"), specified by an order parameter, to construct a mean field equation. This has several degenerate solutions, labelled by different values of the order parameter, and we break the symmetry by picking one solution (e.g., "spin up") instead of averaging over all. Different solutions correspond to different components -- disjoint regions of phase space. The hard step, of course, is finding the order parameter that leads to this decomposition.

The selection of one component in phase space is recognized in rigorous statistical mechanics as a necessary extension of the Gibbs (canonical) prescription.[12,13] Usually known as extremal decomposition, it has been discussed by van Hemmen at this conference. The conventional way of picking a component in that context is by variation of boundary conditions. Note however that this can lead to at most c^{N_s} different components, where N_s is the number of surface sites. One cannot obtain by this method the order c^N different components that would be required to give a finite (order N) intercomponent entropy (or "complexity"[1]). Indeed, c^N components cannot occur in the extended Gibbs prescription,[13] which does not recognize metastable states or $\tau_{obs} < \infty$ broken ergodicity.

Another way of breaking ergodicity is the application of an "ordering" field conjugate to the order parameter. More generally one can modify or bias the Hamiltonian so that the modified system has only one selected component. The bias field can usually be taken to zero after $N \to \infty$.[1]

Finally, the trace in the canonical prescription ($Z = \mathrm{Tr}\exp(-\beta H)$) can be modified directly to include only one component. In ortho/para hydrogen one uses an odd or even J restriction. Young and Kirkpatrick[14] have applied direct restrictions in numerical spin glass work. Ground state expansions (e.g. TAP[10], Walker and Walstedt[15]) also have this effect, by considering only those microstates in the neighbourhood of a particular minimum.

Replica symmetry breaking in the SK model should not go unmentioned. It is conceivable that this symmetry breaking has the effect of picking one component (or TAP solution?), though it is hard to see how. One would expect the components to be $\{J_{ij}\}$ dependent, but $\{J_{ij}\}$ averaging is performed before replica symmetry breaking. It is also worth noting[16] that even the replica symmetric SK solution[7] involves gauge symmetry breaking, without which one has to choose between nonsense ($\langle Z^n \rangle \to 1/2$ as $n \to 0$) and $q = 0$.

In all of the above approaches one has to be able to characterize the components (or "know the order parameter") before proceeding with theory. There is no reason to anticipate otherwise in the spin glass. Furthermore, an order parameter such as the Edwards-Anderson[17] q, which only quantifies the degree of order without distinguishing components, cannot be expected to suffice any more in the spin glass than it does in the ferromagnet. The problem is compounded by the timescale continuum, which suggests the need for breaking ergodicity in a continuously τ_{obs} dependent way.

The first question, then, is how are we to characterize the frozen components in a spin glass, as a function of τ_{obs} and external parameters (T, h, ...)? In terms of figure 1, we must describe quantitatively the typical valley in which the system is stuck for τ_{obs}. This will generally depend upon past history, which must become a theoretical ingredient. Indeed, history dependence is a central feature of spin glass behavior.

I attempt a preliminary characterization of components in section IV below.[1] Meanwhile it is interesting to consider what can be said without a detailed characterization; in particular I examine in section III the general relations between average component values and canonical values. I must assume that ergodicity has been broken, and therefore make an ansatz:

On a given timescale τ_{obs}, phase space Γ can be divided into disjoint components Γ^{α} (with $\Gamma = \bigcup_{\alpha} \Gamma^{\alpha}$) such that

A) the probability of escape from Γ^{α} within τ_{obs} (averaged over initial states within Γ^{α}, and over possible evolutions) is negligible, and

B) within Γ^{α} one may use the usual techniques of equilibrium statistical mechanics, restricted to Γ^{α}.

B amounts to assuming that the components are themselves effectively ergodic. This is true within the extended Gibbs prescription,[13] but is not necessarily a good approximation for finite τ_{obs} broken ergodicity. In A it is essential that "negligible" be used in place of "zero", which would make any finite free energy barrier ineffective. One could demand Prob(escape from Γ^{α}) $\leq p_o$, a small significance level (10^{-3}?).[1] The net effect is to freeze the slow degrees of freedom, and average fully over the fast ones, where "slow" and "fast" are defined with respect to τ_{obs}. The division is artificially sharp, and leads to artificially discrete component trees in section IV, but is nevertheless a sensible first approximation.

III. Component Averaging

This section concerns relations between the canonical prediction Q_c for a quantity Q and the component average

$$\bar{Q} = \sum_{\alpha} P^{\alpha} Q^{\alpha} \tag{2}$$

of its value Q^{α} in each component α. In principle component averaging is undesirable since the physical system remains stuck in a single component. It may nevertheless (a) prove practical where specification of a single component is impractical or unsolved, and (b) yield a typical result when the distribution of Q^{α}'s is narrow, as frequently seems to occur. Indeed, the motivation for component averaging is analogous to that for using statistical mechanics rather than microscopic dynamics. Note that the information required to specify a single component diverges with N in the spin glass.

The canonical prediction Q_c can be computed as an expectation value

$$Q_c = \langle Q(x) \rangle = \mathrm{Tr}[x\epsilon\Gamma]\ \exp(-\beta H(x))/Z \tag{3}$$

if Q is an observable with a value Q(x) in each microstate x. In other cases (e.g. free energy, specific heat, susceptibility), Q_c must be computed from the partition function Z or its derivatives. Similarly, Q^α is computed from the restricted expectation value

$$Q^\alpha = \langle Q(x) \rangle^\alpha = \mathrm{Tr}[x\epsilon\Gamma^\alpha]\ \exp(-\beta H(x))/Z^\alpha \tag{4}$$

where

$$Z^\alpha = \mathrm{Tr}[x\epsilon\Gamma^\alpha]\ \exp(-\beta H(x)) \tag{5}$$

for an observable, or from Z^α otherwise. The probability p^α provides a weight for component α. The real system is described by $p^\alpha = 1$ for $\alpha = \alpha_o$, $p^\alpha = 0$ otherwise, but we are averaging in order to replace such specific knowledge of α_o and its history and parameter dependence. A natural choice is the Gibbs weight

$$p^\alpha = \exp(-\beta F^\alpha)/Z = Z^\alpha/Z, \tag{6}$$

which is correctly normalized because $\Sigma Z^\alpha = Z$ from $\Gamma = \bigcup\Gamma^\alpha$. This is also the least biased choice in an information theoretic sense.[1] It is certainly not always appropriate since, for example, history dependence is eliminated (cf. section IV), but serves as a sensible first guess. An alternative, not discussed further here, is to evaluate Eqn. (6) at a temperature T_o not equal to T.

With the above definitions (2)-(6) it is easy to derive relations between Q_c and $\bar{Q}$ for specific Q's. Some result[1] are:

(i) $\bar{Q} = Q_c$ if Q is an observable. This applies, for example, to energy and magnetization.

(ii) $\bar{F} = F_c + TI$ and $\bar{S} = S_c - I$ for the free energy and entropy, where

$$I = -k \sum_\alpha p^\alpha \ln p^\alpha \tag{7}$$

is the intercomponent entropy, or complexity. $K^* = \exp(I/k)$ is a useful measure of the effective number of components. K^* is less than or equal to the actual number $K = \sum_\alpha 1$, with equality only when the p^α's are identical. F and S are only modified appreciably (i.e. in order N) if K^* is order c^N. The existence of frozen states without long range order implies that I grows with N, since otherwise a fixed amount of information would suffice to describe the frozen order everywhere, amounting to long range order.

Note that $\bar{S} \neq -\partial\bar{F}/\partial T$ in general, because derivatives of I enter. In component averaging p^α is kept fixed during differentiation -- no component jumping is included -- but the canonical prescription does involve $\partial p^\alpha/\partial T$ terms.

(iii) The specific heat and susceptibility obey

$$C_c = \bar{C} + \sigma^2(E^\alpha)/kT^2 \tag{8}$$

and

$$\chi_c = \bar{\chi} + \sigma^2(M^\alpha)/kT \tag{9}$$

where σ^2 means an intercomponent variance computed with the weight P^α. The differences $C_c - \bar{C}$ and $\chi_c - \bar{\chi}$ between canonical and typical values are significant in order N if the relative widths of E^α and M^α are order $N^{-1/2}$, so there is no conflict with experimental reproducibility of energy and magnetization. Relations (8) and (9) imply $C_c \geq \bar{C}$ and $\chi_c \geq \bar{\chi}$. The latter is well known in spin glasses in the form $\chi_{eqm} > \chi_{1\text{-valley}}$.[18] We would only expect an appreciable $\sigma^2(E^\alpha)$ at h = 0, where $C_c > \bar{C}$ might be observable as an increase of specific heat with measuring time. The effect is seen in Monte Carlo simulations.[19,20]

The Sommers[21] order parameter Δ (equal to $\Delta(x=0)$ in the dynamic theory[8]) is given by

$$\Delta = kT\ (\chi_c - \bar{\chi})/N = \sigma^2(M^\alpha)/N \tag{10}$$

and may thus be given a new physical interpretation as the intercomponent magnetization variance.

In some preliminary Monte Carlo simulations I have computed $\bar{\chi}$ and $\sigma^2(M^\alpha)$ for a 2d square lattice of 50x50 Ising spins coupled by nearest neighbour ± J bonds (equal probability). I performed 100 runs each from random starting points at several temperatures with a single $\{J_{ij}\}$ configuration. After aging for 200MCS/s (Monte Carlo Steps per spin), I averaged over τ_{obs} = 1000MCS/s. These run times are intentionally short compared to equilibrium times. Figure 6 shows $T\bar{\chi}$, $\sigma^2(M^\alpha)$, and their sum, taking k = J = 1 and assuming that averaging over 100 random starts approximates component averaging with P^α. χ^α was computed from the magnetization fluctuation. As in other simulations,[22] $\bar{\chi}$ itself has a broad maximum at $T \approx 1.4J$ and rises again at low T. The maximum shifts to lower T at longer τ_{obs}. Apart from the point at T = 0.5J (which was higher in a second run, not shown), the results agree well with the horizontal line

$$T\,\bar{\chi} + \sigma^2(M^\alpha) = NJ^2 \tag{11}$$

with N = 2500, giving the pure Curie law

$$\chi_c = NJ^2/kT \tag{12}$$

from Eqn. (9). This confirms Eqn. 9 and the interpretation of Δ.

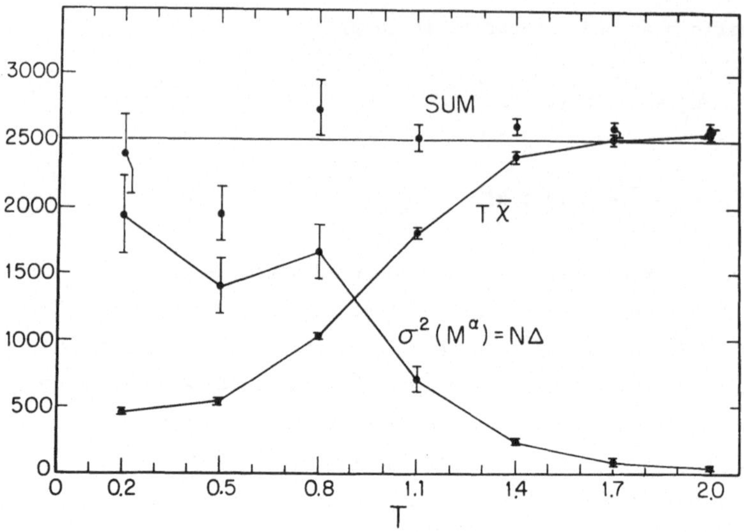

Fig. 6: Monte Carlo results for a 50x50 ±J Ising model, described in the text.

IV. Component Characterization for Spin Glasses

Returning to the central problem of characterizing components in spin glasses, we see that several clues are available. Firstly, as already discussed, there are many relevant timescales; relaxation is far from exponential.

Secondly, there are many possible components, the number depending on system size N, external parameters (including T), and observation time τ_{obs}. The complexity argument following Eqn. (7) implies a growth with N, and may be extended to show that there are order c^N components if the maximum linear extent ξ of local order is finite, as with locally frozen clusters. For then there would be order N/ξ^d independently orientable regions. This case certainly applies to the 2d ± J model, which possesses a network of zero energy contours with $\xi \sim 13$ lattice spacings.[23] Ground state degeneracy estimates[24] also suggest c^N components for this model. It is not clear whether a similar number applies to experimental systems or the SK model. If we identify different TAP solutions as different components[9] we find c^N components in SK too.[25,1] On the other hand an estimate based on correlated clusters of size $\sqrt{N}$ suggests $c^{\sqrt{N}}$ components.[1]

The number of components increases with decreasing T or decreasing τ_{obs}.[1] In terms of Fig. 1, a lower temperature or shorter observation time will allow sticking in a smaller valley, and there are more small valleys than large ones. In the SK model, the number of solutions increases with decreasing T.[25] Indirect experimental and Monte Carlo evidence, mentioned below, also confirms this picture.

A third clue to the component structure is provided by the irreversibility signature of experimental (and simulated[26]) spin glasses. Irreversibility occurs only when the field h is changed within the spin glass phase. At any non-zero fixed h there appears to be a unique "field-cooled" state. Change of h (from $h_1 \neq 0$ to $h_2 \neq 0$, or from $h_1 = 0$ to $h_2 \neq 0$) leads to a different metastable state that relaxes only very slowly towards the field-cooled state at h_2. Similarly a change to $h_2 = 0$ gives the long-lived thermoremanent magnetization. At $h = 0$ there seem to be many different states of similar stability.[26] The special nature of $h = 0$ is also evident in the non-linearity of M(h) there. Fig. 7, based on ref. 27, shows the effect of temperature changes. After turning on h within the spin glass phase ($A \rightarrow B \rightarrow C$), raising T leads to an irreversible increase of M ($C \rightarrow D \rightarrow E$) up to T_c. On the other hand, lowering T at any point below T_c puts the system on a reversible curve (e.g. DF) that can be retraced back to CDE. Thus cooling followed by heating is reversible, but heating followed by cooling is not.

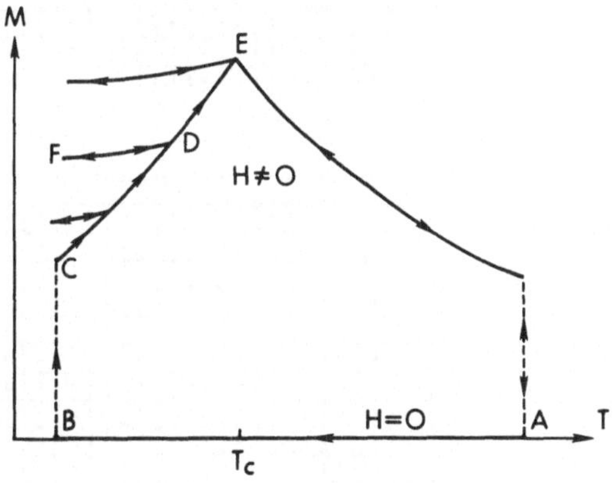

Fig. 7: Idealized magnetization curves for typical spin glasses; the irreversibility signature.

Finally, Monte Carlo simulation allows direct investigation of the nature of the frozen components. I am currently investigating the Monto Carlo properties of spin glass systems at $\tau_{obs} < \tau_{eqm}$, with emphasis on the actual frozen spin configurations. The following results are preliminary. Fig. 8 shows the spins that remained frozen during two Monto Carlo runs of length τ_{obs} = 1280MCS/s from different random starting states on the same lattice. They form frozen clusters, with considerable correlation of the frozen sites between runs. The actual spin directions are not nearly so well correlated as the frozenness -- different clusters can be reversed more or less independently. This gives order c^N components. The frozen cluster locations are also correlated with the underlying frustration distribution, particularly with patches relatively devoid of frustrated plaquettes. The distribution of such patches is in principle calculable.[28] Fig. 9 shows similar data, on the same lattice, for τ_{obs} eight times shorter and longer than Fig. 8. The number of clusters, and hence the number of components, decreases with increasing τ_{obs}. The percolation

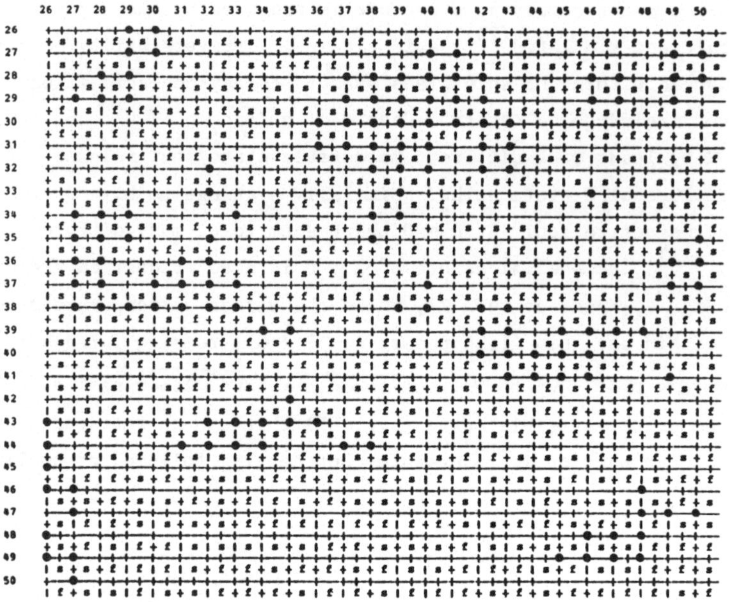

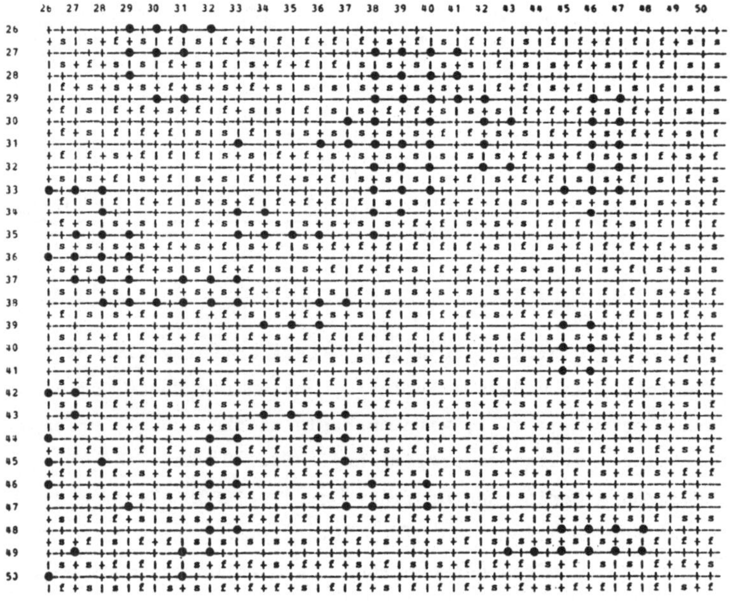

Fig. 8: Frozen Ising spins (●) on ¼ of a random 50x50 ±J lattice in two Monte Carlo runs of length τ_{obs}=1280MCS/s. Each run began 200 MCS/s after choice of a random starting state; the two runs differ only in their starting states. Plus and minus bonds, and frustrated (f) and satisfied (s) plaquettes, are also shown.

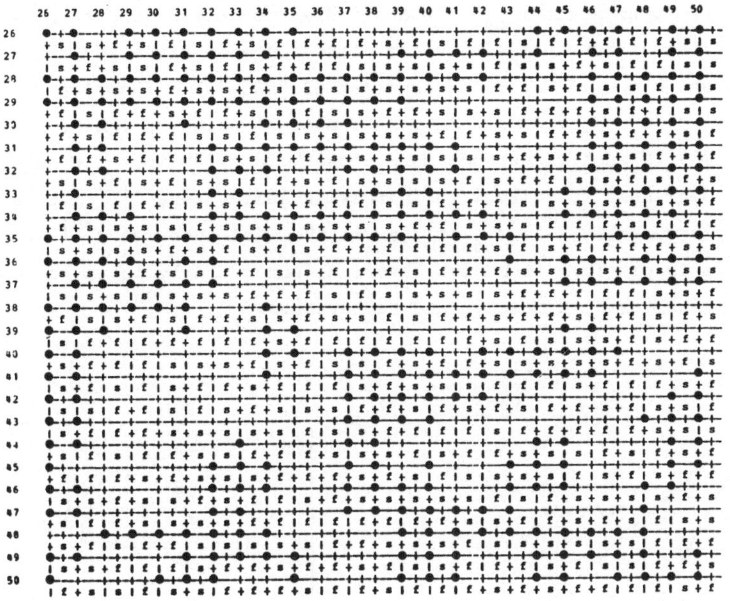

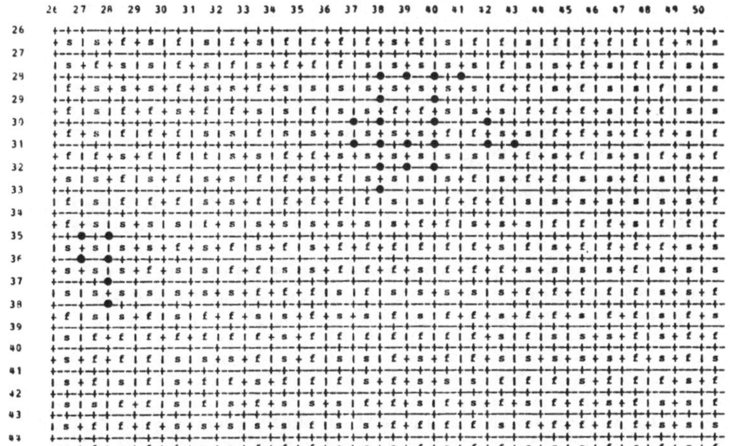

Fig. 9: As Fig. 8, on the same lattice, for τ_{obs}=160MCS/s (upper) and τ_{obs}=10240MCS/s (lower).

of frozenness apparent at τ_{obs} = 160MCS/s does not correspond to a rigidly frozen lattice; different runs are little correlated in spin direction. Fig. 10 is a plot of the average number of spins frozen (out of 2500) as a function of τ_{obs}. There is an approximate $\ell n\ \tau_{obs}$ dependence. The number of spins at least 75% frozen ($|\langle S_i \rangle| \geq 0.75$) is much larger and more closely logarithmic. The number of frozen spins also decreases steadily with increasing temperature.[29] It may perhaps depend on the single variable $T\ \ell n\ \tau_{obs}/\tau_o$, as in the Néel theories.[30] However, the present picture is distinct from such theories, particularly in the dependence of the clusters themselves on timescale and temperature.

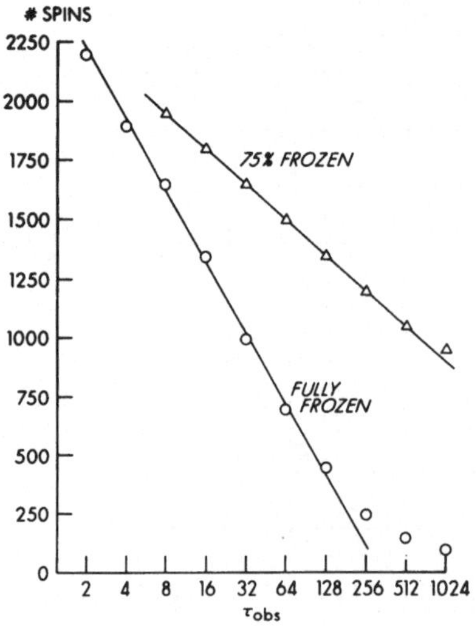

Fig. 10: The number of frozen spins as a function of observation time in a Monte Carlo study. τ_{obs} is in units of 10MCS/s, plotted logarithmically. Straight lines are to guide the eye.

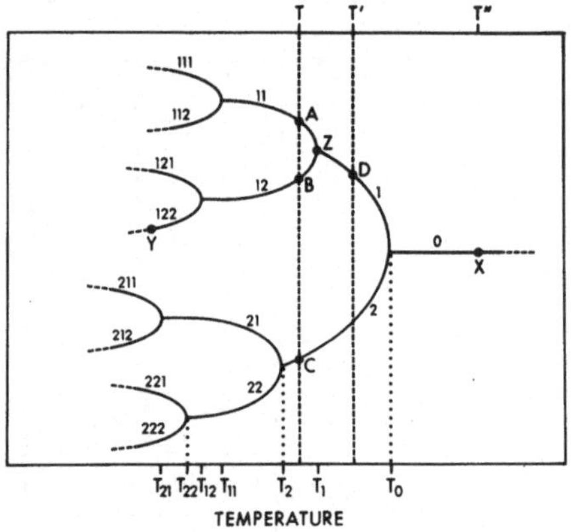

Fig. 11: A component tree arising from repeated branchings.

The preceding clues begin to suggest a phenomenological picture of the component structure in spin glasses. Such a picture is highly desirable, particularly as we have as yet no theory which proceeds satisfactorily from microscopic to macroscopic. Indeed, the discussion of section II suggests that phenomenological understanding must come first. It can also be given a quantitative form in an extension of Landau theory.[1]

Let us first consider h = 0. As we increase T or τ_{obs} the number of components decreases. Presumably this occurs by a merging of, say, subvalleys into valleys, as in Fig. 1. The process continues (valleys to supervalleys, etc.), so we are led naturally to a tree of possible components, as sketched in Fig. 11. The vertical axis serves only to distinguish the components. The horizontal axis shows a small part of the temperature axis, but could also be labelled "τ_{obs}". The sharp bifurcations are of course only an artefact of the sharp ansatz defining components. In reality, more components, fuzzily defined, must arise gradually out of fewer as T is lowered; Fig. 11 represents an extreme idealization of this process. The bifurcation points labelled T_0, T_1, T_2, T_{11}, etc. must become dense on the temperature axis as $N \to \infty$ because the number of components diverges with N (for $T < T_c$).

Krey[31] was the first to suggest a component tree (Fig. 11). Besides the clues already presented, there is other evidence in favour of the picture. Irreversibility occurs at all $T < T_c$, not just at T_c, as is expected from Fig. 11; on raising and re-lowering the temperature the system typically ends up on a different branch (e.g. $A \to D \to B$). The Monte Carlo evidence[32] is quite convincing. An alternative proposal, that all components appear together at T_c, is not viable because it does not give irreversibility below T_c and because (assuming c^N components) it would entail an entropy jump at T_c.[1,13,33] Finally, as a <u>meta</u>theoretical reason for Fig. 11, note that the most successful theories to date (TAP, Parisi, Sompolinsky) have employed a large (diverging with N) number of order parameters. One order parameter is needed for each branching point in the ancestry of a particular state in Fig. 11. Close to T_c the theories actually provide a good description with just a few order parameters, needing more and more for a given accuracy with decreasing T; see especially ref. 34.

Adding a finite field h has two effects. Within each component Γ^α that survives there is a response $\chi^\alpha h + O(h^2)$. More important is the intercomponent effect, which changes the stability of different components. In Fig. 1, some valleys are raised, some are lowered. The component tree may be radically altered, so that at many branching points one branch is preferred over the other. The effect can be modeled by the Ising ferromagnet in a field, transforming Fig. 4 to Fig. 12. On lowering the temperature the upper branch is always taken even though there are two branches (one metastable, but with a lifetime diverging with N) at low T. The metastable branch vanishes at the spinodal point where two extrema of F(M) merge. Application of this model to

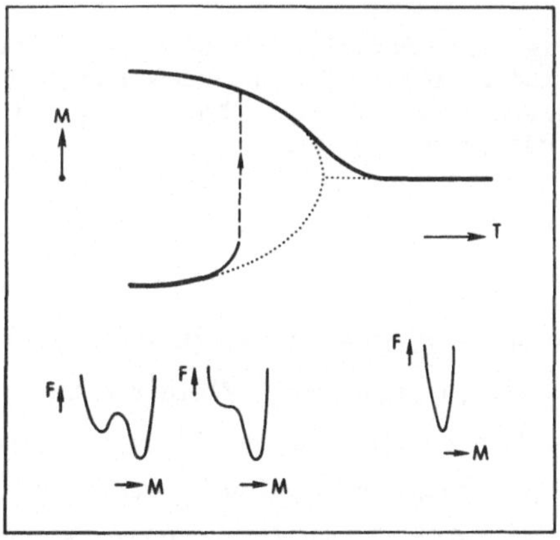

<u>Fig. 12:</u> Component structure of the Ising ferromagnet in a field below T_c. The upper (stable) branch has magnetization M parallel to h, the lower (metastable) one has M reversed. The dashed line is at the spinodal point, which is sharp in mean field theory, but fuzzy (τ_{obs} dependent) in reality. Dotted line is for h=0 (Fig. 4).

Fig. 11 gives Fig. 13. It is crucial to realize however that the vertical ordering of branches is arbitrary. Although for a given h the stable branches can therefore be drawn everywhere upwards for convenience, as in Fig. 13, this will <u>not</u> persist in general if h is changed. The most stable branch will be that with the lowest $F^\alpha(h)$, which depends on h. Fig. 14 is a sketch of some possible $F^\alpha(h)$ curves using

$$F^\alpha(h) \approx F_o^\alpha - hM^\alpha \tag{13}$$

with (not unreasonably)

$$F_o^\alpha = a + b(M^\alpha)^2. \tag{14}$$

We see that different branches become stable with change of h. The parabolic envelope gives (approximately) the canonical average, with a very different susceptibility.[1] Toulouse[35] has proposed a similar picture in terms of F(M). The uniqueness of h = 0 (with a higher degeneracy there) is not yet included in either proposal.

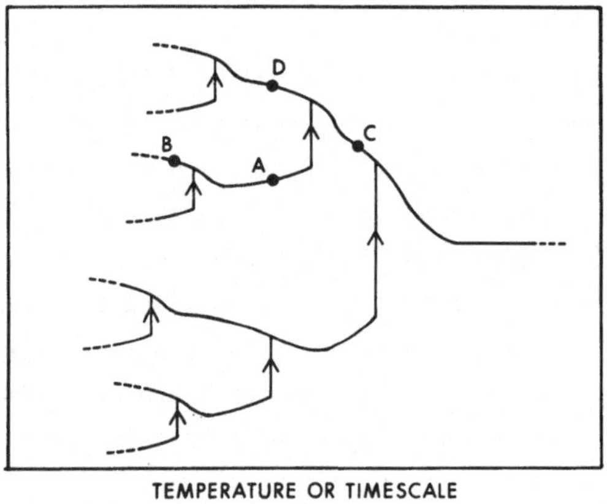

Fig. 13: Modification of Fig. 11 for a finite field, with the stable branch always drawn upwards. This choice is <u>not</u> field independent.

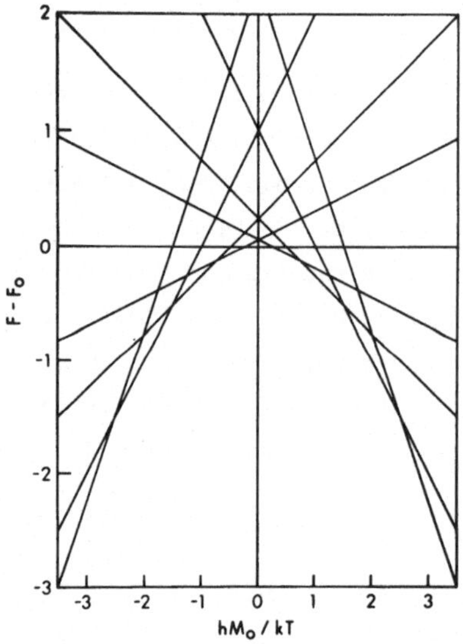

Fig. 14: A simple model for component free energies as a function of field.

Figures 11 ($h \approx 0$), 13 ($h \neq 0$) and 14 give a clear physical explanation of the irreversibility signature. Field cooling is reversible because there is a unique branch followed on cooling in Fig. 13, corresponding to the lowest curve for the chosen h in Fig. 14. Change of h below T_c scrambles the stabilities (Fig. 14), leaving the system on a metastable branch (e.g. point A) in Fig. 13. Thereafter cooling followed by heating is reversible (A → B → A), whereas heating followed by cooling (A → C → D) is not.

The component structure discussed here is only a first qualitative suggestion. Various features, including both the bifurcations and the Fig. 11/13 distinction, have been made artificially sharp. Nevertheless the picture is plausible and fits the clues discussed. A quantitative theory under development will include statistical descriptions of components (distributions of M^{α} etc.) based on the observed frozen clusters, and equations of motion for the probabilities p^{α} as a function of thermomagnetic history.

Acknowledgements

I thank P.W. Anderson, J.L. van Hemmen, J. Jäckle, D. Stein, and G. Toulouse for helpful discussions. Supported in part by NSF grant DMR8011937.

References

1. R.G. Palmer, Adv. Phys. 31, 669 (1982).
2. I. Morgenstern and H. Horner, Phys. Rev. B25, 504 (1982).
3. N.D. Mackenzie and A.P. Young, Phys. Rev. Lett. 49, 301 (1982).
4. R.P. Feynman, Statistical Mechanics (Benjamin, Reading 1972), p. 1.
5. R.B. Griffiths, C.-Y. Weng, and J.S. Langer, Phys. Rev. 149, 301 (1966).
6. F. Mezei and A.P. Murani, J. Magn. Mag. Mat. 14, 211 (1980).
7. D. Sherrington and S. Kirkpatrick, Phys. Rev. Lett. 35, 1792 (1975).
8. H. Sompolinsky, Phys. Rev. Lett. 47, 935 (1981).
9. A.P. Young, J. Phys. C14, L1085 (1981).
10. D.J. Thouless, P.W. Anderson, and R.G. Palmer, Phil. Mag. 35, 593 (1977).
11. I. Morgenstern and K. Binder, Phys. Rev. Lett. 43, 1615 (1979).
12. R.B. Israel, Convexity in the Theory of Lattice Gases (Princeton UP, Princeton 1979).
13. A.C.D. van Enter and J.L. van Hemmen, Phys. Rev. Axx, xxxx (1984).
14. A.P. Young and S. Kirkpatrick, Phys. Rev. B25, 440 (1982).
15. L.R. Walker and R.E. Walstedt, Phys. Rev. Lett. 38, 514 (1977).
16. J.L. van Hemmen and R.G. Palmer, J. Phys. A12, 563 (1979).
17. S.F. Edwards and P.W. Anderson, J. Phys. F5, 965 (1975).
18. F.T. Bantilan Jr. and R.G. Palmer, J. Phys. F11, 261 (1981).
19. J. Jäckle and W. Kinzel, J. Phys. A16, L163 (1983).
20. I. Morgenstern, this conference.
21. H.-J. Sommers, Z. Phys. B31, 301 (1978).
22. A.J. Bray, M.A. Moore, and P. Reed, J. Phys. C11, 1187 (1978).
23. I. Morgenstern, Phys. Rev. B27, 4522 (1983).
24. A.J. Kolan and R.G. Palmer, J. Appl. Phys. 53, 2198 (1982).
25. A.J. Bray and M.A. Moore, J. Phys. C13, L469 (1980).
26. C.M. Soukoulis, K. Levin, and G.S. Grest, Phys. Rev. Lett. 48, 1756 (1982).
27. R.V. Chamberlin, M. Hardiman, L.A. Turkevich, and R. Orbach, Phys. Rev. B25, 6720 (1982).
28. R. Liebmann and H.G. Schuster, J. Phys. C14, 709 (1980).
29. W. Kinzel, Z. Phys. B46, 59 (1982).
30. J.J. Préjean and J. Souletie, J. Phys. (Paris) 41, 1335 (1980).
31. U. Krey, J. Magn. Mag. Mat. 6, 27 (1977).
32 D. Stauffer and K. Binder, Z. Phys. B30, 313 (1978).
33. J. Jäckle (private communication) disputes this conclusion.
34. G. Parisi, J. Phys. A13, L115 (1980).
35. G. Toulouse, in Anderson Localization: Proceedings of the Fourth Taniguchi International Symposium (ed. Nagaoka and Fukuyama, Springer-Verlag, Heidelberg 1982).

SPIN GLASSES AND FRUSTRATION MODELS: ANALYTICAL RESULTS

W.F. Wolff and J. Zittartz
Institut für Theoretische Physik
Universität zu Köln
Zülpicher Straße 77
D-5000 Köln 41

I. Introduction

Spin glasses /1/ are magnetic systems where the interactions are "in competition" with each other due to some disorder in the system and no conventional long-range ferromagnetic or antiferromagnetic order is established. Classical spin glasses consist of dilute magnetic ions in a nonmagnetic metallic matrix at low concentration, e.g. 1% Fe in Au /2/. The RKKY exchange interaction among the spins strongly oscillates with distance and thus, due to the randomness of distances, some of the interactions between spins are ferromagnetic and some are antiferromagnetic. These systems exhibit a "transition" into a state where the spins are more or less frozen in and the susceptibility shows a rather sharp cusp. While the nature of this state and of the freezing transition is still debated, some basic concepts have emerged which are generally accepted. Firstly, although the RKKY exchange is fairly extended, models with short range interactions are considered to give a realistic description. Secondly, it has become clear from the early work of Toulouse /3/ that the concept of frustration is most relevant for spin glasses. This concept essentially describes and quantifies the effect of competition between ferromagnetic and antiferromagnetic interactions. Thus one is led quite naturally to study magnetic models with short range interactions containing both frustration and disorder.

In this survey we shall report on recent theoretical work of the authors on inhomogeneous Ising models where Ising spins, $\mu=\pm 1$, interact via nearest-neighbour couplings varying both in strength and sign. As we were mainly interested in exactly soluble models (numerical work is reviewed by Binder in these proceedings (see also /4/)) we first had to restrict our investigations to two dimensions and secondly we had to restrict ourselves to models with homogeneous couplings in one direction of the lattice and arbitrarily varying interactions only in second direction, namely Ising models of layered structure. These models are defined in Sect. II where we also briefly describe the method of solution and discuss the possible occurrence of phase transitions. Groundstate properties and correlation functions are considered in Sect. III and IV, respectively. In

Sect. V we shall comment on the implications these models have for spin glasses.

In the last Sect. VI we shall describe a different route to the random-bond Ising model. In the general case the nearest-neighbour couplings are randomly distributed according to some probability distribution. So far this model has defied an exact solution. It is possible, however, to formulate a perturbation expansion about the non-random case and to calculate explicitly the first terms in the expansion in two dimensions. This then leads to a conjecture for the scaling function which describes the crossover from the non-random to the random case and determines the critical behaviour.

II. Layered Ising Models

As already mentioned in the introduction we have investigated models which are exactly soluble. It turns out that the most general classes of models one is able to solve so far are two-dimensional Ising models of layered structure. In these models the nearest-neighbour couplings are homogeneous, i.e. translationally invariant, in one direction of the square lattice, but arbitrarily distributed in the orthogonal direction. This then results in two general classes of models, namely models with a horizontally layered structure (HL) and with a diagonally layered structure (DL). Fig. 1 shows the unit cell of different couplings, $K_j=\beta J_j$ for horizontal and $\bar{K}_j=\beta\bar{J}_j$ for vertical couplings with $j=1,\ldots,\nu$.

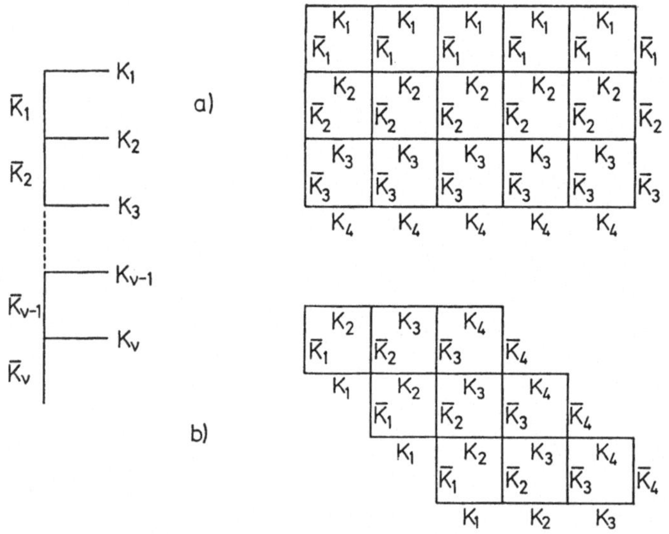

Fig. 1: Elementary cell of a layered Ising model: a) horizontally layered and b) diagonally layered structure.

These unit cells can be stacked together horizontally or diagonally resulting in layers extending in horizontal or diagonal direction. These layers have finite but arbitrary width ν and are then repeated in the orthogonal direction periodically to make up the whole lattice of our HL or DL models. In typical frustration models we consider widths of say $\nu=2,3,4$, while $\nu=1$ is always the homogeneous Ising model. General results, however, apply to arbitrary ν, and, in particular, for $\nu\rightarrow\infty$ we consider models with a random coupling distribution in one direction. These latter models may be viewed as a first approximation to the situation of a spin glass where the couplings, of course, should also vary randomly in the second direction. The restriction is the price we have to pay for being able to derive exact results.

Layered Ising models are described by Hamiltonians

$$\mathcal{H} \equiv \beta H = -\sum_j \left[K_j \sum_{h.n.n.} \mu\mu' + \bar{K}_j \sum_{v.n.n.} \mu\mu' \right] \tag{2.1}$$

where the sums inside the square bracket run over horizontal (h.n.n.) and vertical (v.n.n.) nearest-neighbour pairs and the external j-sum extends over horizontal or diagonal rows with the periodicity condition $K_j = K_{j+\nu}$. Thermodynamics is derived from the partition function $Z = \text{Trace } e^{-\mathcal{H}}$.

As noticed by Toulouse /3/ the Hamiltonian (2.1) is invariant under the following local gauge transformations (acting on both spin and interaction variables):

$$\mu_i \rightarrow s_i \mu_i \;;\quad K_{ij} \rightarrow s_i s_j K_{ij} \quad (j \text{ adjacent to } i) \tag{2.2}$$

for any choice of $s_i = \pm 1$. One is therefore led to introduce the frustration index

$$\phi = \prod \text{sign } K_{ij} = \pm 1 \tag{2.3}$$

where the K_{ij}'s are the four couplings around a plaquette. The thermodynamic properties, i.e. partition function and free energy, of the models defined by (2.1) are gauge invariant and do not depend on the coupling distribution $\{K_{ij}\}$ but rather on the distribution of coupling strengths $\{|K_{ij}|\}$ and of frustration $\{\phi\}$. This means that many different coupling distributions have the same equilibrium thermodynamics. For our layered Ising models it is clear from the definition (2.1) that the frustrated plaquettes with frustration index $\phi = -1$ are arranged in horizontal or diagonal rows.

Horizontally layered Ising models have been considered in many earlier papers /10-15/, with the restriction on only ferromagnetic couplings, i.e. couplings of the same sign. The effect of frustration is then completely missing. Only after Toulouse's introduction of the frustration concept some typical frustration models have been considered /16-20/ which also appear as special cases of our HL or DL models.

The models are solved by using the transfer matrix technique. This method is well known in the horizontal case where one transfers from a configuration $\{\mu\}$ of spins in a horizontal row to a configuration $\{\mu'\}$ of spins in the next row /21/, it is almost unknown /22/ in the diagonal case where one considers configurations of spins in a diagonal row. However, as shown in /6,7/, the transfer matrix technique is most powerful in the case of diagonal transfer, because we can derive all other results from the results of the DL models. By suitably letting couplings go to infinity and to zero we obtain from the DL models /6/: (i) the HL models /5/, (ii) layered models on the triangular (LT) and on the honeycomb lattice (LH), (iii) layered models on mixed lattices, consisting of arbitrary combinations of rows of triangles, squares, pentagons and hexagons.

In the diagonal case the elements of the transfer matrix $T=T(K,\bar{K})$ in a diagonal row of N spins with horizontal couplings K and vertical couplings $\bar{K}$ are given by

$$T_{\mu\mu'} = e^{\sum_{j=1}^{N}\left[K\mu_j\mu'_j + \bar{K}\mu_j\mu'_{j+1}\right]} \quad . \tag{2.4}$$

In contrast to the horizontal case one cannot easily write down an explicit operator representation for this matrix which could then be used for its diagonalization. This, however, can be achieved in an indirect way. One first observes that the transfer matrix T commutes with the 1-d Hamiltonian

$$\mathbb{H} = \sum_{j=1}^{N}\left[\sigma_j^1 + \sinh(2K)\sinh(2\bar{K})\,\sigma_j^3\sigma_{j+1}^3\right] \tag{2.5}$$

where σ_j^α, $\alpha = 1,2,3$, denote Pauli spin matrices in standard notation. However, the Hamiltonian $\mathbb{H}$ is readily diagonalized by using the Jordan-Wigner transformation to fermions and Fourier transformation. This yields the eigenstates of $\mathbb{H}$ which due to the commutation property are also eigenstates of the transfer matrix T. Using these eigenstates in the corresponding eigenvalue equations and transforming back to the spin variables μ one finally obtains the eigenvalues of T in a rather straightforward way.

To solve layered Ising models the operator $\mathcal{T}$ describing the transfer across the whole layer of width ν is first expressed as a product of transfer matrices $T_j=T(K_j,\bar{K}_j)$ transfering from the j-th row of spins to the next one. Although the individual T_j's do not commute in general and thus cannot be diagonalized simultaneously the knowledge of the eigenstates and eigenvalues of the T_j's is sufficient to allow for the diagonalization of the product operator $\mathcal{T} = \prod_{j=1}^{\nu} T_j$. While for the partition function only the largest eigenvalue of $\mathcal{T}$ is needed, the whole information about $\mathcal{T}$, i.e. eigenvalues and eigenstates, is needed for the correlation functions. The free energy per site and the correlation functions are finally given by closed form expressions.

A phase transition is determined by a singularity in the free energy. It turns out that quite generally there can be at most one transition as a function of temperature. At T_c the system goes over from the unique disordered high temperature phase to a 2-fold degenerate low temperature phase whereby the global $\mu \to -\mu$ symmetry of the Hamiltonian (2.1), which is in fact the only symmetry of our models for fixed couplings, is broken. As expected by universality this transition is always of Ising type, i.e. the critical behaviour and critical exponents are the same as in the homogeneous Ising case, in particular the specific heat diverges logarithmically. The ordered low temperature phase is characterized by an ordering of spins which depends on the particular coupling distribution and is in general a complicated spatial arrangement of ferromagnetic and antiferromagnetic substructures.

The transition temperature T_c for DL models is uniquely determined by the remarkably simple formula /6/

$$1 = \prod_{j=1}^{\nu} \frac{\sinh|K_j - s\bar{K}_j|}{\cosh(K_j + s\bar{K}_j)} \Bigg|_{T=T_c} \tag{2.6}$$

which must be satisfied for either s=1 or s=-1, corresponding roughly to predominantly antiferromagnetic or ferromagnetic ordering, respectively. For HL models one obtains the even simpler formula /5/

$$\sum_{j=1}^{\nu} L_j = \left| \sum_{j=1}^{\nu} K_j \right| \tag{2.7}$$

where without loss of generality vertical couplings could be chosen positive, $\bar{K}>0$, and L denotes its dual coupling, $e^{-2L_j} = \tanh \bar{K}_j$. The conditions (2.6,7) have to be fulfilled at some finite temperature T_c in order to have a phase transition. Otherwise we have no transition except possibly at T=0 and the free energy is analytic for all T>0. For instance, this always happens whenever the right hand side in (2.6) or in (2.7) vanishes.

We remark that both conditions (2.6) and (2.7) are independent of the distribution of the couplings within the period of length ν. Interchanging some of the rows or diagonals in the period merely results in a reordering of the various terms in (2.6,7). This property, of course, is present neither in the Hamiltonian (2.1) nor in the free energy.

Of particular interest are the pure frustration models where all couplings have the same strength, $|K_j|=|\bar{K}_j|=K$, but may differ in sign. Pure frustration models with a diagonal distribution of frustrations never show a phase transition, as (2.6) cannot be satisfied, except for $\nu=1$. This is somewhat different from the HL models which have no transition only if the sum of the horizontal couplings in the period ν adds up to zero. Explicitly, (2.7) reduces to

$$L = K \frac{|\nu_+ - \nu_-|}{\nu_+ + \nu_-} \tag{2.8}$$

where $\nu_\pm$ with $\nu = \nu_+ + \nu_-$ denotes the number of horizontal rows with positive and negative couplings within the period ν. The critical temperature vanishes precisely if $\nu_+ = \nu_-$ when we have an equal number of ferro- and antiferromagnetic rows, but arbitrarily distributed.

The pure frustration HL models also include the case of (m,n) - models with two successive layers of width m and n repeated periodically such that these layers are internally free of frustrations but separated from each other by a row of frustrated plaquettes. These (m,n)-models have $T_c=0$ precisely if m=n, i.e. if the two layers have equal widths. Otherwise we have a phase transition at some finite $T_c>0$. What happens is a compensation effect. If the temperature is lowered, the usual ferromagnetic ordering takes place gradually in the individual layers, i.e. local ordering takes place, but the effective interaction between layers, being ordering dependent, diminishes gradually at the same time. This is so, because the sum of vertical couplings through the separating frustrated row adds up to zero, such that there would be no interaction between layers if these would be completely ordered. This means that with decreasing temperature also the effective interaction between layers diminishes at the same time. The global ordering, i.e. ordering of adjacent layers, is therefore aggravated by the internal ordering process; in fact the interaction is maintained only by the disorder of spins along the edges still present at finite T. This might be called "entropy- or disorder-coupling". It then seems to be a purely quantitative effect that for $m \neq n$ a transition at finite temperatures still takes place, while precisely for m=n T_c is reduced to zero.

Precisely at T=0 the layers are perfectly ordered and the effective interaction between layers vanishes giving rise to many groundstates, namely all combinations of up and down layers. The energy barrier between such groundstates is obviously finite, nevertheless we have a transition for $m \neq n$ and no transition for m=n. The argument frequently encountered in the literature that finite energy barriers between groundstates prevent phase transitions, as a freezing in of the thermodynamic state on the neighbourhood of one groundstate in phase space cannot take place, is therefore incorrect in this simple form. At least it must be supplemented by some entropy argument.

Figs. 2 and 3 taken from /5/ show the specific heat for (m,n)-models. For m=n (Fig. 2) there is no transition, but we see that with increasing m the analytic specific heat develops a sharp maximum and approaches rapidly the homogeneous Ising case in the limit $m \to \infty$ with a logarithmic singularity. Below the maximum we have finite ordering within the layers, but no global ordering. For slightly different m and n (Fig. 3) we see again the local ordering taking place as demonstrated by the broad analytic maximum. The global phase transition happens at a much smaller T_c and is indicated by the thin line in Fig.3, representing the Ising logarithmic singularity. The fact that the area under the logarithmic peak is hardly visible (for m=10 for convenience) indicates that the amount of energy required for the global ordering process is much smaller than that required for local ordering.

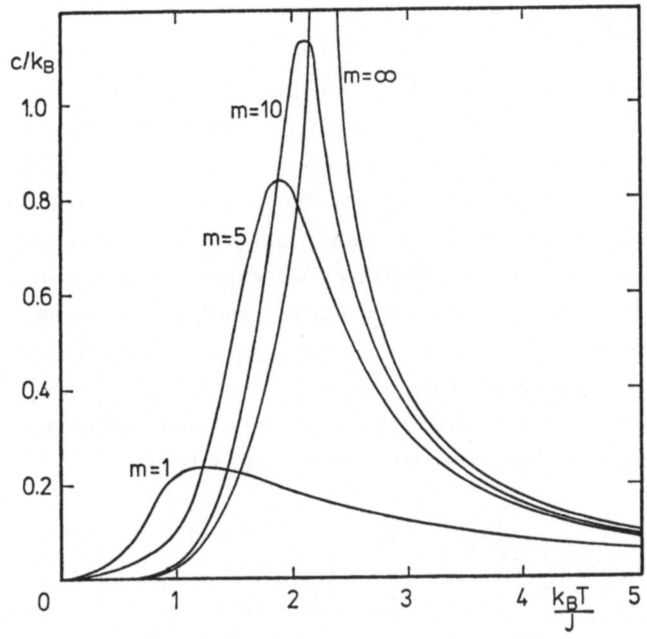

Fig. 2: Specific heat of (m,m)-models for different values of m. m →∞ corresponds to the Ising model.

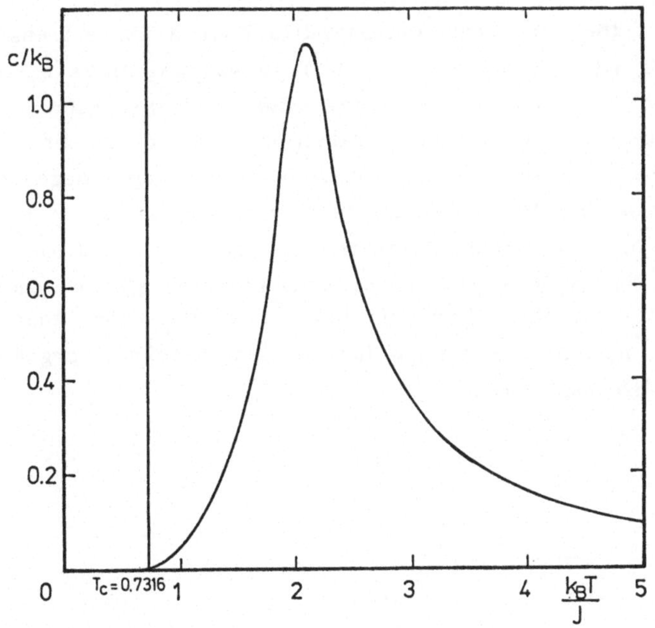

Fig. 3: Specific heat of (m,n)-models for the special case m=10, n=11. The vertical line at T_c indicates the logarithmic singularity.

III. Groundstate Properties

Frustration means that ferro- and antiferromagnetic couplings compete with one another and that not all energy couplings can be satisfied near T=0. Thus there will be loose spins or clusters decoupled from the rest of the system such that one may turn over spins locally without cost in energy. This leads to groundstate degeneracies which might lead to finite rest entropies at T=0. Because of their high symmetry in couplings we expect such degeneracy in particular for pure frustration models with equal coupling strengths. From /5/ we have the result that the rest entropy vanishes, if and only if each row of frustrated plaquettes is separated from the next one at least by one row of non-frustrated plaquettes. In all other cases we have rest entropy. Quantitatively /5/ we have for (m,n)-models the rest entropy per site as

$$s_o = \begin{cases} \frac{G}{\pi} = 0.291 & , \quad m=n=1 \\ (m+n)^{-1} \ln \frac{1+\sqrt{5}}{2}, & m \geqslant 2, \ n=1 \\ 0 & , \quad m,n \geqslant 2 \end{cases} \tag{3.1}$$

where G is Catalan's constant.

We remark that the very same (m,n)-models have a phase transition at some finite T_c only for $m \neq n$. Formula (3.1) then shows that there exists no simple connection between the occurence of a transition and the occurence of a finite rest entropy ; for all combinations, transition and no transition on the one hand and presence or absence of rest entropy (macroscopic degeneracy) on the other hand, are possible. What can be said, however, as shown in /5/, is the following: If for any two groundstates one can be obtained from the other one by a succession of purely local transformations, then the global symmetry $\mu \rightarrow -\mu$ cannot be broken, i.e. then $T_c=0$. If this is not the case, then there might be a transition or not. As seen in the last section the simple argument of finite energy barriers also does not hold.

IV. Correlation Functions

Bulk properties of our layered models can be inferred from the knowledge of the exact free energy and have been described in the foregoing sections, while correlation functions are a suitable tool to gain insight into the microscopic properties, in particular into the details of the ordering process. In /17,25-29/ it has already been shown that correlations in frustrated systems show quite interesting behaviour. One usually studies two-spin correlations where for convenience the two spins are separated by r steps in the direction of homogeneity, i.e. parallel to the layering.

Here we shall describe the main results obtained in /9/. The transfer matrix technique turns out to be also well suited for the calculation of correlations. Formally the correlation $f(r) \equiv \langle \mu_0 \mu_r \rangle$ can be written as a block Toeplitz determinant which in all cases could be reduced to a usual Toeplitz determinant. We then could follow standard treatments described in detail in the book of McCoy and Wu /30/. For a general discussion it is convenient to distinguish the two cases where a) the critical temperature is finite, $T_c>0$, and b) $T_c=0$, more precisely, no transition at finite temperature.

a) $T_c>0$: In models with $T_c>0$ correlations show the usual Ising type critical behaviour. In the disordered high temperature region above T_c f(r) decays exponentially with a correlation length ξ diverging as $|T-T_c|^{-\nu}$ with the Ising exponent $\nu=1$. At T_c we have algebraic decay with $r^{-\eta}$ and $\eta=1/4$. Below T_c f(r) approaches $m(T)^2$ asymptotically, where the spontaneous magnetization or local order parameter m(T) vanishes at T_c with exponent $\beta=1/8$. As far as critical behaviour at finite T_c is concerned, there is thus nothing new to be seen as could be expected from universality. Inhomogeneities and frustrations, however, show their influence in the details.

Below T_c we may have different local order parameters along layers of spins with different couplings. The ordering is, however, always unique, i.e. ferromagnetic or antiferromagnetic and is determined at T>0 by the globally dominating couplings. At zero temperature we may, however, encounter typical one-dimensional chain ordering in a chain with negative couplings while the rest orders ferromagnetically. A typical example for such behaviour is the HL model with period $\nu=2$ (see paper II, ref. /9/). Fig. 4 shows plots of the local order parameter m(T) in the chain with antiferromagnetic couplings (-K) for increasing values of K while the other ferromagnetic couplings are kept fixed. Curve (1) shows m(T) for very small K when the competing effect of (-K) is not yet seen. In curve (2), larger K, m is already substantially reduced, but for $T \to 0$ it crosses over to the perfect ferromagnetic ordering value 1. Curve (3) shows m when K equals the strength of the vertical couplings. Curve (4) shows m for even larger K. While K cannot achieve one-dimensional antiferromagnetic orde-

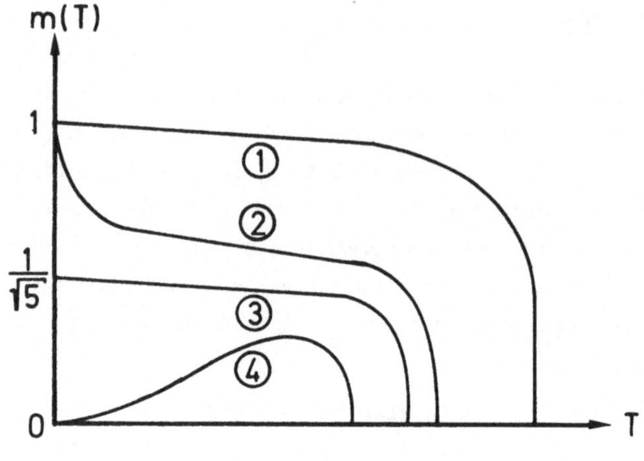

Fig. 4: Spontaneous magnetization in the HL model with period ν=2.

ring in this row at any finite T, it nevertheless is able to reduce the ferromagnetic ordering by competition and ultimately destroys this ordering in the limit $T \to 0$, causing m to vanish at T=0. At T=0 m is discontinuous.

One further effect of frustration is the fact that certain spins may be totally uncorrelated, i.e. f(r)=0, for any temperature $T > T_c$, i.e. in the disordered region. This occurs for instance in the fully frustrated square lattice model (FFS) and in the chessboard model, discussed in detail in paper I, ref. /9/,for spins separated by an odd distance r in diagonal direction.

Due to frustration and competing interactions the decay of correlations in the disordered region above T_c develops in many cases quite interesting oscillatory behaviour superposed on the exponential decay

$$f(r) \simeq e^{-r/\xi} \cos\left[\Theta(T) \cdot r\right] \quad . \tag{4.1}$$

The oscillations may be commensurate with the lattice, in which case $\Theta(T)$ in (4.1) takes on values which are rational fractions of 2π, e.g. 0 for ferromagnetic alignment and π for usual antiferromagnetic alignment resulting in the period 2. But the period 4, or $\Theta = \pi/2$, is also frequently encountered in situations with vanishing correlation for odd distances. Such behaviour is seen for instance in the HL model (paper II, ref. /9/) where in the whole temperature range above T_c we either have ferromagnetic or antiferromagnetic correlations depending on the dominating couplings. However, for the "general square lattice" model (GS) (paper III, ref. /9/) we may have a change from commensurate periods to incommensurate periods whenever the couplings lead to frustration. It is the same behaviour as in the triangular antiferromagnet /25/.

Just above T_c we always have commensurate periods. At some "disorder temperature" $T_D(>T_c)$ correlations decay as a pure exponential, i.e. $f(r) \equiv e^{-r/\xi}$, all r, while above T_D the wavevector $\Theta(T)$ depends continuously on temperature and can assume any value within a finite subinterval of $(0,\pi)$ which depends on the couplings. This change from commensurate to incommensurate oscillations at T_D is seen only in the $r \rightarrow \infty$ limit of the correlation f(r). As all thermodynamic quantities remain analytic, the smooth change at T_D is no phase transition.

b) $T_c=0$: As we have pointed out before frustration may be so effective that the system cannot globally order at any finite temperature. In the disordered high-temperature region, i.e. for $T>0$ in this case, correlations of course decay exponentially and we may encounter similar behaviour as described in a) for $T_c>0$. For instance, in the GS model we also have oscillations with incommensurate periods for all $T>T_D=0$. What then happens at T=0 can best be seen in the spin-spin correlation function. There are three cases to be distinguished:

(i) There is perfect ordering at T=0, i.e. $f(r)=\pm 1$. The ordering is usually one-dimensional chain ordering. As this is expected to be typical for models with layered structure, it is not surprising that the majority of our models with $T_c=0$ show this behaviour, i.e. a first order transition at zero temperature.

(ii) T=0 is a critical point, i.e. $T_c=0$ in the strict sense. In that case the system has a vast groundstate degeneracy and the correlation decays algebraically with $r^{-\eta}$ $(r \rightarrow \infty)$. In this case the critical exponent η is different from the Ising value 1/4 and seems to assume the value $\eta=1/2$ universally in all investigated cases. The first known case is the triangular antiferromagnet (FFT) /25,26/, the second case is the fully-frustrated square lattice (FFS) model /26/ and ref. /9/ paper I. However, we have now found that this also occurs in the general HL model with layer period $\nu=2$ on a whole one-dimensional manyfold in the space of the three couplings of the model and also in the general square lattice (paper III, ref. /9/).

(iii) Even at T=0 the system is disordered. This means that also at T=0 the correlations decay exponentially with a finite correlation length $\xi(T=0) < \infty$. That frustration can be so effective and the groundstate degeneracy so vast in certain models as to cause this fast exponential decay is perhaps the most surprising result. So far we have found two special models with this behaviour, namely the chessboard model (paper I, ref. /9/) which is also - and this is somewhat surprising - that model with the largest rest entropy per site and the fully-frustrated honeycomb model (FFH), recently solved in ref. /7/. The T=0 behaviour of the latter model is thus different from the other fully frustrated lattice models in two dimensions, namely the FFS and FFT models.

V. Implications for Spin Glasses

As we have mentioned before, one of our motivations to study frustration models was to learn something about spin glasses. In particular, we have the popular short range $\pm J$ model in mind where one has to average quantities over the distribution of signs ε with fixed strength such that

$$P(\varepsilon) = p\delta(\varepsilon-1)+(1-p)\delta(\varepsilon+1) \quad (5.1)$$

is the probability and p is the fraction of the positive bonds. Clearly the model is symmetric around p=1/2 and we need to consider only the range $0 \leq p \leq 1/2$.

If we consider our layered pure frustration models as approximations to spin glass models, we must let the period width ν go to infinity. Within the ensemble of frustration distributions or distributions of coupling signs clearly the distribution with an equal, or almost equal, number of rows with positive and negative couplings will dominate because they carry the largest weight. As the right hand side in eq. (2.8) then vanishes, we see that these models which most closely should resemble spin glasses have no phase transition (or T_c is arbitrarily small for large ν). Also in view of the fact that pure frustration DL models never show a phase transition, we must conclude that spin glasses have no transition in the symmetric case, p=1/2 in eq. (5.1), at least in two dimensions. This is in accord with the view recently advanced by many theoretical papers that the spin glass transition as a genuine equilibrium phase transition does not exist, see refs. /1/ and / 4 /, for instance.

In ref. /8/ we have tried to come even closer to spin glasses. With a view to Fig. 1 it is clear that a spin glass is obtained, if we would consider an elementary cell of random couplings extending both in horizontal direction with period ν_1 and vertical direction with period ν_2 and then let both $\nu_{1,2}$ go to infinity. Such models cannot be solved completely. However, in the first paper of ref. /8/ Hoever was able at least to formulate the T_c-condition for arbitrary ν_1 and ν_2 exactly. This condition then had to be solved numerically for arbitrary ν_2 and ν_1=1,2,3,4,..., successively. The results showed for the probability distribution (5.1) that in the range from p=0.5 down to p=0.19, which in fact is an upper bound, there was no transition at finite temperatures. This is in good agreement with other numerical results based on other methods /24/.

In ref. /8/ we also have developed a new high temperature expansion technique proceeding via diagrammatic resummations. This expansion when carried through to third order already then reproduces the exact high temperature expansion in terms of α=tanh K to order 22 and again shows in the symmetric case p=1/2 that the free energy remains analytic down to zero temperature, i.e. no transition.

Many recent spin glass papers, /1, 4/ and references therein, now suggest that the spin glass freezing is a dynamical phenomenon. Due to frustration the random magnetic systems have many groundstates in phase space. Below the freezing temperature the system then should remain fairly long in one of these groundstate valleys and the passing over the barriers of substantial height is quite rare and thus takes a long time. As a result the relaxation times to reach the final equilibrium state without symmetry breaking are very long and comparable to measuring times in realistic experiments.

This metastability of spin glasses below the freezing temperature is also observed in our frustration models. We may for instance consider the (m,n)-models which have no phase transition, but which for increasing m show a very pronounced maximum in the specific heat as shown in Fig. 2 before. Above the maximum which roughly would correspond to a freezing temperature the system is overall disordered while below the maximum all the layers are internally well ordered but only in loose contact among each other because of the separation by the rows of frustrated plaquettes. If a spin within such a layer is picked in up position in some initial state, a flip over is then very improbable as the spin is well frozen in the whole complex of ordered spins within the same layer. Thus the decay of the local magnetization towards the equilibrium value zero takes a very long time.

There are a few more observations which support the view that the spin glass transition is "merely" a dynamical phenomenon. A detailed discussion, however, must be omitted here.

VI. Low-Concentration Expansion

In the foregoing sections we have considered inhomogeneous Ising models with fixed couplings varying only in one direction of the lattice. This restriction is sufficient to allow for an exact solution and the computation of correlation functions. The more general case where the couplings vary randomly also in the second direction is presumably not solvable. One therefore has to use either numerical or approximate treatments.

In the general random-bond model Ising spins on a square lattice interact via nearest-neighbour couplings $K_{ij}=\beta J_{ij}$ randomly chosen with the probability distribution

$$P(K_{ij})=(1-p)\delta(K_{ij}-K)+p\delta(K_{ij}-\alpha K) \tag{6.1}$$

where $-1\leq\alpha\leq1$ can be assumed without loss of generality. These systems may be thought of as Ising models with couplings K and randomly distributed impurity bonds αK with concentration p. Two special cases have received widespread attention, namely the bond-diluted ferromagnet with $\alpha=0$ /31-34/ and the $\pm J$ spin glass model (5.1) with $\alpha=-1$ (see for instance /4/). In the first case a fraction p of interactions K is randomly removed from the system. For small dilution p one still expects a ferromagnetic phase transition at some finite transition temperature $T_c(p)$ as in the pure, nondiluted system. Above the percolation threshold p_c /36/ the transition temperature vanishes due to the disappearance of the largest percolating cluster. The phase diagram has been obtained by using renormalization group calculations /31/. However, there has still remained the problem of determining the nature of the transition at $T_c(p)$.

Similarly, in the $\pm J$ model ($\alpha=-1$) the ferromagnetic phase transition is still present for small enough p /37/. However, with increasing p frustration becomes so effective that the transition is already destroyed for p larger than $p\simeq0.12$ /35/. Though numerical analysis could rule out the existence of a spin glass state /35/, it was not possible so far to clarify the nature of the transition at $T_c(p)$.

Unfortunately the general case $p\neq0$ is presumably not soluble. However, it is possible to investigate the case of small p by performing a concentration expansion about the soluble pure system, p=0. Such an expansion has been recently performed by us /38/. As we have pointed out, it is essential to obtain results at least to second order in p to include impurity correlations and to clarify the difference between quenched and annealed randomness. In two dimensions where the pure Ising model is soluble, all relevant quantities can be explicitly calculated. In particular we have obtained

a) the critical temperature $T_c(p;\alpha)$ to second order in the impurity concentration

b) the exact leading singular terms of the quenched free energy, $\bar{f}=\langle f[K_{ij}]\rangle_{av.}$, to second order in p.

The critical temperature $T_c(p;a)$ determines the phase diagram of the quenched system. For p=0 we have the pure Ising system with a transition at $T_{c,o}=K_o^{-1}=2/\ln(\sqrt{2}+1)$. To second order in p the transition temperature is given by

$$T_c(p;a)/T_{c,o} = 1-s_1p-s_2p^2 \tag{6.2}$$

and the coefficients $s_{1,2}$ can be expressed in closed form in terms of $T_{c,o}$ and the interaction parameter a /38/. The explicit formulas are omitted for brevity. For the bond-diluted case, $a=0$, we obtain $s_1=1.329$, $s_2=0.135$. Approximate numbers have also been calculated in /32,33,34,39/. For the $\pm$J model, $a=-1$, one gets $s_1=3.208$, $s_2=4.399$. As expected and explicitly demonstrated in /38/ the critical temperature $T_c(p;a)$ decreases for increasing concentration p $(a\leq 1)$. Depending on the values of a we encounter two different cases (see Fig. 5). For positive a all impurity bonds are ferromagnetic and only disorder but no frustration is present in the system. Then the critical temperature $T_c(p;a)$ is finite for all values of p until for p=1 it reaches the known value $T_c(1;a)$ for the pure Ising system with couplings aK. On the other hand the transition temperature is expected to vanish above some threshold p_c if $a\leq 0$ and a separate antiferromagnetic phase occurs near p=1 for $a<0$. However, for clarity this phase has been left out of the figure. As can be seen from the explicit formulas /38/, and also from the figure, frustration is more effective in reducing the critical temperature than dilution and the threshold concentration $p_c(a)$ decreases with a.

In Fig. 5 the transition temperature T_c is plotted for several values of a from eq. (6.2) and numerical estimates /31, 35/ are also shown for comparison as dashed curves. For small values p the agreement is very good, only near p_c we have substantial deviations where higher order corrections become relevant.

Our second result is the expansion of the quenched averaged free energy $\beta\bar{f}$ in the vicinity of $T_c(p;a)$, i.e. near the transition. Keeping only the leading terms the singular part of $\beta\bar{f}$ is given up to second order in p as /38/

$$\beta\bar{f}\Big|_{sing.} \simeq -\frac{t^2}{2b_0^2p}\left[y-\frac{1}{2}y^2+\frac{1}{3}y^3+O(y^4)\right] \tag{6.3}$$

with

$$y=p\,b_0^2\,\frac{8}{\pi}\,\ln|t|^{-1} \tag{6.4}$$

and $t\sim T_c(p;a)-T$ goes to zero. The coefficient b_o depends on the interaction parameter a and is given in terms of the critical coupling of the pure system, $K_o=1/2\ln(\sqrt{2}+1)$, as

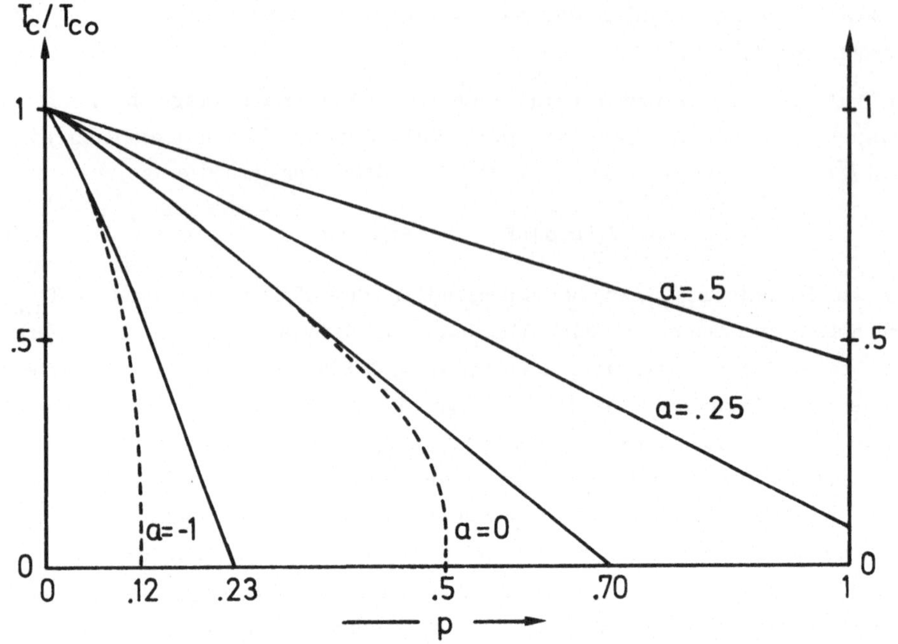

Fig. 5: The critical temperature $T_c(p;\alpha)$ obtained from (6.2) is shown for various interaction parameters α. For clarity the antiferromagnetic phase which occurs for $\alpha < 0$ in the lower right has been omitted from the picture. Numerical estimates for $\alpha=0$ /31/ and $\alpha=-1$ /35/ (dashed curves) are also shown for comparison.

$$b_o = \tanh\left[(\alpha-1)K_o\right]\cdot\left\{1+\frac{1}{\sqrt{2}}\tanh\left[(\alpha-1)K_o\right]\right\}^{-1} \quad . \tag{6.5}$$

The expression (6.3) suggests that the expansion in brackets defines some universal scaling function G(y) which describes the crossover from the pure Ising model, in which case p=0 and $G(y)\Rightarrow y$, to the random model. In the latter case p is nonzero and the critical behaviour of the free energy, as $t\to 0$, is then determined according to (6.3) by the asymptotic behaviour of G(y) for $y\to\infty$. As all p and α dependence is absorbed in the one variable y, apart from the unimportant amplitude in (6.4), this strongly suggests that we have universally the same critical behaviour along the whole T_c curve for all concentrations p and also for all values of the parameter α. This would mean that frustration, which occurs for $\alpha<0$, in particular for $\alpha=-1$, would have no qualitative influence on the type of phase transition, when we compare this situation with the case of a simple mixture of ferromagnetic bonds, $0<\alpha<1$, where we have no frustration.

The exactly known terms in the expansion (6.3) allow to check recent suggestions in the literature on their validity. (6.3) rules out Fisher renormalization /40/ which holds for the annealed case /41/, but was also suggested for the quenched case /32,33/; it also rules out a power law singularity $t^{-\alpha(p)}$ with an exponent continuously varying with p, as suggested in /42/. In both cases the free energies do not fit the correct expansion (6.3). However, quite recently it has been suggested /39/ on the basis of an approximate mapping of the random-bond Ising model onto a zero-component Gross-Neveu model that the scaling function G(y) should be given as

$$G(y) = \ln \left(1+y\right) \tag{6.6}$$

whose expansion obviously fits with (6.3). We believe that (6.6) is essentially correct. In the limit $t \to 0$ $(y \to \infty)$ it leads to the universal critical behaviour

$$\beta \bar{f}\,\Big|_{sing.} \simeq -\frac{t^2}{2b_0^2 p}\, \ln\big|\ln|t|\big| \, . \tag{6.7}$$

Most remarkable is the fact that the critical behaviour, i.e. the type of singularity, is the same for all values of the interaction parameter a, which is definitely true to second order in p. This would mean that frustration, which occurs for $a<0$, does not qualitatively change the criticalities and thus the nature of the transition as compared to the case of $a>0$, where we have no frustration, but only disorder.

Next-to-leading terms in the full expansion of $\bar{f}$ to second order in p have also been calculated by us /38/. However, they modify our conclusions in an unessential way and merely lead to a redefinition of the amplitude and the variable y such that the full expansion is consistent with

$$\beta \bar{f}\,|_{sing.} \simeq -t^2 A(p;a) \ln\left[1+B(p;a)\ln|t|^{-1}\right] \tag{6.8}$$

with two amplitudes A,B depending both on p and a. The form of the scaling function (6.6) and the critical behaviour (6.7) are both left unchanged.

References:

/1/ For a recent review see: Fischer, K.H.,Phys. Stat. Solidi (b) 116, 357 (1983)

/2/ Binder, K.: Festkörperprobleme XVII, 55 (1977)

/3/ Toulouse, G.: Commun. Phys. 2,115 (1977); Vannimenus, J., Toulouse ,G.: J. Phys. C10, L 537 (1977); Fradkin , E., Huberman, B.A., Shenker,S.H.: Phys. Rev. B18, 4789 (1979)

/4/ Binder, K., Kinzel, W., in: Lecture Notes in Physics 149, Berlin-Heidelberg-New York, Springer 1981

/5/ Hoever, P., Wolff, W.F., Zittartz, J.: Z. Phys. B - Condensed Matter 41, 43 (1981); Wolff, W.F., Hoever, P., Zittartz, J.: Z. Phys.B - Condendensed Matter, 42, 259 (1981); Hoever, P., Zittartz, J.: Z. Phys. B - Condensed Matter, 44, 129 (1981)

/6/ Wolff, W.F.; Zittartz, J.: Z. Phys. B - Condensed Matter, 44, 109 (1981); Wolff, W.F.: Dissertation, Köln (1981)

/7/ Wolff, W.F., Zittartz, J.: Z. Phys. B - Condensed Matter, 49, 139 (1982)

/8/ Hoever, P.: Z. Phys. B - Condensed Matter, 48, 137 (1982); Hoever, P., Zittartz, J.: Z. Phys. B - Condensed Matter, 49, 39 (1982)

/9/ Wolff, W.F., Zittartz, J.: Z. Phys. B - Condensed Matter, 47, 341 (1982); 49, 229 (1982); 50, 131 (1983)

/10/ McCoy, B.M., Wu, T.T.: Phys. Rev. 176, 631 (1968); 188, 1014 (1968); B2, 2795 (1970)

/11/ Fisher, M.E.: J. Phys. Soc. Jpn. Suppl. 26, 87 (1969)

/12/ Barouch , E.: J. Math. Phys. 12, 1577 (1971)

/13/ Au-Yang, H., McCoy, B.M.: Phys. Rev. B10, 886, 3885 (1974)

/14/ Hahn, H.: J. Mag. Mag. Mat.: 7, 209 (1978): Decker, I., Hahn, H.: Physica 83A, 143 (1976); 89A, 37 (1977); 93A, 215 (1978)

/15/ Rander, K., Hahn, H.: Phys. Lett. 53A, 287 (1975); Rander, K.: Z. Naturforschung 31A, 1465 (1976)

/16/ Villain, J.: J. Phys. C10, 1717 (1977)

/17/ Villain, J., Bideaux, R., Carton, J.-P., Conte, R.: J. Phys. (Paris) 41, 1263 (1980)

/18/ André, G., Bideaux, R., Carton, J.-P., de Seze, L.: J. Phys. (Paris) 40, 479 (1979)

/19/ Longa, L., Oleś, A.M.: J. Phys. A13, 1031 (1980)

/20/ Bideaux, R., de Seze, L.: J. Phys. (Paris) 42, 371 (1981)

/21/ Schultz, T.D., Mattis, D.C., Lieb, E.H.: Rev. Mod. Phys. 36, 856 (1964)

/22/ Stephen, M.J., Mittag, L.: J. Math. Phys. 13, 1944 (1972); Temperley, H.N.V., in: Domb, C., Green, M.S.: Phase Transitions and Critical Phenomena, Vol. I, London, Academic Press 1972

/23/ Binder, K., Kinzel, W., in: Lecture Notes in Physics 149, Berlin-Heidelberg-New York, Springer 1981; Binder,K.: Z. Phys. B, to be published

/24/ Ono, I.: Phys. Soc. Jap. 41, 345 (1976); Kirkpatrick, S.: Phys. Rev. B16, 4630 (1977); Vannimenus, J., de Seze, L.: J. Appl. Phys. 50, 7324 (1980); Morgenstern, I., Binder, K.: Phys. Rev. B22, 288 (1980); Bieche, I., Maynard, R., Rammal, R., Uhry, J.P.: J. Phys. A13, 2553 (1980); Domany, E.: J. Phys. C12, L 119 (1979); Horiguchi, T., Morita, T.: J. Phys. A15, L 75 (1982)

/25/ Stephenson, J.: J. Math. Phys. 5, 1009 (1964); 7, 1123 (1966); 11, 413, 420 (1970)

/26/ Forgacs, G.: Phys. Rev. B22, 4473 (1980)

/27/ Gabay, M.: J. Physique Lettres 41, 427 (1980)

/28/ Sütő, A.: Z. Phys. B - Condensed Matter 44, 121 (1981)

/29/ Peschel, I.: Z. Phys. B - Condensed Matter 45, 339 (1982)

/30/ McCoy, B.M., Wu, T.T.: The Two-Dimensional Ising Model, Harvard University Press, Cambridge, Mass. 1973

/31/ Tsallis, C., Levy, S.V.: J. Phys. C13, 465 (1980)

/32/ Harris, A.B.: J. Phys. C7, 1671 (1974)

/33/ Tamaribuchi, T., Takano, F.: Progr. Theor. Phys. 64, 1212 (1980): Tamaribuchi, T.: Progr. Theor. Phys. 66, 1575 (1981)

/34/ Domany, E.: J. Phys. C11, L 337 (1978)

/35/ Morgenstern, I., Binder, K.: Phys. Rev. Lett. 43, 1615 (1979); Phys. Rev. B22, 288 (1980)

/36/ Stauffer, D.: Phys. Rep. 54, 1 (1979)

/37/ Avron, J.E., Roepstorff, G., Schulman, L.S.: J. Stat. Phys. 26, 25 (1981)

/38/ Wolff, W.F., Zittartz, J.: Z. Phys. B - Condensed Matter, to be published

/39/ Dotsenko, V.S., Dotsenko, V.S.: JETP Lett. 33, 40 (1981); J. Phys. C15, 495 (1982)

/40/ Fisher, M.E.: Phys. Rev. 176, 631 (1968)

/41/ Kasai, Y., Syozi, I.: Progr. Theor. Phys. 50, 1182 (1973)

/42/ Suzuki, M.: J. Phys. C7, 255 (1974)

A Study of Short-Range Spin Glasses

H. Sompolinsky
Dept. of Physics, Bar-Ilan University,
Ramat-Gan, Israel*
and
A. Zippelius
Institut für Festkörperforschung
Kernforschungsanlage Jülich GmbH,
D-5170 Jülich, Federal Republic of Germany

Abstract: Spatial fluctuations in spin glasses are studied by an expansion around the dynamic mean field theory. We discuss the properties of the low temperature phase, in particular the stability and consistency of mean field theory and identify the most divergent fluctuations and the resulting special dimensionalities of the model. The dynamic critical behaviour is studied within an ε expansion around the upper critical dimension $d_u=6$.

I. Introduction

So far most theoretical effort has concentrated on a Mean Field Theory (MFT) of the Spin Glass (SG) phase, using the replica trick, a dynamic approach or other means[1]. Now it is well known from the theory of phase transitions in uniform systems that MFT is incorrect for the following reasons:

a) Below the upper critical dimension d_u, fluctuations modify the critical behaviour.

b) Below the lower critical dimension d_ℓ, fluctuations destroy the order completely for any finite temperature.

This last point is of particular interest from a theoretical point of view, since many experiments seem to be consistent with a truly static phase transition[2], whereas high temperature series expansions[3] as well as careful Monte Carlo simulations[4] predict $d_\ell=4$.

As a first step towards a more realistic description of SG we have undertaken a study of fluctuations in systems with short range exchange. Thereby we have focussed on two problems:

A) What are the properties of the low temperature phase and

B) what is the critical behaviour in models with short range interaction?

*present address: Bell Laboratories, Murray Hill, NJ 07974, USA

II. Low temperature phase

Our aim is to go beyond MFT and study the properties of spatial fluctuations in the short range model. This has been done with the help of an expansion around the full marginally stable MF solution[5]. In this section we are going to discuss and analyze such an expansion on the level of a Gaussian approximation, i.e. cubic as well as higher order interactions between fluctuations are neglected. Thus the results of this section are not concerned with the critical properties near the transition temperature, rather we want to learn more about the low temperature phase in the short range model. In particular we would like to know:

- Is MFT consistent and stable for all temperatures and time scales?
- Does the spectrum of relaxation times of MFT pertain to nonlocal fluctuations?
- Which are the most important long wavelength fluctuations?
- What are the resulting special dimensions?
- And in particular: What is the lower critical dimension d_ℓ, such that for $d<d_\ell$ fluctuations destroy the order completely at any finite temperature?

We study the soft-spin version of the Edwards-Anderson SG model given by the Hamiltonian

$$H = -\sum_{<ij>} J_{ij}\sigma_i\sigma_j + \sum_i V\{\sigma_i\} \tag{1}$$

where the interaction J_{ij} between each spin σ_i and its z nearest neighbors are random Gaussian variables with zero average and mean square fluctuations $[J_{ij}^2]=\tilde{J}^2/z$. The potential V provides the local constraint on the length of the spins. Purely relaxational dynamics is introduced by the equations of motion,

$$\Gamma_o^{-1}\partial_t\sigma_i(t) = -\frac{\delta(\beta H)}{\delta\sigma_i} + \phi_i(t) + h_i(t) \tag{2}$$

where, as usual, the variance of the Gaussian noise ϕ is related to the kinetic coefficient Γ_o via the FDT. This model has been previously[5,6] studied in the MF limit of z=N,N being the number of spins in the system. The dynamic MFT deals mainly with the properties of the ensemble averaged local susceptibility and correlation,

$$\chi(t) \equiv [\delta<\sigma_i(t')>/\delta h_i(t'-t)] \tag{3}$$

$$C(t) \equiv [<\sigma_i(0)\sigma_i(t)>]$$

where < > refers to average over the noise ϕ and [ ] over the quenched disorder J_{ij}.

Our study of nonlocal fluctuations focusses on the time dependent, nonlocal SG correlation

$$C_{SG}(k,t) = \frac{1}{N}\sum_{ij} \ell^{i\vec{k}\cdot\vec{R}_{ij}}[<\sigma_i(0)\sigma_j(t)>^2] \tag{4}$$

and the frequency dependent nonlocal SG susceptibility

$$\chi_{SG}(k,\omega) = \frac{1}{N}\sum_{ij} \ell^{i\vec{k}\cdot\vec{R}_{ij}}\left[\left(\frac{\partial<\sigma_i(\omega)>}{\partial h_j(-\omega)}\right)^2\right] \tag{5}$$

These have been calculated[7] for the dynamic model defined in Eq.(1) by expanding up to quadratic order in deviations from MFT. For $\chi_{SG}(k,\omega)$ we find a simple k dependence at all T and small k

$$\chi_{SG}(k,\omega) = \frac{\chi^2(\omega)}{1+ck^2-\beta^2\tilde{J}^2\chi^2(\omega)} \tag{6}$$

where $\chi(\omega)$ is the result of MFT for the average local susceptibility, c is a constant and we limit ourselves for simplicity to the fixed-length limit $\sigma_i^2=1$. For $T>T_c=\tilde{J}$ and small ω, $\chi(\omega)\approx 1+i\omega/\Gamma(\omega)$ with $\Gamma(\omega)\approx(T-T_c)f(\omega/(T-T_c)^2)$ where $f(0)$=const., and $f(z\to\infty)\propto z^{1/2}$. Then $\chi_{SG}(k,\omega)$ has the following scaling behavior above T_c,

$$\chi_{SG}(k,\omega) = \xi^{2-\eta}g(k\xi,\omega\xi^{z^+}) \tag{7}$$

with $\xi=(T-T_c)^{-\nu^+}$, $\nu^+=1/2$, $\eta=0$, $\gamma=1$ and $z^+=4$. For $\omega=0$, Eq.(7) yields $\chi_{SG}^{-1}(k)\simeq T-T_c+ck^2$ which agrees with previous results[8]. Above T_c, $C_{SG}(k,t)$ can be obtained from χ_{SG} by the FDT.

Below T_c, it was found in the MFT[5] that both $\chi(t)$ and $C(t)$ have a spectrum of 'macroscopic' relaxation times denoted by $\tau_x, x\varepsilon[0,1]$ where τ_1 and τ_0 are respectively the finite time and the infinite time limits and $\tau_{x'}<<\tau_x$ if $x'>x$. Only the non-equilibrium parts of χ and C obey the FDT whereas the long-lived parts serve as the order parameters $q(x)\equiv C(\tau_x)$ and $\chi(x)\equiv\chi(\omega_x=\tau_x^{-1})$. In zero external field, $q(x\to 0)=0$ and $\chi(x)-T/\tilde{J}\propto -q^2(x)$ for $x \ll 1$ at all $T\leq T_c$. At long but finite time $(t<\tau_1)\chi(t)$ varies as $t^{-\mu(T)-1}$ where $\mu(T)\simeq \frac{1}{2} -O(T_c-T)^6$.

Expanding around the MF solution below T_c we find that (1) the nonlocal propagators relax to their equilibrium values with the same spectrum of relaxation times $\{\tau_x\}$ as that of the order parameters; (2) Not only q(x) but also the non-local correlations $<\sigma_i(0)\sigma_j(t)>$ decay completely to zero (in zero field) as $t\to\tau_0$; (3) The continuum of time scales gives rise to a distribution of correlation lengths

$$\xi_x = (\chi(0)-\chi(x))^{-1/2}\approx 1/q(x) \tag{8}$$

characterizing the spatial extent of non-local fluctuations at time scale τ_x. The correlation length ξ_x diverges at the longest time scales, $x\to 0$, at all $T\leq T_c$ reflecting the marginal nature of the SG phase. More specifically, using Eq.(5) with $\omega=\omega'=\omega_x$ we find

$$\chi_{SG}(k,\omega_x) \approx \xi_x^2 f^{(1)}(k\xi_x) \tag{9}$$

where $f^{(1)}(0)=\text{const.}$ and $f^{(1)}(z\to\infty) \propto z^{-2}$. Thus at true equilibrium $\chi_{SG}(k,0) \propto k^{-2}$ for all $T\leq T_c$.

For the SG correlation we find

$$C_{SG}(k,\tau_x) \sim k^{-4} f^{(2)}(k\xi_x) \tag{10}$$

where $f^{(2)}(0)=\text{const.}$ and $f^{(2)}(z\to\infty)\sim z^{-2}$. This means that the Fourier Transform of $[<\sigma_i(0)\sigma_j(\tau_x)>^2]$ diverges as k^{-4} for $k\to 0$ at all $T\leq T_c$. This is true for the equal time correlations $[<\sigma_i(t)\sigma_j(t)>^2]$ as well. Integrating Eq.(10) over k one finds for the local correlations

$$[<\sigma_i(0)\sigma_i(\tau_x)>^2] \propto \begin{cases} q^2(x), & d>6, \\ q^{d-4}(x), & 4<d<6 , \\ \infty, & d\leq 4 . \end{cases} \tag{11}$$

Since the local correlations are necessarily finite, we conclude that the expansion around the MFT breaks down below d=4 even far below T_c. This is certainly consistent with four being the lower critical dimensionality of the SG transition as suggested by previous work[3]. However, it is possible that below d=4 there is still a SG phase but with properties which are different from the MFT at all temperatures below T_c. Such a modification of the MF results at low temperature is particularly plausible in a phase which remains critical below T_c. Equation (11) also indicates that 6 is the upper critical dimensionality since in MFT $[<\sigma_i(0)\sigma_i(\tau_x)>^2]$ is indeed proportional to $q^2(x)$. This conclusion can also be derived from the hyperscaling relation as applied to $\chi_{SG}(k)$ above T_c as well as from the anomalous dimensionality of the cubic vertex[8] which is present in the model.

So far we have discussed the fluctuations on macroscopic time scales. In the dynamic regime, $\omega \gg \omega_1$, Eq.(6) yields

$$\chi_{SG}(k,\omega)\sim\xi_1^2 g_-(k\xi_1,\omega\xi_1^{z-}) \tag{12}$$

with $\xi_1=1/q(1)$ is finite and $z^-(T)=2/\mu(T)$ which also implies that $\nu^-=1$, $\gamma^-=1$ and $\eta^-=0$. In this regime $C_{SG}(k,t)$ is related to Eq.(12) by the FDT.

The results we have presented so far support the conclusion that the dynamic

MFT is marginally stable at all temperatures. More precisely, $\chi_{SG}(k,\omega)$ and $C_{SG}(k,t)$ are special cases of the more general four spin correlation, which is the dynamic analog of the stability matrix. We have calculated it for the longest timescales, but have not yet completed the calculation of the full matrix. Nevertheless, since correlation lengths grow with time scale it is unlikely that the more general correlation functions will exhibit stronger singularities than those which were found here.

Furthermore our analysis can equally well be applied to other MF solutions to check their stability. In the solutions of Sherrington and Kirkpatrick[9] and Sommers[10] frozen correlations, i.e.q, remain finite on the longest timescales. Then $\chi_{SG}(k,\omega)$ is still of the same form as Eq.(6) except that $\chi^2(\omega)$ is replaced by $[\chi_{ii}^2(\omega)]_{MF}$ where χ_{ii} is the unaveraged MF local susceptibility. In the Sherrington and Kirkpatrick solution, $\chi_{ii}(0)=1-\langle\sigma_i\rangle^2$, one then obtains $\chi_{SG}^{-1}(\vec{0},0)\propto 1-\beta^2\tilde{J}^2[(1-\langle\sigma\rangle^2)^2] \simeq -\frac{4}{3}(T_c-T)^2$ which clearly signals instability. In Sommers solution, there is a frequency scale ω_1 such that $\chi_{ii}(\omega\geq\omega_1)-1+\langle\sigma_i\rangle^2=0$ and $\chi_{ii}(\omega\ll\omega_1)-1+\langle\sigma\rangle^2>0$. For $\omega\gg\omega_1, \chi_{SG}^{-1}(k=\vec{0})\propto 1-\beta^2\tilde{J}^2[(1-\langle\sigma\rangle^2)^2]\simeq\frac{4}{3}(T_c-T)^2$. Therefore this solution is stable in the finite frequency regime as was indeed found in the dynamic MF analysis. However for $\omega\ll\omega_1, \chi_{SG}^{-1}(k=\vec{0})\propto 1-\beta^2\tilde{J}^2[\chi_{ii}^2(0)]\simeq -\frac{4}{3}(T_c-T)^2$, implying an instability of Sommers solution in the static limit, in agreement with the stability analysis in the replica formalism[11].

III. Critical Dynamics

In uniform systems the most successful description of critical behaviour has been an expansion around the upper or lower critical dimensionality. In the SG case no low temperature expansion exists so far, whereas several attempts have been made to study fluctuations via an expansion around the upper critical dimension: Harris et al.[12] used a renormalization group approach to calculate critical exponents in an expansion in $\varepsilon=6-d$. Subsequently Bray & Roberts[13] extended their work to finite fields in order to study the critical behaviour along the Almeida-Thouless[14] line. Surprisingly they were unable to locate a stable fixed point below 6 dimensions. Another surprising result was obtained by Chen & Lubensky[15], who considered a model with competing ferromagnetic and antiferromagnetic ordering. They found complex exponents for XY and Heisenberg spins at the ferromagnetic-spinglass multicritical point. Neither of these results appears to be understood.

We have investigated the dynamic critical behaviour of an Ising spinglass for $T\geq T_c$ and $h=0$[16]. Our starting point is the purely relaxational dynamic model of Eq. (1). The dynamic quantities of interest are the local responsefunction $\chi(\omega)$ as well as the dynamic spinglass susceptibility $\chi_{SG}(k,\omega)$. Above T_c we expect that time dependent fluctuations decay exponentially for long times, so that it is meaningful to define characteristic relaxation rates

$$\nu_{local}^{-1} = i\chi(\omega=0)\frac{\partial}{\partial\omega}\chi^{-1}(\omega=0) \tag{13}$$

$$\nu_{nonlocal}^{-1} = i\chi_{SG}(k=0,\omega=0)\frac{\partial}{\partial\omega}\chi_{SG}^{-1}(k=0,\omega=0) \tag{14}$$

Before discussing the effect of fluctuations, we briefly recall the predictions of MFT for the dynamic critical behaviour[6]. In the limit of low frequency and long wavelength, both local

$$\chi^{-1}(\omega) = 1+\frac{i\omega}{\Gamma_o r_o} \tag{15}$$

as well as nonlocal fluctuations

$$\chi_{SG}^{-1}(k,\omega) = r_o+k^2+\frac{i\omega}{\Gamma_o r_o} \tag{16}$$

are characterized by two parameters: the deviation $r_o=T-T_c^o$ from the MF transition temperature $T_c^o=\tilde{J}$ and the bare relaxation rate Γ_o, which remains finite as $T\to T_c^o$. So above T_c the decay in time is indeed exponential. As $T\to T_c^o$ the local and nonlocal relaxation rates diverge as $\nu_{local}\sim\frac{1}{\chi_{SG}}\sim|T-T_c|$ and $\nu_{nonlocal}\sim\frac{1}{\chi_{SG}^2}\sim|T-T_c|^2$ respectively.

Now we would like to know: How are these quantities modified if fluctuations are taken into account. We follow the standard procedure: First a coupling constant w is formally introduced to characterize the strength of the interactoin of fluctuations. We then calculate perturbatively the effect of fluctuations with wavenumbers $\frac{1}{b}<k<1$ and rescale the remaining degrees of freedom $\sigma'=b^{a/2}\sigma$, $k'=bk$ and $\omega'=b^z\omega$. The resulting dynamic correlations are rather complicated functions of wavenumber and frequency. However, in the hydrodynamic limit $k\to0$ and $\omega\to0$ the simple form of MF theory prevails

$$\chi^{-1}(\omega) = 1+\frac{i\omega}{\Gamma r} \tag{17}$$

$$\chi_{SG}^{-1}(k,\omega) = r+k^2+\frac{i\omega}{\Gamma r} \tag{18}$$

with renormalized parameters r and Γ. The recursion relations for the latter read:

$$r' = b^{d-2a}(r+2w^2(A(o)-2rK_d\ \ell n\ b)) \tag{19a}$$

$$w' = b^{d-3a}w(1-2w^2K_d\ \ell n\ b) \tag{19b}$$

$$\Gamma' = b^{z-2d+4a}\Gamma \tag{19c}$$

Here $K_6 = 2/\Gamma(3)(4\pi)^3$, $A(o) = \frac{K_6}{2}(1-b^{-2})$ and a is related to η via $d-2a=2-\eta$. Above 6 dimensions the recursion relations have a stable fixed point $w^*=0$, $r^*=0$, $\eta^*=0$, and $z^*=4$, corresponding to critical MF behaviour. Below 6 dimensions a new fixed point becomes stable with $(w^*)^2=\frac{\varepsilon}{2K_6}$, $r^*=-\frac{\varepsilon}{2}$, $\eta^*=-\frac{\varepsilon}{3}$ and dynamic exponent $z^*=2(2-\eta^*)=2(2+\frac{\varepsilon}{3})$. The results for the static susceptibility and the coupling constant agree with previous static calculations using the replica trick[12].

To summarize: MFT correctly predicts the dynamic critical behaviour above six dimensions. Below 6 dimensions the kinetic coefficient Γ remains finite, so that critical slowing down is solely determined by the divergence of the order parameter susceptibility. In other words: van Hove theory is shown to be correct to lowest order in $\varepsilon=6-d$, implying for the dynamic exponent $z=2(2-\eta)$.

References

/1/ For a recent review, see R. Rammal and J. Souletie in Magnetism of Metals and Alloys, Ed. M. Cyrot, North Holland 1981, p. 342; K. Fischer, phys. stat. sol. (b) 116, 357 (1983)
/2/ B. Barbara, A.P. Malozemoff, and Y. Imry, Phys. Rev. Lett. 47, 1852 (1981); P. Monod and H. Bouchiat, J. Phys. Lett. (Paris) 43, L45 (1982); A. Berton, J. Chaussy, J. Odin, B. Rammal, and R. Tournier, J. Phys. Lett. (Paris) 43, L-153 (1982); R. Omari, J.J. Préjean, and J. Souletie, J. Phys. (Paris) in press
/3/ R. Fisch and A.B. Harris, Phys. Rev. Lett. 38, 785 (1977)
/4/ I. Morgenstern and K. Binder, Phys. Rev. Lett. 43, 1615 (1979); Phys. Rev. B22, (1980); Z. Phys. B39, 227 (1980)
/5/ H. Sompolinsky, Phys. Rev. Lett. 47, 935 (1981)
/6/ H. Sompolinsky and A. Zippelius, Phys. Rev. Lett. 47, 359 (1981); Phys. Rev. B25, 6860 (1982); J.A. Hertz, A. Khurana, M. Puoskari, Phys. Rev. B25, 2065 (1982)
/7/ H. Sompolinsky and A. Zippelius, Phys. Rev. Lett. 50, 1297 (1983)
/8/ A.J. Bray and M.A. Moore, J. Phys. C12, 79 (1979); E. Pytte and J. Rudnick, Phys. Rev. B19, 3603 (1979)
/9/ D. Sherrington and S. Kirkpatrick, Phys. Rev. Lett. 35, 1792 (1975)
/10/ H.J. Sommers, Z. Phys. B32, 173 (1979)
/11/ C. De Dominicis and T. Garel, J. Phys. (Paris) 40, L576 (1979)
/12/ A.B. Harris, T.C. Lubensky and J.H. Chen, Phys. Rev. Lett. 36, 415 (1976)
/13/ A.J. Bray and S.A. Roberts, J. Phys. C13, 5405 (1980)
/14/ J.R.L. de Almeida and D.J. Thouless, J. Phys. A11, 983 (1978)
/15/ J.H. Chen and T.C. Lubensky, Phys. Rev. B16, 2106 (1977)
/16/ A. Zippelius, to be published

The spin glass transition: a comparison of Monte Carlo simulations of nearest-neighbor Ising Edwards-Anderson models with experiments

K. Binder and W. Kinzel
Institut für Festkörperforschung, Kernforschungsanlage Jülich,
D-5170 Jülich, W.-Germany

Abstract: Numerical studies of Ising square lattices with random bonds ($J_{ij}=\pm J$ or drawn from a gaussian distribution) are reviewed. Particular attention is paid to the temperature- and field dependence of the equilibrium magnetization $M(H,T)$. While for a symmetric bond distribution the zero-field susceptibility trivially follows a Curies law $\chi_o \propto T^{-1}$, the nonlinear susceptibility $\chi_{n\ell}$ shows a dramatic temperature-dependence, which can nearly be mistaken for a power-law divergence at a freezing temperature T_f. These findings are compared in detail with corresponding experimental data, including possible "scaling" representations. We relate this behavior to the onset of long-range Edwards-Anderson order as $T\to 0$, as measured by the correlation function $g_{EA}(\vec{r}_{ij})=[\langle S_i S_j\rangle_T^2]_{av}$.

We then discuss time-dependent quantities: the spin-spin autocorrelation function and the time-dependent Edwards-Anderson order parameter $q(t)$, dynamic susceptibility $\chi(t)$ etc.; also the onset of irreversible behavior at critical magnetic fields $H_c(t)$ is emphasized, and again compared to experiments. A possible explanation of this behavior in terms of the free energy barriers separating the various "valleys" in configuration space is indicated.

I. Introduction

Spin glasses are magnets (diluted or otherwise disordered) where conventional long-range ferro- or antiferromagnetic order does not occur, but nevertheless a transition occurs to a state where the spin directions are frozen-in. First it was thought that these phenomena could simply be explained in terms of the "Neél-model", which describes the "blocking" of the reorientation of the magnetic moments of superparamagnetic clusters [1,2], or variants thereof [3-5]. However, the observation of very sharp cusps in the AC-susceptibility in small fields [6] was an indication that small clusters well isolated from each other - which do nicely account for materials such as $Eu_xSr_{1-x}S$ at concentrations $x\approx 0.05$ far below the percolation treshold [7] are not enough to explain typical spin glasses, and rather collective phenomena must be involved. This idea is manifest in the Edwards-Anderson model [8] which has the hamiltonian

$$\mathcal{H} = -\sum_{i\neq j} J_{ij}S_iS_j - H\sum_i S_i \ , \ S_i = \pm 1 \text{ (Ising spins)} \quad . \tag{1}$$

It is now well established that this model exhibits a new kind of phase transition in the limit of infinitely weak infinitely long-range forces [9-13]: While for a symmetric bond distribution $P(J_{ij})=P(-J_{ij})$ the zero-field susceptibility χ_o is described by the Curie law $\chi_o=1/T$ for temperatures T larger than the freezing temperature T_f, χ_o has a cusp at T_f described by the onset of Edwards-Anderson order. This order is not described by a single order parameter but rather the Parisi [10] order parameter function q(x), with

$$\chi_o = (1-\bar{q})/T, \ \bar{q} = \int_0^1 q(x)dx \propto (1-T/T_f)^\beta, \ \beta = 1 \quad . \tag{2}$$

This behavior is due to the fact that below T_f the model is strongly non-ergodic [11-13], there are many "valleys" in phase space separated by infinitely high barriers between them. While in a time average one would sample only from a single valley, by statistical mechanics-averaging one samples all the valleys and hence finds rather different properties. Above T_f this transition shows up in the susceptibility χ_{EA} defined as

$$\chi_{EA} \equiv \sum_i g_{EA}(\vec{r}_{ij}) \ , \ g_{EA}(\vec{r}_{ij}) = [\langle S_iS_j\rangle_T^2]_{av} \ , \tag{3}$$

which diverges at T_f as

$$\chi_{EA} \propto (T/T_f-1)^{-\gamma} \ , \ \gamma = 1 \quad . \tag{4}$$

For $T < T_f$ this suceptibility becomes $\chi_{EA}=q_{EA}^2 N$, where - implying that a weak symmetry-breaking field is applied and is taken to zero after the thermodynamic limit $N\to\infty$ is taken - $q_{EA}=[\langle S_i\rangle_T^2]_{av}$ is the order parameter originally defined by Edwards and Anderson, with $q_{EA} \propto (1-T/T_f)^\beta$ also.

This transition signalling a breakdown of ergodicity and hence the onset of truly irreversible behavior (remanence, etc.) in the dynamics occurs also for finite magnetic field, at the Almeida-Thouless (AT) instability line [14], which near T_f is described by

$$H_c(T)/T_f \propto (1-T/T_f)^{3/2} \quad . \tag{5}$$

For $H < H_c(T)$ the magnetization is essentially temperature-independent [15].

For more convential phase transitions it is well-known, of course, that systems with more realistic interactions of shorter range may behave quite differently from the infinite range case (see e.g. [16]). This mean-field transition is qualitatively correct (i.e. the critical exponents have their mean-field values) for spatial dimensionalities d exceeding the upper critical dimension d_u. For spin glasses one believes $d_u=6$ [17]. For $d_\ell<d<d_u$ one has nontrivial critical behavior (exponents γ,β different from mean-field values), while for d less than the lower critical dimension d_ℓ the fluctuations destroy the order completely at any finite temperature and the critical temperature is zero.

A central question for the spin glass transition hence is to find d_ℓ. Most theoretical predictions suggest $d_\ell=4$ [18-23], while initially this question was far from being clear [24-26]. But a number of careful experiments of the field cooled magnetization {which is widely believed to yield the equilibrium magnetization although there may be problems with time-dependence, too [27]} have been interpreted as being consistent with a truly static phase transition at a nonzero T_f {[28-32]; see also [33-40]}. While the exponents β,γ extracted from this analysis are rather different from their mean-field counterparts [28-32], the critical magnetic field where irreversible behavior sets in seems still to be described by Eq.(5).

In the present review talk numerical simulation results will be discussed which are pertinent to this question. Initially the Monte-Carlo results both for d=2 [41-43] and d=3 [44] were interpreted as being consistent with a static transition at a nonzero T_f and a nonzero order parameter q_{EA} for $T<T_f$. More careful analysis has revealed ample evidence that $T_f=0$ at least for d=2 [19,21,45-52]. From the very beginning it has been noted that many results are in surprisingly close qualitative agreement with experiment [41,43,50], and hence we shall compare the simulation results to experimental data in detail. More complete reviews discussing also other models can be found in [53].

II. Model and numerical techniques

All work described below will concern the Hamiltonian Eq.(1) with nearest-neighbor interaction on a square lattice, with distribution $P(J_{ij})$ being either

$$P(J_{ij}) = \frac{1}{2}\{\delta(J_{ij}-J)+\delta(J_{ij}+J)\}, \quad (\text{"}\pm J\text{ model"}) \quad , \tag{6a}$$

or

$$P(J_{ij}) \propto \exp[-J_{ij}^2/2(\Delta J)^2] \quad , \quad (\text{"gaussian model"}) \quad . \tag{6b}$$

The numerical techniques discussed are two-fold: (i) A recursive exact calculation of partition functions $Z_{\{J_{ij}\}}$ [and susceptibilities, correlation functions $<S_iS_j>^2_{T,\{J_{ij}\}}$, etc.] for a chosen set $\{J_{ij}\}$ of random couplings. The results are then numerically averaged over typically 30-100 sets $\{J_{ij}\}$. Although this "simple random sampling" in the general case has to be considered with precautions [54], in the cases of interest to us here the errors are well under control [55], and sufficiently accurate results have been obtained [21] for lattice sizes up to 18x18 lattice spacings, with periodic boundary conditions in one direction and free boundary conditions in the other (in which then the decay of correlation $<S_iS_j>^2_T$ is measured). (ii) The second method is the standard importance sampling Monte Carlo method [53], which can be interpreted as a realization of a dynamic relaxation process described by a Markovian master equation for the probability $P(\vec{X},t), \vec{X} \equiv (S_1,\dots,S_N)$,

$$\frac{d}{dt}P(\vec{X},t) = -\sum_{\vec{X}'} W(\vec{X}\to\vec{X}')P(\vec{X},t) + \sum_{\vec{X}'} W(\vec{X}'\to\vec{X})P(\vec{X}',t) \quad . \tag{7}$$

The "time" t is proportional to the (sequential) label of configurations $\vec{X}$ generated by the transitions $\vec{X}\to\vec{X}'$ in the Monte Carlo process. These transitions are carried out with transition probability (per unit time) W,

$$W(\vec{X}\to\vec{X}') = (1/2\tau)\{1-\tanh([\mathcal{H}(\vec{X}')-\mathcal{H}(\vec{X})]/2k_BT)\} \quad . \tag{8}$$

Eq.(8) satisfies a detailed balance condition with the probability $P_{eq} \propto \exp(-\mathcal{H}/k_BT)$ in thermal equilibrium, and hence it is ensured (at least in a finite system which necessarily is ergodic) that equilibrium is reached for $t\to\infty$. The parameter τ is arbitrary and sets the time-scale; we eliminate it by measuring time in the units of Monte Carlo steps (MCS) per spin (=attempted single spin flips per spin).

At this point already the question arises to what extent (if at all) this model relates to the real world; and if so, to estimate conversion factors for the scales of magnetic moment, magnetic field, temperature, length, time to the scales of these quantities for real systems. Unfortunately, a convincing quantitive answer to these questions can not be given. We argue, however, that at least a qualitative relation should exist: imagine that a sort of coarse-graining is performed, where we average out the properties on very short length scales (which depend on detailed material parameters and the detailed character of disorder-dilution or structural disorder, absence or presence of chemical clustering, etc.-). Each Ising spin of the model then already corresponds to a whole cluster of strongly correlated spins of the real system and the conversion factor for magnetic field and magnetic moment obviously must depend on the nature of the (ferro-or antiferromagnetic) short-range order in the cluster. Only on this coarse-grained level it makes sense to describe the interactions between the Ising spins by a strictly symmetric bond disorder as done in Eq.(6), which implies for the standard correlation function

$$g_F(r_{ij}) \equiv [\langle S_i S_j \rangle_T]_{av} = \delta_{ij} , \tag{9}$$

i.e. the absence of any ferro-or antiferromagnetic correlation between the Ising spins. Even for the canonical spin glasses such as CuMn this absence of magnetic correlations is not true on the level of the magnetic moments of single Mn atoms [56]. Moreover on this atomic level it is clear that one has Heisenberg rather than Ising spins, with some weak anisotropy. On the coarse-grained level, however, a model such as Eqs.(6-9) should be qualitatively reasonable, as the degrees of freedom which have been averaged over in a dynamical sense act as a heat bath on the remaining ones. This heat bath then induces flips of the Ising spins, on a much larger time-scale, which is the process modelled by the master equation, Eq.(7). This coarse-graining, interpreting the Edwards-Anderson model as a model for randomly interacting clusters of magnetic moments (for more detais see [42]), is similar in spirit to the introduction of time-dependent Ginzburg Landau Wilson Hamiltonians, which are so successful in describing the universal features of static [16] and dynamic [57] critical phenomena at ordinary phase transitions.

As the standard Monte Carlo averaging means time-averaging, we define a time-dependent Edwards-Anderson order parameter by

$$q(t) = (1/N) \sum_{i=1}^{N} (\int_0^t S_i(t')dt'/t)^2 , \tag{10}$$

which in zero field is related to a time-dependent susceptibility defined via the fluctuation relation

$$\chi(t) = \frac{N}{k_B T} \{\overline{M^2}(t) - [\bar{M}(t)]^2\} = \frac{1}{k_B T} [1-q(t)], \tag{11}$$

where

$$\bar{M}(t) = \frac{1}{N} \sum \int_0^t S_i(t')dt'/t , \quad \overline{M^2}(t) = \int [\frac{1}{N} \sum_i S_i(t')]^2 dt'/t . \tag{12}$$

Other averages are defined similarly. We here anticipate already a central result of the numerical studies of short range spin glasses: while at low temperatures the description of the spin glass in terms of many valleys in phase space is similar to the infinite range model, the essential difference is that the free energy barriers separating the various valleys are finite rather than infinite [21,22,45-48]. While for short times the system stays in a valley and thus looks well ordered, q(t) being distinctly nonzero, for large times the system explores more and more valleys by jumping over these barriers, q(t) thus slowly decaying to zero for very large times. The freezing temperature T_f where these barriers buildup is thus itself time-dependent, albeit only very weakly [48]: the temperature - and field - dependence of these barriers can probably be linked [52] to the gradual increase of the correlation length ξ_{EA}, which describes the decay of $g_{EA}(\vec{r}_{ij})$ [Eq.(3)] with distance. This purely static correlation length diverges as $T\to 0$ for $H=0$ [21], as well as for $H\to 0$ if $T=0$ [50-52]. This interplay of the truely static gradually increasing correlations, which are still rather short range at T_f [21,51], with the barrier heights seen only in dynamic quantities, is still incompletely understood [52].

III. The magnetization process in a field: static linear and nonlinear susceptibilities

There have been many experimental indications that the equilibrium magnetization $M(T,H)$ can be obtained, if one cools the sample slowly enough from a high temperature to the desired temperature T (near T_f), keeping the applied field H fixed during the cooling (see e.g. [27,28]). This procedure to obtain a field-cooled magnetization and "susceptibility" $\chi(H,T)\equiv M(T,H)/H$ is readily simulated by Monte Carlo [43,52]. In a particular case it was also checked by the recursive partition function calculation that the resulting magnetization is indeed the equilibrium magnetization for the model chosen [58]. Fig.1 shows typical data obtained by simulating the field cooling procedure (cooling from $kT/\Delta J=1.6$ to $T=0$ at a rate $dT/dt=6.25\times 10^{-5}$ J/kMCS, and averaging over 5 to 30 runs) for the two-dimensional gaussian Edwards-Anderson spin glass, for a 60x60 lattice with periodic boundary conditions.

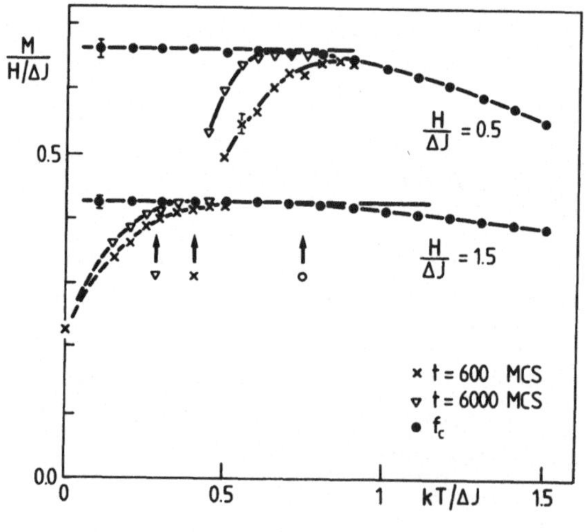

Fig.1: Field-cooled susceptibility (full dots) plotted vs. temperature at two values of the field {$H/\Delta J=0.5$ and $H/\Delta J=1.5$}. Other symbols denote the susceptibility obtained by cooling in zero field and then applying at the desired temperature a field for an observation time $t=600$MCS/spin (crosses) or $t=6000$MCS//spin (triangles) [52].

Fig.1 shows that the susceptibility increases with decreasing temperature (for $H\to0$ it is described by a Curie law $\chi_o=\Delta J/kT$ [21]), until a boundary is reached, (denoted by an arrow with a circle in Fig.1) where it saturates to a temperature-independent "plateau" value. For $H/\Delta J\gg1$ M saturates and thus $M/H\propto H^{-1}$ trivially; but for $0.1\lesssim H/\Delta J\lesssim1.0$ we observe a nontrivial power law (Fig.2)

$$M/H\propto H^{-x},\quad x\approx0.28\pm0.05\quad. \tag{13}$$

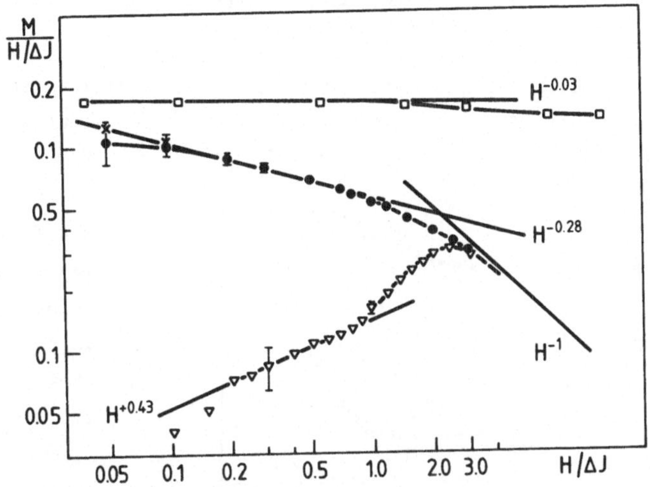

Fig.2: Log-log plot of "susceptibility" $M(T\to0)/(H/\Delta J)$ plotted vs. field. Full circles denote field cooled magnetization for $|dT/dt|=2.5\times10^{-4}\Delta J/kMCS$, crosses for $|dT/dt|=6.25$ and $1.5\times10^{-5}\Delta J/kMCS$. Open squares denote experimental data of [28] for AgMn10.6% {on arbitrary scales}. Triangles are the magnetization obtained from systems cooled to T=0 without a field [52].

Thus there is no contradiction between the plateaus seen in Fig.1 and the fact that the zero-field susceptibility for this model is divergent at T=0. Unfortunately, we cannot estimate the nontrivial exponent x in Eq.(13) with higher precision by studying smaller fields, however: (i) at smaller fields a still smaller cooling rate would be required, since at too high cooling rates the "plateau" value observed for M is too small (cf. full circles at $H/\Delta J=0.05$ in Fig.2) (ii) while we estimate that the correlation length ξ_{EA} is much smaller than the lattice linear dimensions in the regime of T,H on which Fig.2 is based [21,51], for smaller H also larger lattice sizes would be needed.

Fig.3 now shows corresponding experimental data for AgMn10.6% due to Monod and Bouchiat [28]. A qualitatively similar increase of the susceptibility is found, which smoothly merges (at a boundary denoted by arrows) into temperature independent plateau values, which slowly increase with decreasing field. On a quantitative level, there are several discrepancies: (1) The experimental values for H/T_f in Fig.3 are much smaller than corresponding ones in Fig.1. However, due to the unknown conversion factors for the scales of temperature, field, etc. involved in the coarse-graining mentioned above a comparison of these numbers is not meaningful. Thus, this discrepancy cannot serve to criticize the relevance of the theoretical models, as sometimes done by experimentalists. (2) A more important difference is that the experimental data [28] behave like $M/H \propto H^{-0.09}$ for intermediate fields while the effective exponent x crosses over to $x \approx 0.03$ (Fig.2). The question arises whether this crossover

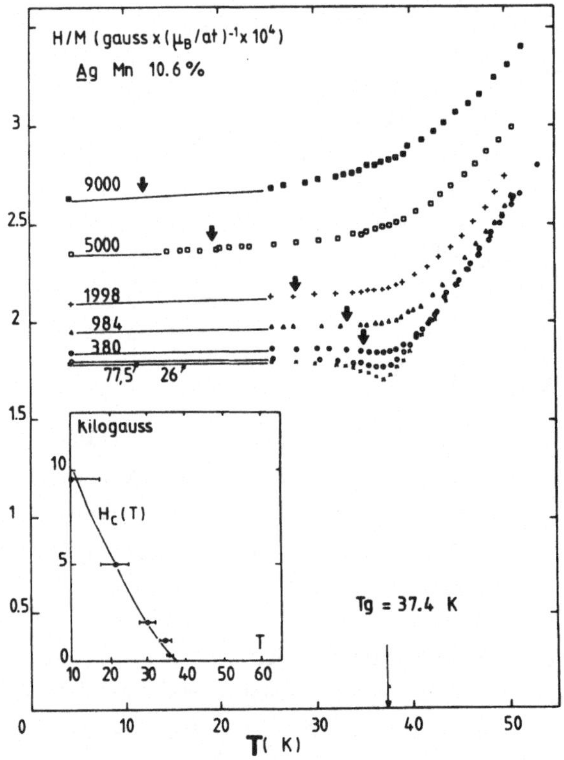

Fig.3: Inverse of the apparent susceptibility H/M for AgMn10.6% as a function of magnetic field as indicated on each set of point symbols (in gauss). The inset shows the critical field $H_c^{eq}(T)$, as estimated by the arrows in the H-T plane [28].

reflects equilibrium properties or (as the full dots in the simulation, Fig.2) too large cooling rates at the smallest fields. In fact, after cooling $AuFe_{8\%}$ in small fields the magnetization still has not relaxed to its equilibrium value see [27]. Careful experiments on this question (and at still smaller fields) are needed, as the exponent x must be strictly zero for H=0 if the dimensionality d is at (or above) d*, while a nonzero value of x is evidence for d<d*. If d*=4, we would expect that x is smaller for d=3 (experiment) than for d=2 (simulation).

(3) A qualitative distinction between Fig.1 and Fig.3 is that the experimental

equilibrium susceptibility develops a small peak in the smallest fields before the plateau sets in. It is possible that simulations would show this phenomenon also, if still smaller fields could be studied with meaningful accuracy. It is more likely, however, that this phenomenon is strictly absent in our strictly symmetric model, Eq.(6) [it also does not occur in mean-field], but rather is due to a bit of ferromagnetic short range order. In fact, in systems with more ferromagnetic short range order such as $Eu_xSr_{1-x}S$ [59] or AlGd37% [30] this susceptibility peak is much more pronounced. In addition, simple theoretical arguments show that even the static zero-field susceptibility must have a peak [50] if the system is close to a re-entrant ferromagnetic phase boundary in the temperature-concentration phase diagram.

(4) Another qualitative distinction between experiment and simulation seems to be the critical field $H_c^{eq}(T)$ which in Fig.3 is extrapolated smoothly to a nonzero T_f in the H-T plane, while the simulation shows (Fig.4) that $H_c^{eq}(T)$ bends backwards and approaches $T_f=0$ for $H_c^{eq}=0$. For small $H_c^{eq}(T)$ we have

$$H_c^{eq}(T) \propto T^{\Delta} \quad ; \tag{14}$$

the exponent Δ in this law can be related to the exponent x in Eq.(13) as $x=1/\Delta$ if we impose a scaling relation [50,52]

$$\frac{M(T,H)}{H} = \frac{1}{T}\tilde{\chi}(H/T^{\Delta}) \quad , \tag{15}$$

which describes the crossover from the Curie law [for small values of the scaling variable $\zeta=H/T^{\Delta}$ we have $\tilde{\chi}(\zeta)\approx\tilde{\chi}(0)=\text{const}$] to temperature-independent plateaus [for large ζ we have $\chi(\zeta)\propto\zeta^{1/\Delta}$].The scaling law Eq.(15) will be studied in more detail below.

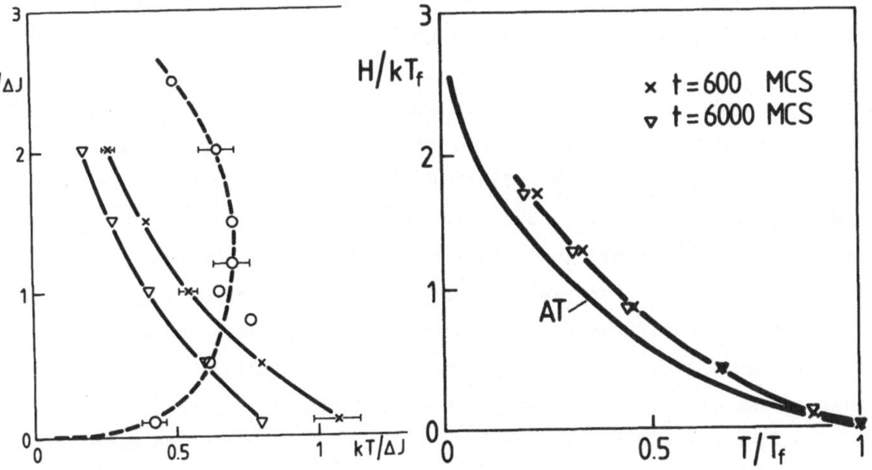

Fig.4: A) Static critical field $H_c^{eq}(T)$ [open circles] and dynamic critical fields $H_c(t)$ for t=600MCS (crosses) and t=6000MCS (triangles) plotted vs. temperature [62] B) Normalized dynamic critical field plotted vs. normalized temperature.

At this point, we emphasize that perhaps there is no contradiction between the qualitative behavior of $H_c^{eq}(T)$ in Figs.3,4: at the smallest fields in Fig.3, which were not included in the insert to Fig.3, the onset of the plateau seems again to occur at slightly lower temperatures. This decrease of the "effective freezing temperature" with decreasing field has been seen in AuFe8% [27], CuMn4,6% [30] and AlGd37% [30,60]. If one tries to interpret the data for AlGd37% [60], see Fig.5, according to Eq.(14), one finds an exponent Δ of the order 10^2 [60]: this result

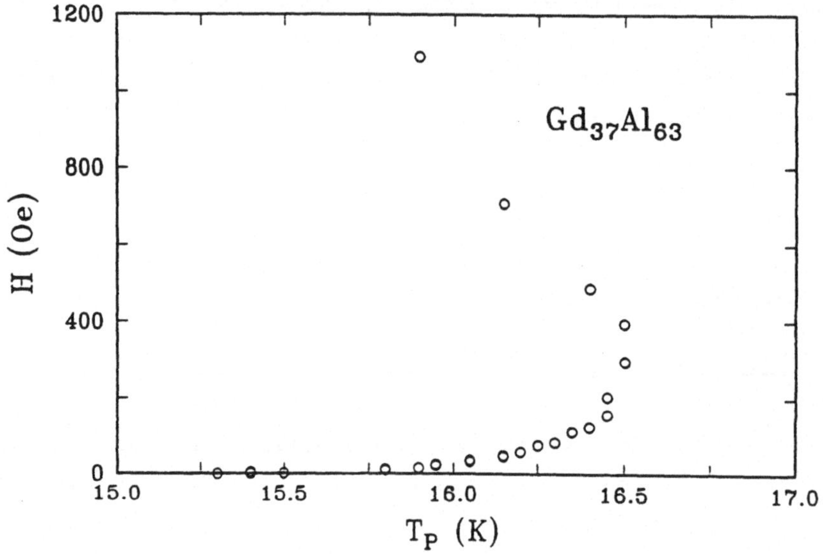

Fig.5: Static critical field $H_c^{eq}(T)$ from AlGd37% estimated from the position of the maximum of the field-cooled magnetization [60].

would be consistent with the smallness of the exponent x extracted from the AgMn10.6% data [28], see Fig.2. If these values really are equilibrium results, they might indicate that rather than the power laws Eqs.(13), (14) one has logarithmic laws {e.g. $M(T\to 0)/H \propto |\log H|$, $H_c^{eq}(T) \propto 1/|\log T|$}. Such laws would be expected if d=3 were the lower critical dimension{which would imply that either $d_\ell=4$ is incorrect for the short range case, or long range forces such as dipolar interactions could decrease d_ℓ for spin glasses}. This is not the interpretation adopted by Malozemoff et al. [29,30,60], who insist on a finite nonzero T_f. But their data [60] clearly indicate that the precise location of T_f, if it exists, is highly ambiguous [61].

In spite of the fact that it is hard to locate T_f in zero field with meaningful accuracy, there have been numerous attempts to estimate critical exponents for spin glasses {e.g. [28-32],[60]}. One of the first attempts to do this was based on the data shown in Fig.3 near T_f, replotted versus field (Fig.6). At small

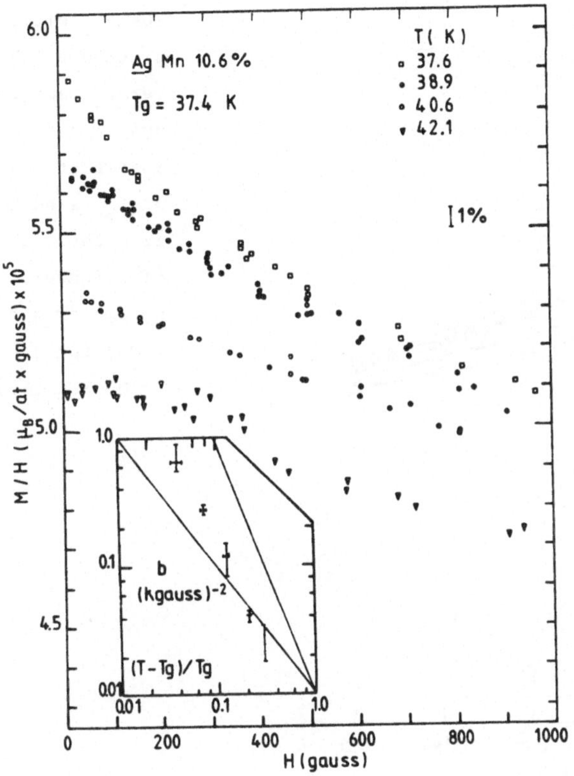

Fig.6: M/H for AgMn10.6% plotted vs. field at four temperatures above T_f (assumed to be T_f=37.4K). Insert shows the temperature variation of the nonlinear susceptibility $\chi_{n\ell}$ {Eq.(16)} [28].

fields the variation of M/H with H in the paramagnetic region must be quadratic,

$$\frac{M}{H} = \chi_o(T)-H^2\chi_{n\ell}(T)+O(H^4) \quad , \tag{16}$$

which hence defines the nonlinear susceptibility $\chi_{n\ell}(T)$. The data of Fig.6 are interpreted in terms of a power law divergence,

$$\chi_{n\ell}(T) \propto (T/T_f-1)^{-\gamma} \quad , \tag{17}$$

with $\gamma\approx1.6$ {while other workers, e.g. [29,30,60], find $\gamma\approx3.5$}. Indeed the critical parts of $\chi_{n\ell}$ {Eq.(16)} and χ_{EA}{Eq.(3)}are proportional [50], for symmetric models {Eq.(6)} we have simply [measuring H in units of J of ΔJ]

$$\chi_{n\ell}(T) = T^{-3}(\chi_{EA}(T) - 2/3).\qquad(18)$$

Replotting [50] Monte Carlo simulation data [41,62] as in Fig.6 one again notes striking qualitative similarity to experiment, Fig.7. One notes that for T exceeding T_f

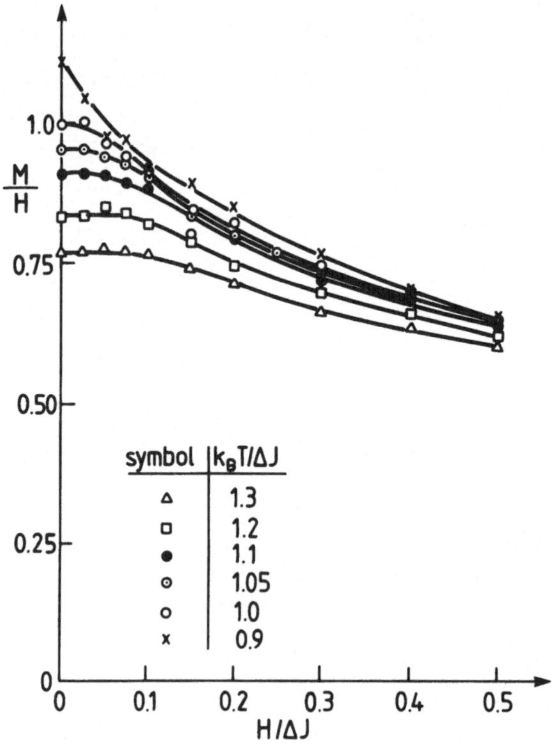

Fig.7: Susceptibility M/H plotted vs. H for several temperatures near $T_f \approx \Delta J/k_B$ [50]. Note that this value of T_f is estimated from the appearance of a nonzero order parameter q(t) { Eq. (10)} for t=2000MCS [41].

symbol	$k_B T/\Delta J$
△	1.3
□	1.2
●	1.1
◉	1.05
○	1.0
x	0.9

the variation of M/H with H indeed is quadratic at small H, while for $T \lesssim T_f$ a cusp-shaped behavior occurs. The nonlinear part of M/H is analyzed further in Fig.8. The data at not to small fields would indeed be consistent with a behavior

$$\frac{M}{H} = \chi_o(T) - H^{a(T)-1}\chi_1(T)$$

with an exponent $a(T) \approx 1.7$ at (and below) T_f, similar to experiment [29]. Of course, since in our model a static $T_f > 0$ does not exist, for H small enough we must always find $a(T) \equiv 3$ independent of T. Again this plot is rather similar to experimental work [32], Fig.9, where more decades in H are accessible and the crossover to $a(T)=3$

Fig.8: Log-log plot of the nonlinear part of M/H vs. H (left part) and temperature variation of $\chi_{n\ell}(T)$ (right part) [50].

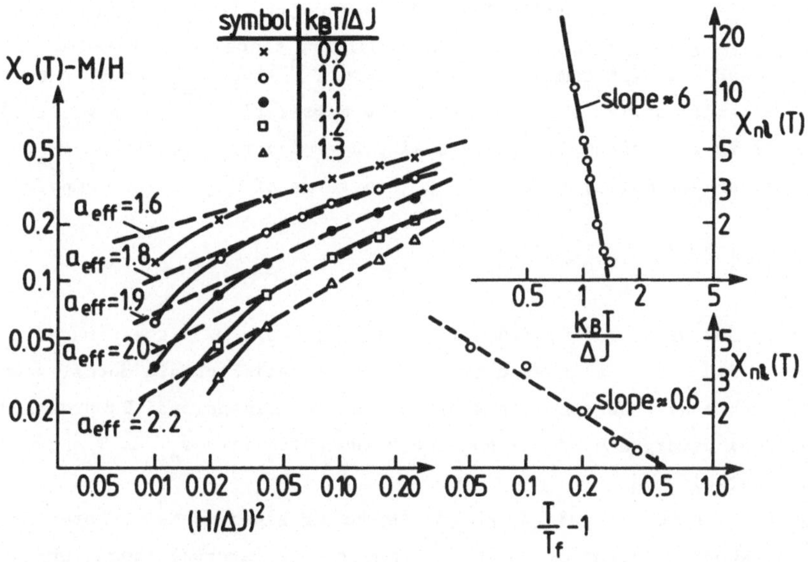

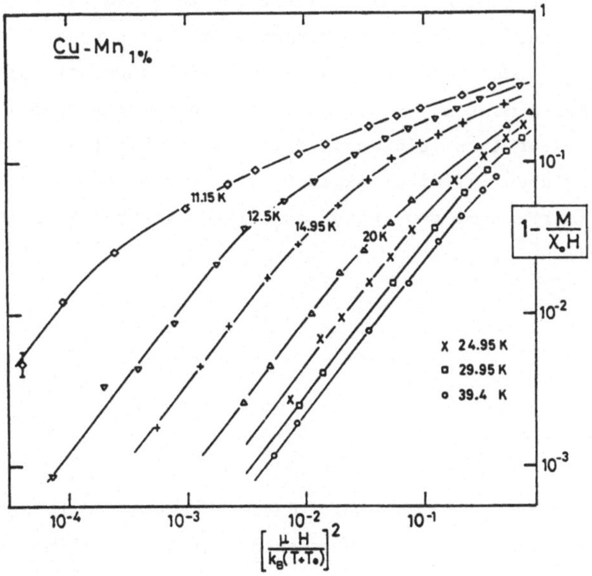

Fig.9: $\mathrm{Log}(1-M/\chi_o H)$ vs. log H^2 for CuMn1% [32]

is hence demonstrated more clearly. If we nonetheless fit Eq.(16) to the data shown in Fig.8, $\chi_{n\ell}(T)$ is obtained as also shown in Fig.8 (right part). The data for $\chi_{n\ell}(T)$ above T_f could be fitted to the power-law divergence, Eq.(17), with $\gamma\approx 0.6$, in the temperature interval shown. But in our case this clearly is an artefact of an inappropriate data analysis: the same data plotted in log-log form versus T itself rather than T/T_f-1 also yield a straight line, thus suggesting a divergence at zero temperature only: for $T\to 0$ we have $\chi_{n\ell}(T) \propto T^{-y}$, with $y\approx 6$. From the scaling law Eq. (15) we now infer that the exponent y can again be expressed in terms of Δ as $y=1+2\Delta$. Using $\Delta=1/x$ and Eq.(13), we find $\Delta=3.5\pm0.5$ and hence $y=8\pm1$. It is not surprising that the direct estimate from Fig.8 is somewhat smaller, since it is based on not so low temperatures, and for high temperatures $\chi_{n\ell}(T) \propto T^{-3}$ since χ_{EA} tends to unity.

Since the transfer matrix calculations of the correlation function $g_{EA}(\vec{r}_{ij})$ has shown [21] that it is well approximated by a simple exponential function, $g_{EA}(r) \propto \exp(-r/\xi_{EA})$, we can use Eq.(3) to express $\chi_{n\ell}$ in terms of ξ_{EA} at low temperatures:

$$\chi_{n\ell} \propto T^{-3}\chi_{EA} \propto T^{-3}\int d\vec{r}\, g_{EA}(\vec{r}) \propto T^{-3}\xi_{EA}^{d} \tag{19}$$

For dimensionality d=2 we have [21,49,50] $\xi_{EA} \propto T^{-\nu}$, where ν is estimated as $\nu\approx 2$, and hence $y=3+d\nu\approx 7$. It is seen that all these estimates are mutually well consistent with each other. We conclude that for the two-dimensional Edwards-Andersonmodel the gradual divergence of the nonlinear susceptibility $\chi_{n\ell}$ as T approaches zero is now well established.

Comparing now once more Fig.8 to the corresponding experiments we note that $\chi_{n\ell}$ increases only by about a factor of 10 in the temperature interval shown, while in the experiment of Omari et al. [32] it increases by a factor 10^3 in a similar interval. Thus the experiment is much closer to a true divergence of $\chi_{n\ell}$ at a nonzero T_f than the simulation, and hence it was concluded that there is a true phase transition at T_f [32]. Conversely, the experiment may still be consistent with a transition occurring only at T=0 as well, but then it necessarily implies that the exponents y,ν,Δ must be very large or even infinite (exponential divergence rather than power law divergence). This is the same conclusion as already reached from the consideration of the critical field $H_c^{eq}(T)$ above.

IV. Scaling behavior

If a phase transition at a nonzero T_f occurs, one expects that the nonlinear part of the magnetization (i.e., the quantity $1-M/\chi_0 H$) should satisfy a scaling hypothesis of the form [62,32]

$$\left(1-\frac{M}{\chi_o H}\right)\left(1-\frac{T_f}{T}\right)^{-\beta} = \tilde{m}\left\{\left(\frac{H}{T}\right)^2 \left(1-\frac{T_f}{T}\right)^{-\gamma-\beta}\right\}, \qquad (20)$$

$\tilde{m}$ being the appropriate scaling function. Malozemoff et al. [29] and later Omari et al. [32] found that their experimental data are indeed nicely consistent with Eq.(20) see e.g. Fig.10. The authors of these works consider this agreement with scaling as their strongest

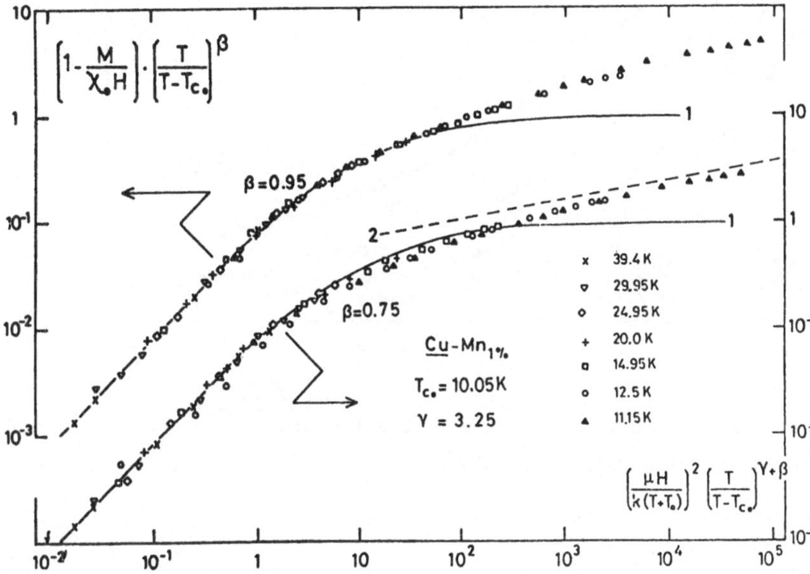

Fig.10: Scaling plot of the magnetization data for CuMn1% for two choices of the exponent β and γ = 3.25 [32].

evidence for a phase transition.

In order to check this scaling property also from the point of view of the simulation extensive data on the field cooled magnetization over wide regimes of temperature and field were generated [52], Fig.11. Again there is a striking qualitative similarity to corresponding experimenta data of Ref.32. Thus it is perhaps not too surprising that the simulation data, within their statistical scatter, are consistentwith Eq.(20) as well: of course, fixing T_f at $k_BT_f/\Delta J=1$ (Fig.12) we have to include data only in the regime $T/T_f \geq 1.1$ in our case, as there is no singularity at all at T_f. It just appears that the three adjustable parameters of Eq.(20), namely γ,β and T_f, are enough to scale such a smoothly varying family of functions as shown in Fig.11. In fact, scaling plots of similar qualitaty are obtained as well for other choices of T_f between $k_BT_f/\Delta J=1$ and $T_f=0$. Only the scaling with $T_f=0$, where Eq.(20) reduces to Eq.(15) and hence there is only one adjustable exponent is physically meaningful, of course. Fig.13 shows the success of this scaling representation appropriate for a zero temperature phase transition. Clearly, it would be desirable to analyze experimental data along similar lines.

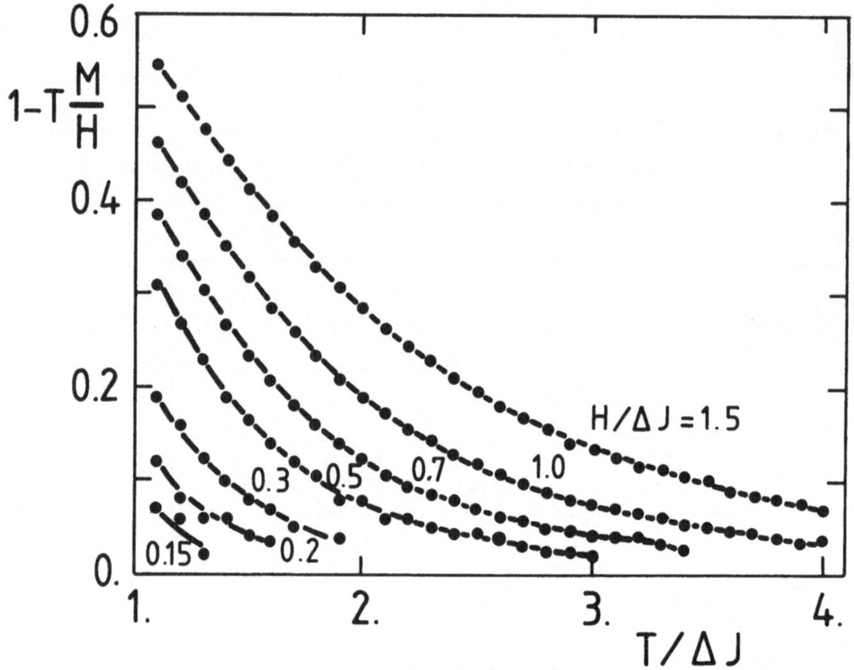

Fig.11: Nonlinear part of the magnetization of the two-dimensional nearest neighbor gaussian Edwards-Anderson model plotted vs. temperature for various fields [52].

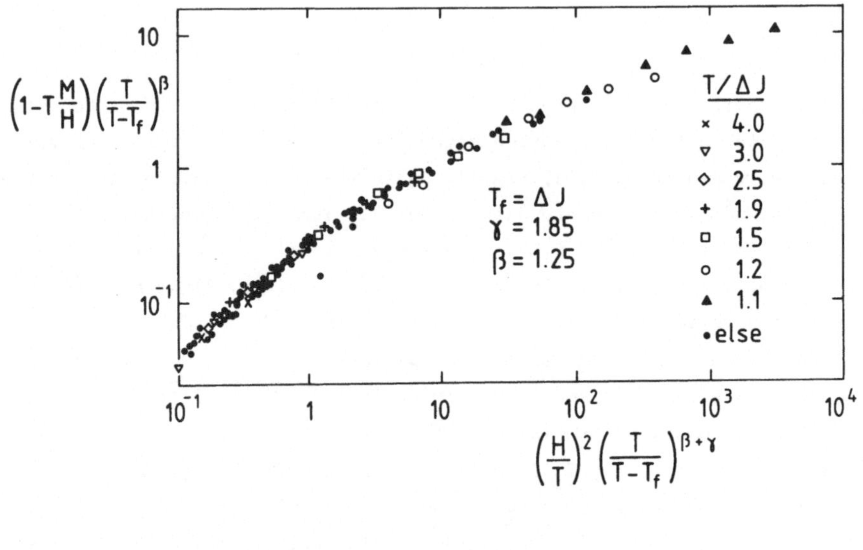

Fig. 12: Scaling plot of the magnetization data of the simulation (Fig.11). In the axis labes we have chosen units $k_B=1$, $\Delta J=1$ [52].

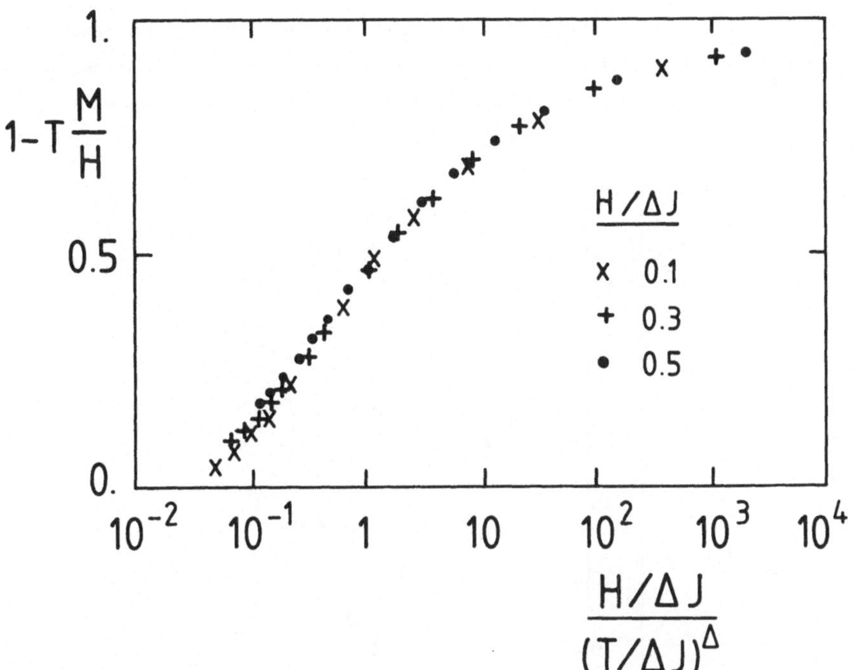

Fig. 13: Scaling plot of the magnetization data of the simulation (Fig. 11) appropriate to a transition at zero temperature {Eq.(15)} [52].

V. Dynamic behavior

While the field-cooled susceptibility at slow enough cooling rates does not show any time-dependent effects and hence yields the magnetic equation of state in equilibrium, stronger time-dependent effects show up in the zero-field cooled case. Fig.1 includes simulation data where the magnetization was measured which is obtained when a static field is applied to a zero-field cooled system for a time of 600 or 6000 MCS/spin, respectively. While at high temperatures the magnetization thus obtained agrees with the field-cooled magnetization, for each time t and field H there exists a temperature (indicated by arrows for $H/\Delta J=1.5$ in Fig.1) where the zero-field cooled magnetization falls below the field-cooled one. In the H-T plane these temperatures form a family of curves $H_c(t)$ {Fig.4}. While for high fields this fall-off starts in the plateau region, for small fields it starts already well before the plateau is reached. Therefore $H_c(t)$ intersects the curve $H_c^{eq}(T)$ and terminates at H=0 at a finite temperature $T_f(t)$, which we may identify as a time-dependent freezing temperature {Fig.4}. As there is no static T_f in our model, $T_f(t)$ tends to zero as $t\to\infty$, albeit very slowly. The general shape of these critical field curves is again strikingly similar to critical fields $H_c(t)$ identified experimentally by various procedures (e.g. Fig.14). As observed also experimentally [36-39], these critical field curves can

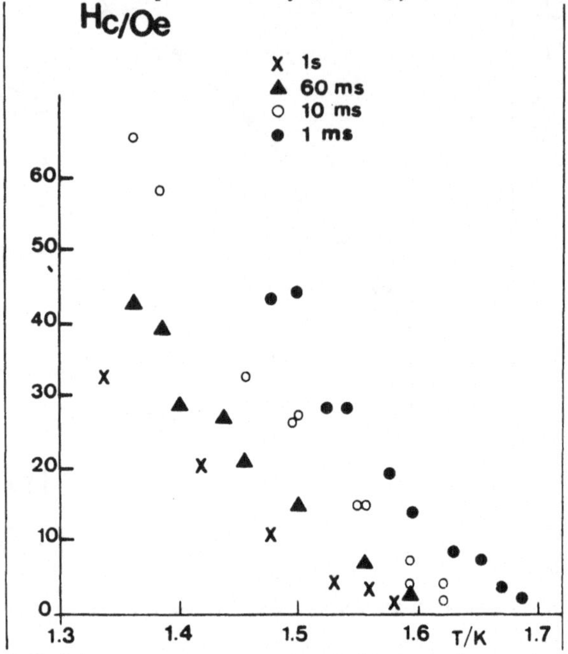

Fig.14: Experimental values of $H_c(t)$ versus temperature T in $Eu_{0.4}Sr_{0.6}$ for four different time-scales, as obtained from the decay of the Faraday rotation angle [38].

(at least approximately) be fit on a single curve by adjusting the axes, i.e. measuring both T and H in units of $T_f(t)$ in our case (Fig. 4B). It is a really surprising coincidence that the resulting curve is very similar to the AT-line (Fig. 4B). Of course, there is no phase transition and no broken replica symmetry etc. in our model. We do not see any reason why the AT-line should have any significance in our short-range model below the lower critical dimension. In fact, at least a speculative explanation of the shape of this curve can be given [52] which is based on very different ideas. Thus we feel it would be completely wrong to count this coincidence (which anyway is not really perfect) as a success of mean field theory, which clearly fails to describe all the other properties of our short range models. This observation should be a waining signal to experimentalists concerning any "fits" or "misfits" they encounter with the various theories.

Rather than associating these critical fields $H_c(t)$ with the AT-line, it is much more useful to interpret them as contours of constant (average) relaxation time $\tau_{av}(T,H)=t=const$ in the H-T-plane. In fact, measuring a relaxation time from the time-integral of the spin autocorrelation function, Young [51] has obtained quite similar results (for the ±J-model) for $H_c(t)$ as shown in Fig.4A (for the gaussian model). It is clear that $\tau_{av}(T,H)$ at low T,H is dominated by processes where the system moves from one "valley" in phase space over a free energy barrier to another valley. With increasing field the number of valleys is reduced (ultimately only one remains), and thus it is plausible that the field has a tendency to reduce the barrier heights, and therefore the curve $H_c(t)$ must decrease to lower temperatures as H in creases.

Of course, in the absence of any phase transition $\tau_{av}(T,H)$ must be analytic in H, and since no sign is preferred, it must have an expansion of the same type as Eq.(16), i.e.

$$\tau_{av}(T,H) = \tau_{av}(T,0)-H^2\tau'(T)+O(H^4) \quad . \tag{21}$$

This implies that the curves $H_c(t)$ in Fig.4 must start out at H=0 with infinite rather than zero slope! Just as Eq.(16) for the nonlinear part of the magnetization is not observed at low temperatures and there is an apparent cusp if one uses fields which are not small enough (Fig.7), there is also an apparent cusp in $\tau_{av}(T,H)$ if one uses fields which are not small enough. This claim is proven by the data of Young [51] taken for $\tau_{av}(T,H)$ at higher temperatures. A clear experimental evidence that the curves $H_c(t)$ come down to H=0 with vertical tangent has very recently been obtained in the $Eu_xSr_{1-x}S$ system [40,64].

The behavior of other dynamic quantities has already been reviewed elsewhere [53,65] and hence will here be mentioned only briefly. For instance, the remanent magnetization was found to exhibit an apparent decay with time as $M_r(t) \propto t^{-\alpha(T,H)}$, the exponent $\alpha(T,H)$ depending on temperature [41] and field [43], Fig.15, in analogy with experimental data, e.g. Fig.16. Similarly, the spin autocorrelation function $\langle S_i(0)S_i(t)\rangle$ is found [41] to decay exponentially fast with time at temperatures much larger than T_f, while near T_f a broad spectrum of relaxation times contributes and below T_f the decay even is logarithmic over intermediate time-scales (Fig.17). This behavior is reminiscent of corresponding neutron scattering

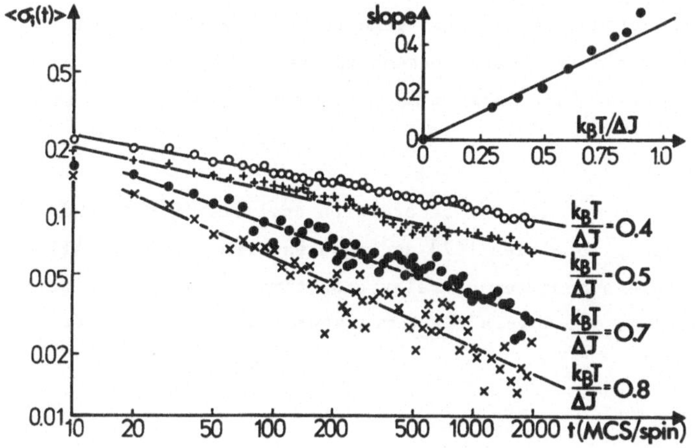

Fig.15: Decay of the remanent magnetization with time in the two-dimensional gaussian Edwards-Anderson model for various temperatures. Insert shows temperature variation of the apparent exponent $\alpha(T,0)$

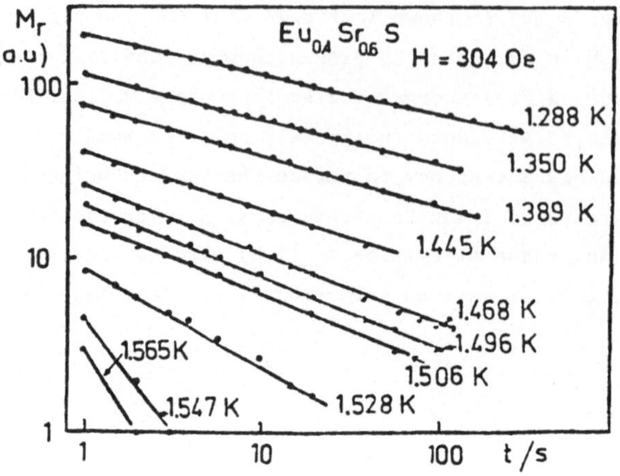

Fig.16: Decay of the remanent magnetization in $Eu_{0.4}Sr_{0.6}S$ with time at various temperatures [33]

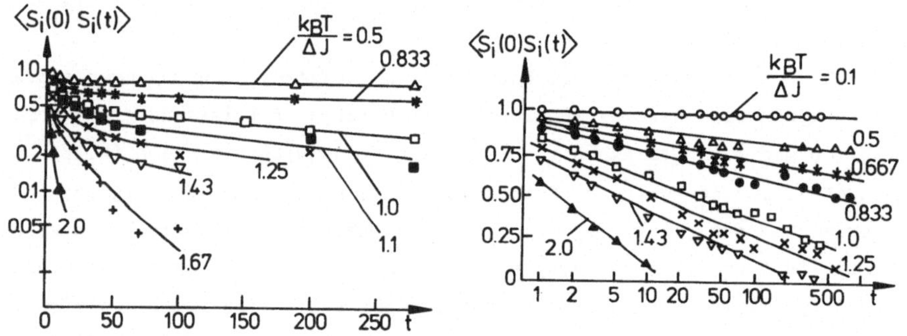

Fig.17: Decay of the self-correlation function with time for the two-dimensional gaussian Edwards-Anderson model at various temperatures. Left part has ordinate scale logarithmic, right part shows same data when abscissa scale is logarithmic [41]

results [66], Fig.18. It is also in qualitative accord with analysis of real and imaginary parts of dynamical susceptibilities [67] and muon experiments [68]. All these data suggest a picture where a spectrum of relaxation times exists which gradually -but strongly- broadens when one lowers the temperature; below $T_f(t)$ this spectrum exceeds the observation time t and hence is no longer clearly resolved.

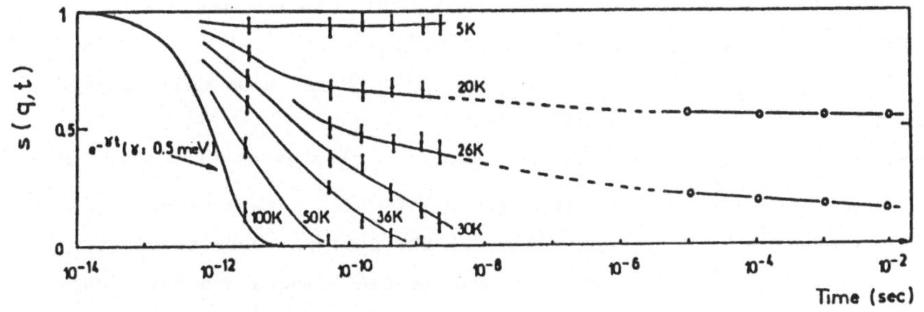

Fig.18: Spin relaxation in CuMn5% at various temperatures. Data points at times shorter than 10^{-8} sec are directly measured dynamic structure factors S(q,t) at $q=0.093Å^{-1}$ via the neutron spin echo method, these beyond 10^{-6} sec were calculated from a.c. susceptibilities [66].

A very direct evidence for such a picture of the dynamical freezing process has earlier been obtained by Kinzel [47], sampling the distribution $P_{t_o}(|\langle S_i\rangle|)$, $\langle S_i\rangle \equiv \int_t^{t+t_0} S_i(t')dt'/t$. At $T \gg T_f$, P_{t_0} is sharply peaked at $\langle S_i\rangle=0$; near T_f P_{t_0} broadens, while below T_f a fraction P of spins during t_0 has not changed its orientation at all, Fig.19. Now the dynamic Edwards-Anderson

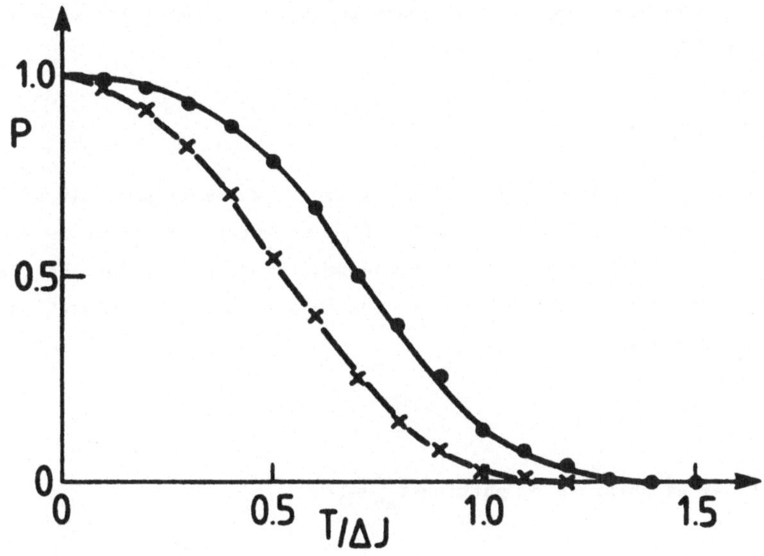

Fig.19: Fraction P of frozen spins plotted vs. temperature for t_0=400MCS/spin (crosses). Dots denote fraction of spins with $0.9 \leq |\langle S_i\rangle| \leq 1$. [47]

order parameter is $q(t_o)=1$ for these slow spins, while $q(t_o)\approx 0$ for the other spins. Therefore $\chi(t_o)\approx(1-P)/T$, which approximation accounts for the peak of the time-dependent susceptibility rather well [47]. At $T \lesssim T_f$ P is still rather small, and thus the slow spins must occur in small clusters well isolated from each other only. This statement is also verified by direct inspection of the spin configurations [47]. At lower T these clusters of slow spins grow and finally "percolate" at $T_p(t_o=400$ MCS/spin$)\approx 0.5\Delta J$. Of course, $T_p(t_o)$ tends to zero as $t_o\to\infty$: there is no well defined cluster percolation transition in this model either.

It is tempting to identify the typical size of a cluster of slow spins as being proportional to ξ_{EA} (for a given t_o). The reorientation of such a cluster of slow spins can then be considered as a description of the process where the system passes a saddle point from one "valley" to another. Due to the temperature-dependence of ξ_{EA} we then expect the free energy barrier to be strongly temperature-dependent. As a result, we propose that the frequency dependence of $T_f(\omega)$ is no longer that of a simple Arrhenius law:

$$T_f^{-1} \propto |\ell n(\tau_o\omega)| \tag{22a}$$

as in the Neél model [1,2], but much weaker

$$T_f^{-1} \propto |\ell n(\tau_o\omega)|^{1/\nu z}, \qquad \nu z>1 \quad . \tag{22b}$$

Kinzel [48] estimated this dynamic exponent for d=2 as $\frac{1}{\nu z}=\frac{1}{3}$, but it is nontrivial to relate the barrier height for overturning a slow cluster to its size because of the many low lying states in spin glasses [52], and hence we do not propose a particular value for z here. In any case Eq.(22b) seems to us more natural for this problem than the familiar Vogel-Fulcher law [3-5]

$$(T_f-T_o)^{-1} \propto |\ell n(\tau_o\omega)| \quad , \tag{22c}$$

which again involves another characteristic temperature T where it probably breaks down, while Eq,(22b) possibly holds down to $T\to 0$. If $\frac{1}{\nu z}$ is small the curvature seen in experimental plots of T_f^{-1} vs. $\ell n\ \omega$ [33,34,69] could well be accounted for. Eq.(22b) can also be justified by assuming power laws for the logarithm of the relaxation time ("generalized dynamic scaling")

$$\ell n(\tau_{av}/\tau) \propto \xi_{EA}^z \propto T^{-\nu z} \tag{23}$$

and assuming that $T_f(\omega)$ is obtained by putting $\tau_{av} \propto 1/\omega$.

Of course, this suggestion is a tentative speculation only. The alternative suggestion for explaining data for $T_f(\omega)$ such as shown in Fig.20 is [50] that the static susceptibility in zero field has a (rounded) peak due to the proximity of a reentrant ferromagnetic phase boundary in the temperature-concentration phase diagram.

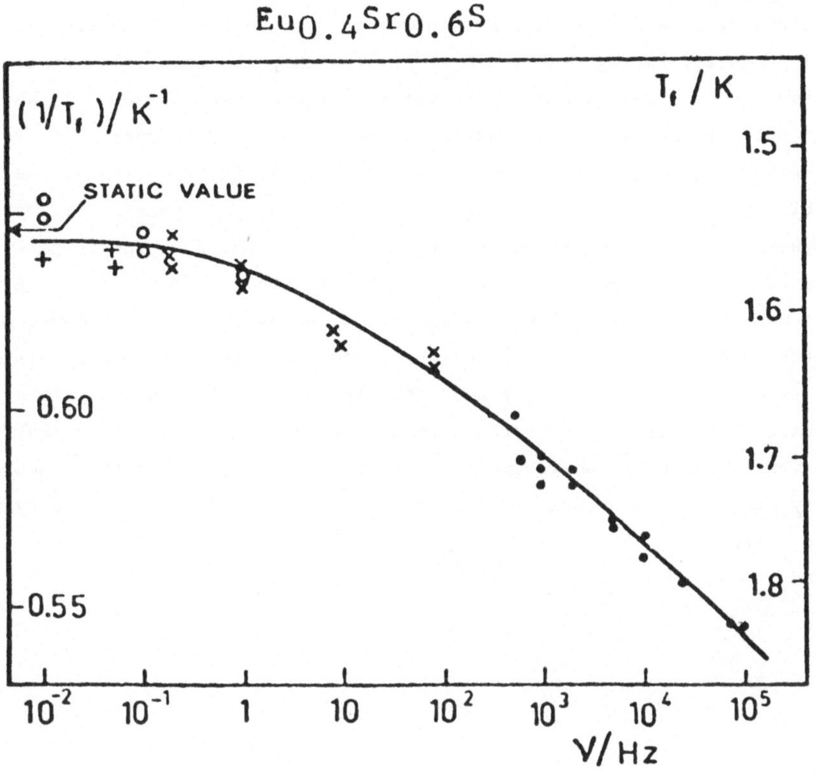

Fig. 20: Inverse freezing temperature T_f^{-1} for $Eu_{0.4}Sr_{0.6}S$ plotted vs. measurement frequency [33].

VI. Conclusions

Ample evidence has been presented to show that the two-dimensional nearest neighbor gaussian Edwards-Anderson model is qualitatively similar to real spin glasses. In particular, it accounts well for the dynamic properties above freezing, which are not well accounted for by mean-field theory (which predicts simple critical slowing down described by a relaxation time diverging with a power law at T_f). While the model has no phase transition at a nonzero T_f, there is a zero-temperature transition where correlation length ξ_{EA}, Edwards-Anderson susceptibility χ_{EA}, linear and nonlinear susceptibility χ_o, $\chi_{n\ell}$ and the average relaxation time τ_{av} diverge. By a static scaling hypthesis, whose validity was demonstrated, all static exponents have been expressed in terms of a single exponent Δ($\simeq$3.5±0.5 in two dimensions). If one tries to interpret experimental data along similar lines, which is logical because of the prediction that the lower critical dimension is four, one finds the surprising result that Δ is very large ($\Delta \sim 10^2$ or so). The implication of this result is as yet unclear. Clearly the necessity of further careful simulations of three-dimensional rather than two-dimensional models emerges, as well as of further careful measurements in very small fields paying particular attention to the effects of varying the cooling rate there.

Acknowledgements: We are greatly indebted to J. Souletie and A.P. Malozemoff for fruitful discussions, correspondence, and the permission to reproduce some of their experimental data (Figs. 9,10 and Fig.5, respectively). We thank P. Monod for the permission to reproduce data shown as Fig. 3,6; N. Bontemps for the permission to reproduce data shown as Fig.14, H. Maletta for the permission to reproduce data shown as Figs.16,20 and F. Mezei for the permission to reproduce data shown as Fig.18.

References

[1] L. Néel, Ann. Geophys. 5, 99 (1949)
[2] J.L. Tholence and R. Tournier, J. Phys. (Paris) 35, C4-229 (1974)
[3] J.L. Tholence, J. Appl. Phys. 50, 7310 (1979); Solid State Comm. 35, 113 (1980)
[4] E.P. Wohlfarth, Physica 86-88B, 852 (1977); J. Phys. F10, L241 (1980); Phys. Lett. 70A, 489 (1979); E.P. Wohlfarth and S. Shtrikman, Phys. Lett. 85A, 467 (1981)
[5] M. Gyrot, Solid State Comm. 39, 1009 (1981)
[6] For a review, see J.A. Mydosh, J. Mag. Magn. Mat. 7, 237 (1978)
[7] G. Eiselt, J. Kötzler, H. Maletta, D. Stauffer and K. Binder, Phys. Rev. B19, 2664 (1979)
[8] S.F. Edwards and P.W. Anderson, J. Phys. F5, 965 (1975)
[9] D. Sherrington and S. Kirkpatrick, Phys. Rev. Lett. 35, 1792 (1975)
[10] G. Parisi, J. Phys. A13, L115 (1980); Phil. Mag. B41, 677 (1980)
[11] H. Sompolinsky, Phys. Rev. Lett. 47, 935 (1981); H. Sompolinsky and A. Zippelius, Phys. Rev. B25, 6860 (1982); J. Hertz, J. Phys. C16, 1219 (1983); 1233 (1983)
[12] A.P. Young, J. Phys. C14, L1085 (1981); A.P. Young and S. Kirkpatrick, Phys. Rev. B25, 440 (1982); A. Houghton, S. Jain and A.P. Young, preprint; N.D. Mackenzie and A.P. Young, Phys. Rev. Lett. 49, 301 (1982), and preprint; A.P. Young and C. de Dominicis, preprint
[13] R.G. Palmer, Adv. Phys. 31, 669 (1982)
[14] J.R.L. De Almeida and D.J. Thouless, J. Phys. A11, 983 (1978)
[15] G. Parisi and G. Thouless, J. Phys. Lett. (Paris) 41, L361 (1980); G. Toulouse, M. Gabay, T.C. Lubensky and J. Vannimenus, J. Phys. Lett. 43, L109 (1982)
[16] M.E. Fisher, Rev. Modern Phys. 46, 597 (1974)
[17] A.B. Harris, T.C. Lubensky, and J.H. Chen, Phys. Rev. Lett. 38, 765 (1976)
[18] R. Fisch and A.B. Harris, Phys. Rev. Lett. 38, 785 (1977)
[19] A.J. Bray and M.A. Moore, J. Phys. F7, L333 (1977); A.J. Bray, M.A. Moore, and P. Reed, J. Phys. C11, 1187 (1978)
[20] A.J. Bray and M.A. Moore, Phys. Rev. Lett. 41, 1068 (1978); J. Phys. C12, L441 (1979); see also H. Sompolinsky and A. Zippelius, Phys. Rev. Lett. 50, 1294 (1983)
[21] I. Morgenstern and K. Binder, Phys. Rev. Lett. 43, 1615 (1979); Phys. Rev. B22, 288 (1980)
[22] I. Morgenstern and K. Binder, Z. Phys. B39, 227 (1980)
[23] J.R. Banavar and M. Cieplak, Phys. Rev. Lett. 82, 832 (1982)
[24] W. Kinzel and K.H. Fischer, J. Phys. C11, 2115 (1978), and references therein
[25] P.W. Anderson and C.W. Pand, Phys. Rev. Lett. 40, 903 (1978)
[26] D. Stauffer and K. Binder, Z. Physik B30, 313 (1978); B34, 97 (1979)
[27] L. Lundgren, P. Svedlindh, and O. Beckmann, Phys. Rev. B26, 3990 (1982)
[28] P. Monod and H. Bouchiat, J. Phys. Lett. (Paris) 43, L45 (1982)
[29] B. Barbara, A.P. Malozemoff, and Y. Imry, Phys. Rev. Lett. 47, 1852 (1981); A.P. Malozemoff, B. Barbara, and Y. Impry, J. Appl. Phys. 53, 2205 (1982)
[30] B. Barbara, A.P. Malozemoff, and Y. Imry, Physica 108B+C, 1289 (1981)
[31] A. Berton, J. Chaussy, J. Odin, B. Rammal, and R. Tournier, J. Phys. Lett.(Paris) 43, L-153 (1982)
[32] R. Omari, J.J. Prejean, and J. Souletie, J. phys. (Paris) in press
[33] J. Ferré, J. Rajchenbach and H. Maletta, J. Appl. Phys. 52, 1967 (1981)
[34] M. Guyot, S. Foner, S.K. Hasanain, R.P. Guertin, and K. Westerhold, Phys. Lett. 79A, 339 (1980)
[35] M.B. Salamon, K.V. Rav, and Y. Yeshurun, J. Appl. Phys. 52, 1687 (1981)
[36] R.V. Chamberlin, M. Hardiman, L.A. Turkevich and R. Orbach, Phys. Rev. B25, 6720 (1982)
[37] Y. Yeshurun and H. Sompolinsky, Phys. Rev. B26, 1487 (1982)
[38] N. Bontemps, J. Rajchenbach, and R. Orbach, J. Phys. Lett. (Paris) 44, L47 (1983)
[39] M.B. Salamon, and J.L. Tholence, J. Appl. Phys. 53, 7684 (1982); J. Magn. Mag. Mat., in press
[40] J. Hamida, C. Paulsen, S.J. Williamson, and H. Maletta, preprint
[41] K. Binder and K. Schröder, Phys. Rev. B14, 2142 (1976)
[42] K. Binder, Z. Phys. B26, 339 (1977)

[43] W. Kinzel, Phys. Rev. B19, 4594 (1979)
[44] K. Binder and D. Stauffer, Phys. Lett. 57A, 177 (1976)
[45] Note that finite energy barriers between different ground states need not rule out a phase transition, see P. Hoever, W.F. Wolff, and J. Zittartz, Z. Phys. B41, 43 (1981); B42, 259 (1981); P. Hoever and J. Zittartz, Z. Phys. B44, 129 (1981)
[46] I. Morgenstern, J. Appl. Phys. 53, 7682 (1982); Phys. Rev. B27, 4522 (1983); see also I. Morgenstern, and H. Horner, Phys. Rev. B25, 504 (1982)
[47] W. Kinzel, Z. Phys. B46, 59 (1982)
[48] W. Kinzel, Phys. Rev. B26, 6303 (1982)
[49] K. Binder and I. Morgenstern, Phys. Rev. B27, 5826 (1983)
[50] K. Binder, Z. Phys. B48, 319 (1982)
[51] A.P. Young, Phys. Rev. Lett. 50, 917 (1983)
[52] W. Kinzel and K. Binder, Phys. Rev. Lett. 50, 1509 (1983); and to be published
[53] K. Binder and D. Stauffer, in Monte Carlo Methods in Statistical Physics (K. Binder, ed., Springer, Berlin 1979) p.301; K. Binder, J. Phys. (Paris) 39, C6-1527 (1978); K. Binder, in Ordering in Strongly Fluctuating Condensed Matter Systems (T. Riste, ed., Plenum Press, New York 1979) p.423; K. Binder and D. Stauffer, in Monte Carlo Methods in Statistical Physics II (K. Binder, ed., Springer, to be published).
[54] H. Hilhorst and B. Derrida, J. Phys. C14, L539 (1981)
[55] J.L. van Hemmen and I. Morgenstern, J. Phys. C15, 4353 (1982)
[56] J.A. Mydosh, J. Phys. Soc. Jpn. 52, Suppl. p.85 (1983)
[57] P.C. Hohenberg and B.I. Halperin, Rev. Mod. Phys. 49, 435 (1977)
[58] I. Morgenstern, Phys. Rev. B25, 6067 (1982)
[59] H. Maletta, J. Phys. (Paris) C6, 115 (1980), and references therein
[60] B. Barbara, and A.P. Malozemoff, Proc. 16th Rare Earth Research Conf., Florida State Univ., April 1983 to be published
[61] It should be noted that the physical significance of the shift of the susceptibility maximum at very small fields can be questioned on a various physical grounds: (i) it may be an artefact due to macroscopic sample inhomogeneities (H. Alloul, private communication) (ii) it may be an effect specific for Heisenberg spins, just as there exists for isotropic Heisenberg antiferromagnets in a field a cusp-shaped umbilicus of the critical temperatures $T_c(H),T_c(-H)$, which merge at the T-axis in a bicritical point (G. Toulouse, private communication). In Ising spin glasses for $d<d_\ell$ none of these alternative explanations can apply, however.
[62] These data were obtained in [41] by stepwise decrease of H at constant T, and thus for small H near T_f are slightly affected by observation time effects. The scaling analysis of Sec.IV hence uses more accurate recent simulation data [52] obtained by field-cooling as in Figs.1,2.
[63] J. Chalupa, Solid State Commun. 24, 429 (1977); M. Suzuki, Progr. Theor. Phys. 58, 1151 (1977); K. Binder, Festkörperprobleme 17, 55 (1977)
[64] N. Bontemps and J. Rajchenbach, to be published
[65] K. Binder and W. Kinzel, in Lecture Notes in Physics 149, p.124 (ed. by C. di Castro) Springer Berlin, Heidelberg and New York, 1981.
[66] F. Mezei, in Recent Developments in Condensed Matter Physics, ed. J.T. Devreese (Plenum, Press, New York 1981) Vol.1, p.679
[67] D. Hüser, L.E. Wenger, A.J. van Duyneveldt and J.A. Mydosh, Phys. Rev. B, in press
[68] D.E. Mac Laughlin, L.C. Gupta, R.H. Heffner, M. Leon, and M.E. Schillaci, to be published
[69] A.P. Malozemoff and Y. Imry: Phys. Rev. B24, 489 (1981)

NUMERICAL SIMULATIONS OF SPIN GLASSES

I. Morgenstern
Institut für Theoretische Physik, Universität Heidelberg,
Philosophenweg 19, D-6900 Heidelberg, FRG

Abstract

In this contribution to the proceedings I carried out the full chronological line in numerical simulations to give clear evidence against a phase transition in short-range spin glass models at least in two dimensions. Furtheron I consider the observation time necessary to obtain equilibrium states at low temperatures - it exceeds 100 years. Then I provide the theory of a new phenomenon: the time dependent specific heat. It is hoped that this effect helps experimentalists to provide convincing evidence in favour or against a transition.

Introduction

Here I want to present numerical results of various short-range Ising spin-glass models. In particular I provide calculations of the Ising-type models, as their properties are easily evaluated by the Transfermatrix (TM) method. I compare these results with those of the dynamical approach of Monte-Carlo (MC) simulations, as long as MC runs are not too time consuming. I deal with the following Hamiltonian:

$$-\beta \mathcal{H} = \sum_{<ij>} J_{ij} \, S_i S_j \tag{1}$$

where we have Ising spins $S_i = \pm 1$ on a two or three-dimensional lattice coupled via the random bonds J_{ij}. The first suggestion about the J_{ij}'s was by Edwards and Anderson /1/ - they introduced a Gaussian distribution as a truncation of the RKKY-couplings - more because of the sake of computational simplicity than of obvious physical reasons. Lateron it was found by Toulouse et al. /2/ that the behaviour of spin glasses is heavily dominated by the frustration effect and therefore by the sign of the couplings J_{ij}. As an abstraction of the Gaussian case they suggested the "$\pm$ J-model" where only the sign of the random bonds survives, i.e. we have 50% + J and 50% -J. As shown by my own calculations, this model contains all the necessary information to provide a good "spin-glass model". In particular we have the famous freezing effects due to the existence of valleys and hills in phase space.
I want to repeat the main line of my numerical calculations here in brief. Furtheron I give a short description of the numerical methods I mainly used to obtain these results. For a not too far advanced reader, I would like to give the essen-

tial information of the basics of the numerical mystery in Appendix A. To point out here: I mainly used two methods: firstly the Transfermatrix (TM) approach and secondly the usual "Monte-Carlo" (MC). My own invention: TM is able to calculate "numerically exact" the partition function of a given, not too large lattice. In two dimensions I was restricted to 18 x 18 and in three to 4 x 4 x 10. But with the further development of computers - mainly the Cray's - I should be able to cope with larger systems but as we will see later, in 2-dim it is not quite necessary.

The second method is the standard MC approach following a one-spin dynamic. The most useful results I obtained in comparing results of both methods, being able to decide unambigously between equilibrium and non-equilibrium effects.

The outline of my contribution to the proceedings of the Heidelberg Colloquium is the following:
I want to present the complete line of results and observations to give unambigous evidence against a phase transition, at least in two dimensions. I want to include further results in 3-dim to provide arguments following the 2-dim case also against a transition. Furthermore I look at the dynamical aspects of the freezing effect in spin glasses. The observation times necessary to obtain equilibrium effects in experiments are estimated for low temperatures T - they exceed the lifetime of most physicists - more than 100 years at low enough T. At the end I will present a new effect: the time-dependent specific heat. Considering this effect experimentalists should have a tool to decide in favour or against a phase transition in real spin glasses.

I) First Evidence against a Phase Transition
Order Parameter ψ^2 and Correlation Function $\langle S_o S_R \rangle^2$

In chapter I) I present the chronological line of my considerations about the transition problem. The first and most important work was done in collaboration with my doctor father Prof. K. Binder /3-5/ in Jülich. We used the TM approach to calculate the order parameter ψ^2 for the $\pm$ J-model and lateron for the Gaussian case. ψ^2 is the projection on a ground-state of the system, i.e. we have

$$\psi^2 = \frac{1}{N^2} \sum_{ij} \langle S_i S_j \rangle_o \langle S_i S_j \rangle_T \qquad (2)$$

We denote by $\langle \ldots\ldots \rangle_T$ the thermal average at temperature T. $\langle \ldots\ldots \rangle_o$ corresponds to the ground-state. We had to choose this form of Binder's order parameter to avoid the pathological case $\langle S_i \rangle_T \equiv 0$ for finite systems. The outcome was quite a surprise- at that time people in Jülich were quite in favour of a transition. But after averaging over a lot of lattices – about 25-100 – considering different system sizes, we obtained a decrease of ψ^2 with the system size as shown in fig. 1. Comparison to the Mattis-type of spin glass where we have a transition shows quite obviously the different behaviour and the absence of a transition as a further decay of ψ^2 is to be expected as the system size is increased. Fig. 1 shows ψ^2 against tempera-

Symbol	L
×	6
○	12
△	16

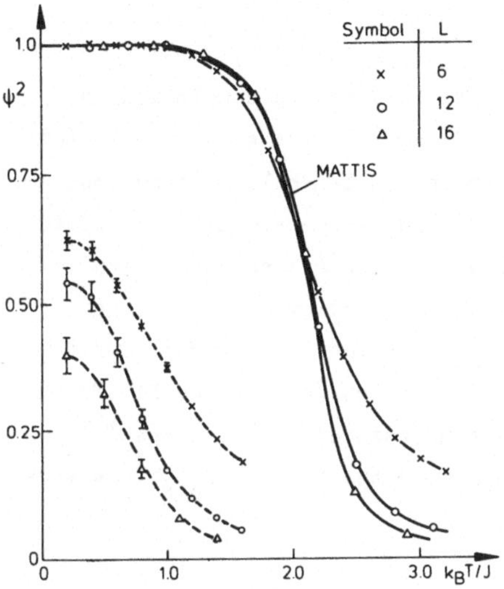

Fig. 1. Average spin-glass order parameter $\overline{\psi^2}$ plotted vs temperature for several L. Error bars are calculated from averages 100 realizations $\{J_{ij}\}$ for L = 6, 40 realizations for L = 12, and 25 realizations for L = 16. Full curves denote exact results for $\pm$ J Mattis spin glasses of the same size (note $k_B T_c$ Mattis/J $\simeq$ 2.27).

ture T for various system sizes. The full curve is the Mattis, i.e. via gauge transformation the ferromagnetic Ising case. A further striking result is the behaviour of ψ^2 in MC. The system is frozen in below T_f while the exact solution is definitely smaller. Fig. 2 shows this behaviour. In the left part we plot the Edwards-Ander-

Fig. 2. Spin-glass order parameters q(t) left part) and ψ^2 (right part) plotted vs temperature, as obtained from Monte Carlo and exact calculation, with use of L = 16 and a realization $\{J_{ij}\}$ which has particularly small ψ^2. Various observation times are shown (data for t = 2000 are the results for L = 80 of Ref. 9; these data and the full circles have random spin configurations as initial condition, while the others have a ground state as initial condition).

Symbol	t
×	20000
○	10 000
□	5000
●	20000
+	2000

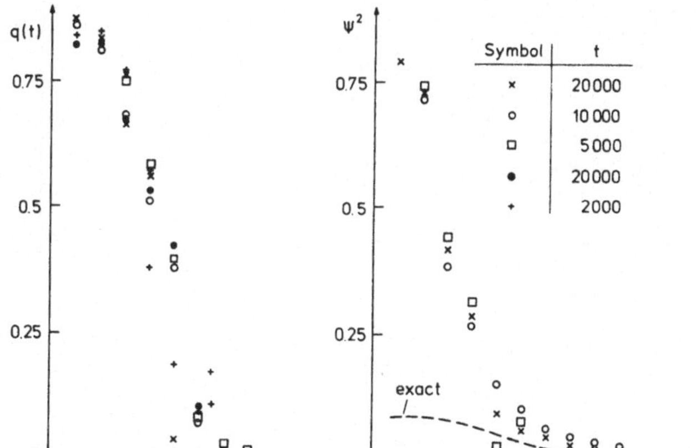

son order parameter q(t) where

$$q = \frac{1}{N} \{ \sum_i <S_i>^2_T \}_{av} \tag{3}$$

$\{....\}_{av}$ means averaged over different realizations of random bonds. q(t) here means the value obtained after the observation time t measured in Monte-Carlo steps per spin (MCS/spin).

The different behaviour of MC and TM order parameters lead to the conclusion that the freezing of spin glasses is a metastable effect - equilibrium shows $\psi^2 = 0$ for all finite temperatures. These results gave rise to the valley-hill picture considered later. But a lot of people were not quite convinced by the arguments provided by the ψ^2 calculation. They objected the possiblity that the states could turn in phase space in such a way that they give only a small projection on one of the ground states. But further calculations destroyed their arguments. We considered the behaviour of the correlation function $<S_o S_R>^2$ (two spins in distance R). We obtained a clear exponential decay with R at finite temperatures for both the $\pm$ J (fig. 3) and the Gaussian model (fig. 4). In both fig. 3 and fig. 4 $<S_o S_R>^2$ is shown

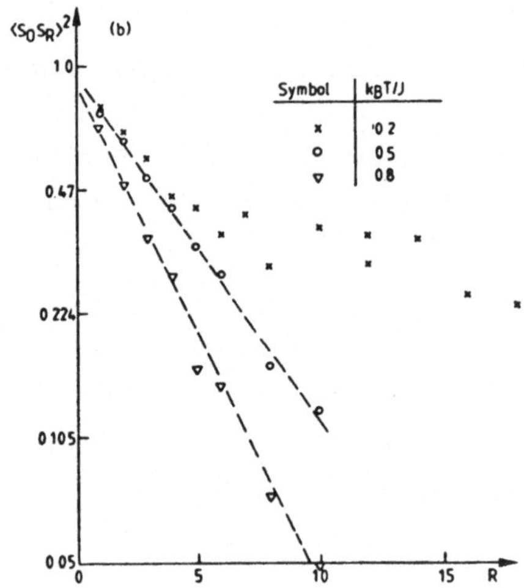

Figure 3

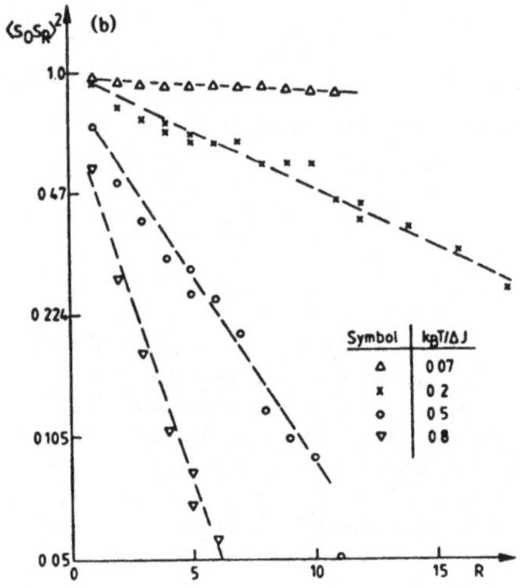

Figure 4

in logarithmic plots against distance R for various temperatures. A different behaviour is seen at T = 0. We concluded for the ± J-model an algebraic decay while the Gaussian case yields a constant value for the correlation function. Thus we have a transition only at T = 0. This means the correlation length only diverges at T = 0. On the other hand we obtained a ground-state entropy of S(0) = 0.075 ± 0.05 for the ± J-model suggesting a huge degeneracy at T = 0 giving further indications for the nature of phase space. In three dimensions we were only able to provide relatively small lattices: max. 4 x 4 x 10. We also calculated the beh aviour of $<S_0 S_R>^2$, R in the longer direction. The behaviour of the correlation function is quite similar to that in 2-dim. We always have an exponential decay at finite temperatures. This is seen in the following two figures. First the ± J (fig. 5) and then the Gaussian case (fig. 6). I should mention that we also did not find any qualitative

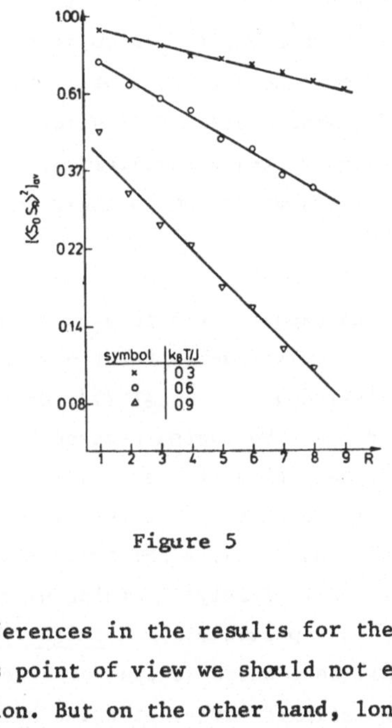

Figure 5

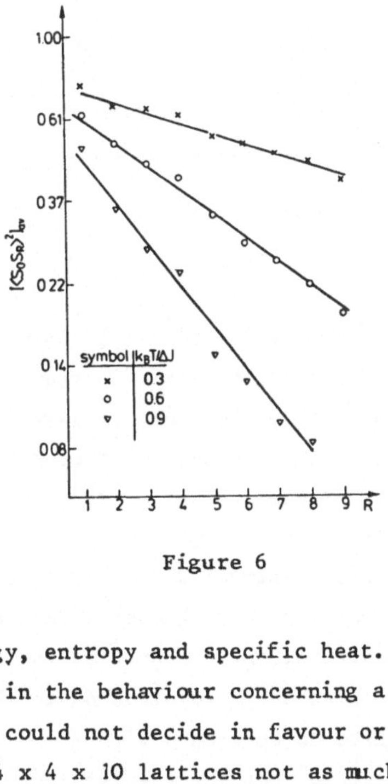

Figure 6

differences in the results for the internal energy, entropy and specific heat. From this point of view we should not expect a change in the behaviour concerning a transition. But on the other hand, long-time MC runs could not decide in favour or against the transition. One should consider the 4 x 4 x 10 lattices not as much too small as MC shows even for 4 x 4 x 4 lattices the usual freezing behaviour combined with irreversibility effects,the main indications for typical spin-glass behaviour. So far the results obtained in the very fruitful collaboration with K. Binder.

II) Valley-Hill Picture
Energy and Entropy Barriers

On the basis of the results of chapter I) I invented in collaboration with H. Horner the low-temperature picture of the "± J-model" /6/. First I would like to explain the valley-hill picture mentioned above several times. In fig. 7 I show in a relatively abstracted scheme the behaviour of the energy barriers E_B in phase space. P denotes some coordinate in phase space: i.e. an N-component vector containing all Ising-spins. The energy barriers are only seen in MC because they are a consequence of the 1-spin dynamic. An MC run floats at higher temperatures above the summit of the hills and lowering the temperature the system is suddenly trapped in one valley. This happens at the freezing temperature. At lower temperatures the probability is relatively small to escape from a valley and to reach the next one. Such considerations will be viewed lateron giving a crude estimation of the time necessary to reach equilibrium, i.e. to visit all valleys according to their Boltzmann weights.

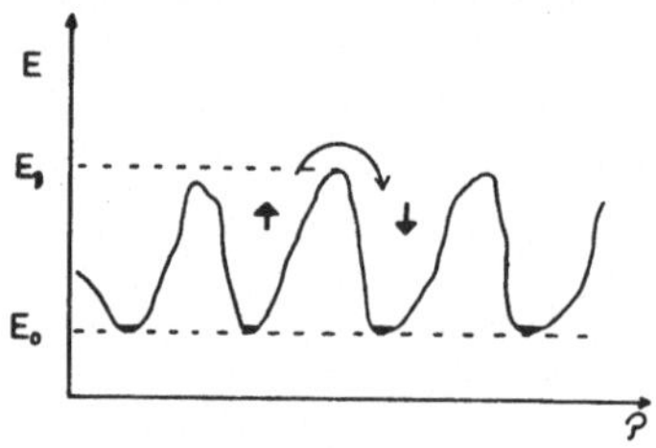

Figure 7

From the picture of fig. 7 it is clear that one has to expect infinitely high energy barriers in the case of a transition, as the system only reaches for $t \to \infty$ the next valley. This picture is seen in the Sherrington-Kirkpatrick model /7/. On the other hand there exists another possibility to avoid that the system escapes from a valley - even at finite energy barriers. This fact can arise if there exists a vanishing probability to find a way out, i.e. an infinite entropy barrier. In ± J-models with restricted distributions of the J's Wolff and Zittartz /5/ found this phenomenon (see also these proceedings). But to point out clearly: A vanishing probability, i.e. the probability to find a way out is only possible if an <u>infinite</u> number of spins is involved as in the Zittartz et al. case where stripes of size L x M have to be turned with $M \to \infty$ and L finite. An infinite number of spins have to be turned, but there exists a finite number of ways out; therefore we have zero probability while on the other hand the energy barrier is finite according to the other length involved - here L. I will return to this argument lateron.

First I will show the finiteness of the energy barriers E_B. We consider one given realization of a 16 x 20 ± J-lattice. TM calculates the correlation function $\langle S_0 S_R \rangle_0^2$ where S_0 is a spin on a free boundary of our lattice with cylindric boundary conditions. We have at T = 0 a power law after averaging over a lot of realizations.

$$\{<S_0S_R>_0^2\}_{av} = C \cdot R^{-p} \tag{4}$$

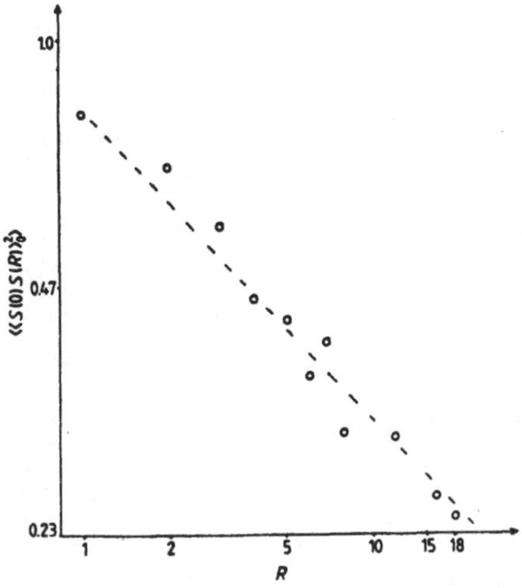

Figure 8

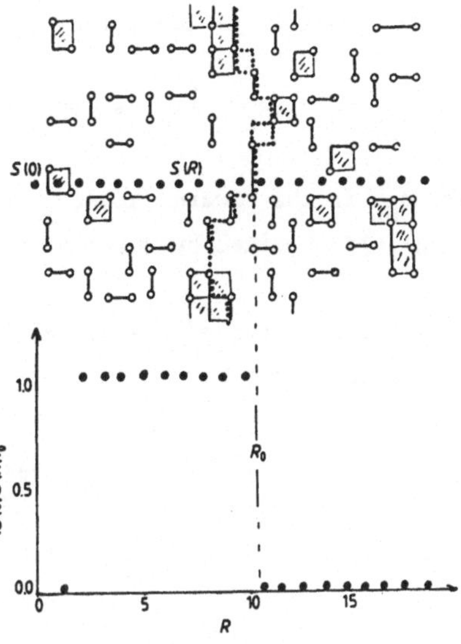

Figure 9

This behaviour is seen in fig. 8. In contrast we have for one realization a clear breakdown at a characteristic distance R_0. This is shown in fig. 9. Averaging over different lattices this effect is smoothed out as R_0 varies from lattice to lattice. A study of the ground-state structure clarifies the effect. Connecting frustrated plaquettes /2/ by the right number of strings leading to the ground state energy known from TM. All other ground states are now reached by flipping those spins surrounded by a zero energy surface ("loop"). We find "trivial loops" only surrounding single spins and larger regions built up by connecting "trivial loops". "Trivial loops" are "seen" by MC - their dynamic due to mutual dependence effects is accessible by the 1-spin-dynamic even at low temperatures. But there still exist "nontrivial" loops not only composed by "trivial" loops. They surround a large number of solidary spins (called "shells") not affected by the 1-spin-dynamic. In our small lattices the "nontrivial loops" surround the whole cylinder. A comparison with $\langle S_0S_R\rangle_0^2$ shows that the breakdown effect just takes place at R_0 where the "nontrivial loop" crosses the lattice. The resulting two parts of the lattice can be turned against each other (concerning TM) with no cost of energy. The correlation function typically tends to very small values (see fig. 9) affected by the nature of the network of "trivial loops" and their connections to the considered "nontrivial loops". Taking the mutual dependence effects into account, it is clear that

$$<S_0S_R>_0^2 = \{\frac{Z(+) - Z(-)}{Z(+) + Z(-)}\}^2 \tag{5}$$

where Z(+) denotes the number of possible states connected to the "nontrivial loop" with $S_0S_R = 1$ and Z(-) analogous. As the spin S_0 on the free boundary cannot be surrounded by a trivial loop, i.e. it is fixed in MC we have

$$\{<S_0S_R>^2\}_{MC} = \{<S_R>^2\}_{MC} = q_{MC}(T \to 0) \tag{6}$$

where q_{MC} is the Edwards-Anderson order parameter obtained by MC, while $\{.....\}_{MC}$ analogously obtained by MC.

As the "trivial loops" are not affected by longer distances R we therefore conclude that (4) takes the form

$$\{<S_0S_R>_0^2\}_{av} = \{<S_0S_R>_0\}_{MC} \cdot \{<S_0S_R>_0^2\}_{\text{nontr. loops}} \tag{7}$$

Taking (4) and (7) together we get:

$$C = q_{MC}(T \to 0) \tag{8}$$

in fair argument with different MC publications /9-11/. Furthermore we have very small values for $\{<S_0S_R>_0^2\}_{\text{nontr. loops}}$ after crossing the R_0-distance. Therefore we can give an upper boundary for the existence of a "nontrivial loops" between S_0 and S_R.

$$P\,\{\text{nontrivial loop}\} \leqslant 1 - R^{-p} \tag{9}$$

leading to the described decay (4) of $\{<S_0S_R>_0^2\}_{av}$. The solidary spins inside the loop lead to the energy barrier height E_B. It is clear that $E_B \propto R_0$. Therefore we have for the probability:

$$P\{E_B\} \propto E_B^{-p} \tag{10}$$

This means our shells, i.e. the region surrounded by the "nontrivial loops", were put into a box of radius R_0. We have now an estimation of the size of this box as the probability for an infinite size tends to zero as $p > 0$. Therefore it is impossible to have a) an infinite energy barrier height and b) as only a finite number of spins is contained in a shell, we also have a <u>finite</u> entropy barrier.

Furthermore we can show by a very nice indirect proof that $<S_0S_R>^2 \to 0$ for $R \to \infty$ for the expected network of "nontrivial loops" in larger lattices. The number of states Z(+) + Z(-) in (5) depending on the "nontrivial loops" tends to infinity supposing an infinite correlation. But as seen from symmetry arguments and the finiteness of the shells the difference Z(+) - Z(-) at least has to stay finite.

Thus, the correlation as seen from (5) has to tend to zero for infinite distances as already supposed by extrapolating the TM results.

To conclude: We have constructed a box surrounding one shell. Neglecting dependence effects between the shells, i.e. setting $\{\langle S_0S_R\rangle^2\}_{\text{nontr.loops}} = 0$ for $R > R_0$ the box grows faster than the shell as $R \to \infty$. But the probability to find an infinite box is zero. The qualitative picture following the above considerations is seen in fig. 10. In the above part of fig. 10 we notice again the qualitative valley-hill picture for the $\pm$ J-model and analogously for all short-range models. The ground states are located at the bottom of the valleys with energy $E = E_0$. We have energy barriers E_B between the valleys. P is a parameter in phase space. This is an abstracted picture - the real thing looks more like the Rocky Mountains or the Alps. But as we are only interested in a qualitative estimation, the picture contains the vital information of barriers between the valleys. Then below we have the connected groundstate structure. The thick lines denote zero energy surfaces that surround the shells of solidary spins. For example, to turn the shaded shell (denoted by ↑) in MC runs the system has to climb over the energy hill E_B separating the ↑ and ↓ valley in the upper part of the figure.

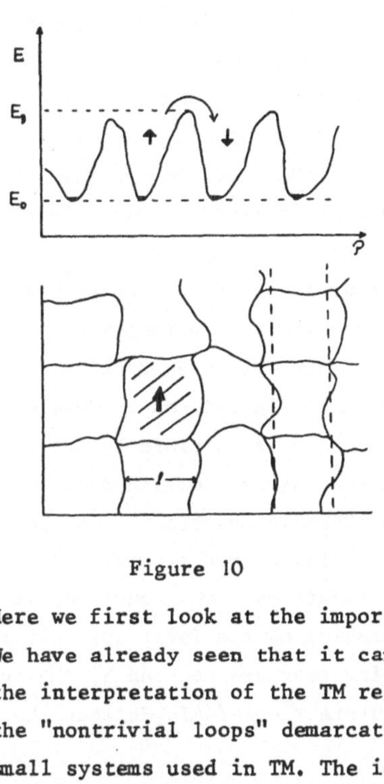

Figure 10

Here we first look at the important question: What is the average size of a shell? We have already seen that it can be shown that the size has to be finite following the interpretation of the TM results for $\langle S_0S_R\rangle_0^2$. I already mentioned above that the "nontrivial loops" demarcating the shells surround the whole cylinder in the small systems used in TM. The idea is now that we consider our TM lattice as a segment of the whole quite larger lattice; as seen in fig. 10 we cut out a stripe and perform our calculations. The existence of the "nontrivial loops" is connected to an even number of spins in let's say horizontal direction. In the case of an odd number we do not get any of the "nontrivial loops". This means: We calculate the zero temperature entropy S(0) by TM. S(0) is connected to the number of ground states which is increased by a factor 2 when a loop crosses the stripe. Thus we are able to determine the number of loops in a stripe of length L by looking at the difference of the entropies of odd and even stripes. We then obtain an average length between successive loops. But this length depends on the width of the stripe. To have a self-consistent problem the width and the average length have to be equal. We carried out different even and odd stripe widths. An extrapolation leads to an average self-consistent length $\lambda \sim 13 \pm 1$ following the formula

$$S(\text{even}) = S(\text{odd}) + \ln(2^{L/\lambda} - 1) \tag{11}$$

The error bars are relatively large as the entropies of different samples differ strongly.

III. Long-time Behaviour at Low Temperatures

Here I follow the basis of the above results to obtain results about the long-time behaviour of spin glasses, i.e. the observation time to reach equilibrium. But first I would like to return to the question whether the considered "± J-model" is the best choice. To carry out our investigation it is necessary to choose a model which is on one hand still to be handled by our computational techniques, but on the other hand contains all necessary physical properties. In most spin glasses (CuMn, EuSrS, etc.) we deal with Heisenberg spins interacting by an RKKY coupling. Walstedt and Walker /12/ introduced an anisotropy as found in experiments in an RKKY-Heisenberg spin glass /13/. They showed that the spins behave Ising-like at low temperatures. As the RKKY-interaction decreases very rapidly with distance, it is appropriate to choose Gaussian nearest neighbour interactions as originally introduced by Edwards and Anderson/1/. Considering only the long-time behaviour it is possible to obtain a further abstraction: our "± J-model". In this model - as already outlined above - only the frustration effects survive from the original RKKY interaction. It is now accepted that the frustration effects dominate the spin glasses qualitatively /2/ and - as we are only interested in more qualitative estimations - the model should be sufficient. K. Binder has given a further interpretation of the Ising spin glass models: He considers mainly clusters of Heisenberg spins that are coupled by the relative strong short-distance components of the RKKY interactions /14/. Between the clusters we only have nearest neighbour interactions.

We now return to the valley-hill picture of fig. 10. We do not only have energy but also entropy barriers between the valleys, i.e. we have free energy barriers F_B. Here we consider the possibility P(T) that the systems climbs from the bottom of one valley over the hill to the next:

$$P(T) = \frac{1}{\mathcal{N}} \exp F_B/T \tag{12}$$

F_B denotes the free energy barrier defined by

$$F_B = E_B + T \cdot S(T) \tag{13}$$

with the energy barrier E_B and the temperature-dependent entropy S(T) related to the considered valley. The normalization factor $\mathcal{N}$ also contains p.e. temperature independent influences as the number of possible ways out of the valley. This number is fairly small - as we will see later - compared to the entropy effect, and we therefore neglect it. Considering low temperatures we make use of the fact that a MC run is mainly located at the bottom of a valley at $T \lesssim 0.3$ (when $J/k_B \equiv 1$). We are therefore able to restrict ourselves to calculate the transition probability from the bottom of a valley to the next one. We turn first to the energy barriers.

a) Energy barriers

Here we consider one shell. I looked at different system sizes L x L using a new approach to calculate E_B. The highest energy hill the shell has to climb over is to turn it to its inverse. Therefore we consider a ground state and try to find a minimum path through phase space to turn all spins by the 1-spin-dynamic. As we know the ground-state energy E_o from TM it is quite easy to obtain a state by cooling MC samples. First the zero-energy loops are identified. Then the numerical procedure tries to find the minimum path. The program is based on the following principles: In the ± J-model we have always a gap of 4·J between different energy levels. Therefore it is not too difficult to identify a minimum path. We have three kinds of spins in a ground state: spins with energy surface 0, 4·J and 8·J. Turning one spin with 4·J or 8·J we create new strings, i.e. new spins with surface 0 or even negative surface. These spins are now easily turned by MC - in most cases the originally flipped spin is turned too and we fall back to the ground state. My procedure now keeps the special first spin fixed and then allows the others to arrange according to the new situation. Then the next spin is turned and kept fixed; the others arrange again. After a sufficient number of spins are turned and fixed successively the whole system turns over and reaches the inverse picture. Carrying out different successions of these special spins, i.e. following different paths through phase space, it is not too difficult to obtain the minimum path. It is some kind of computer game to sit at the screen and look for a minimum path. Investigating about 50 different samples from 8 x 8 to 20 x 20 size I obtained on the average

$$E_B = (\ell + 2) \cdot J \tag{14}$$

where ℓ is the diameter of the particular shell. For a ferromagnet we would get $E_B = 2 \cdot (\ell + 2) \cdot J$.

As seen above the average size of a shell is $\{\ell\}_{av} \equiv \lambda = 13 \pm 1$. Therefore we have an average energy barrier $E_B = (15 \pm 1) \cdot J$ and taking the energy gap of $4 \cdot J$ into account we see that the "± J-model" is governed mainly by 16·J barriers.

b) Entropy

The crucial point to determine the free energy barrier F_B is the calculation of the entropy S(T). As outlined above we restrict ourselves to the transition from the bottom of a valley to the next one. Therefore it is sufficient to consider the zero-temperature entropy S(0) of the valley. TM calculates exactly the partition function of the system and hence exactly S(0) considering the typical representation of the valleys - 13 x 13 lattices. Neglecting nontrivial loop effects by the oddness of the system and subtracting the inverse states we obtain for the zero-temperature entropy per spin

$$s(0) \simeq 0.0685 \pm 0.0005 \qquad (15)$$

Finally the averaged free energy barrier $F_B = E_B + T \cdot S$ is given with

$$S = \lambda^2 \cdot s(0)$$

and

$$E_B = (\lambda + 2) \cdot J. \qquad (16)$$

Considering equation (12) we see that the highest probability P(T) is related to the process that two spins with an energy surface of 8·J flip immediately one following the other to overcome the mainly 16·J barriers not allowing the system to relax to a state of lower energy in the meantime. The latter process is seen to have a lower transition probability.

c) Observation times

Using the knowledge of the free energy barriers it is now possible to determine the observation time t which is necessary to obtain the equilibrium state of the system at a fixed temperature $T_f(t)$. From (12) we get

$$t = t_o \cdot \exp F_B/T_f(t) \qquad (17)$$

Increasing the observation time t the system will show a lower $T_f(t)$, i.e. the system is able to climb over the hills during time t. $T_f(t)$ separates the paramagnetic from the "spin-glass" phase and is therefore identical to the freezing temperature in MC or experiments. Fig. 11 shows this time dependence of the measured T_f. The upper part represents a qualitative experimental picture given by Mydosh /13/ inspired by /15,16/.

We notice a similar qualitative behaviour of the curves. In the limit of very long observation times no significant decrease of T_f is seen. In this region T_f is below T = 0.3 where the starting condition holds that the system is mainly located

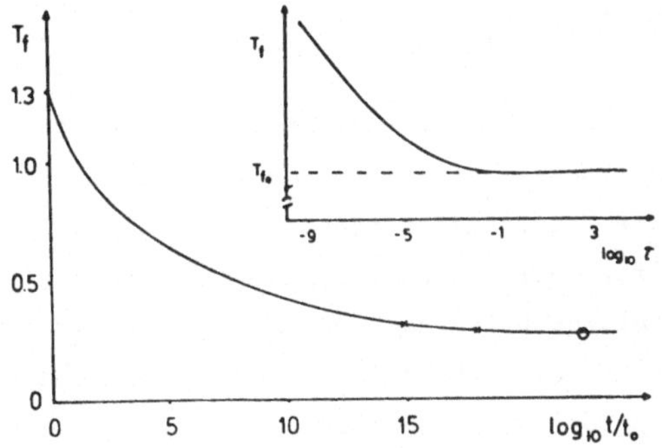

Figure 11

in ground states. Fitting $t_o = 10^{-13}$ (as in most experiments /15,16/), we see that in the region where Malozemoff and Imry /15,16/ found only a single T_{fo} (6 min $\leqslant$ t $\leqslant$ 2400 min) the theoretical curve is already very flat (region denoted by two crosses in fig. 11). Increasing the observation time to 100 years (open circle) we see that the new T_f is just at about 80% of the T_{fo} measured in /15/. Taking these facts into account we conclude that it is hopeless to expect the nonequilibrium effects mentioned above in experiments at low temperatures. In MC the case is even worse as we have to take $t_o \simeq 10^3$ MCS/spin. Therefore spin glasses have to be considered as non-ergodic at low temperatures.

d) Conclusion

Here I would like to repeat the very important above result again. Spin glasses are non-ergodic at low temperatures. The absence of phase transition is only seen in mathematics. The barrier heights lead to huge observation times to reach equilibrium states. It is therefore impossible to obtain the nonequilibrium effects - as predicted by our static theories - in experiments. In particular it is impossible to see the decay of the freezing temperature T_f down to zero. Furthermore, as we have such huge times, it is quite obvious that experimentalists seem to obtain only a single T_{fo} in their measurements. I should point out here that the 1/lnt behaviour seen in the analysis of the $\pm$ J-model is only a qualitative estimation of the decay of the freezing temperature - it is some sense a lower boundary for T_f, i.e. one has at least to wait that times, at some temperatures at least 100 years. A 1/lnt plot against T_f is therefore no check for this theory. And it is clear that the theory of finite energy and entropy barriers in spin glasses cannot be proved or disproved by the long-time experiments. A way out is shown in the next chapter, where we consider the time dependence of the specific heat which is related to finite barriers.

IV) Time-Dependent Specific Heat

In this last chapter I look in more detail at the ground-state valleys. With the help of a low-temperature series expansion I calculate the number of possible states at the lowest levels. The knowledge of these numbers allows to calculate the time-dependence of the specific heat and as a byproduct I am able to give some preliminary results about the dynamics below the freezing temperature. But first the time dependence of the specific heat - this dependence is related to the existence of finite energy and entropy barriers and therefore we can expect from experiments evidence in favour or against the finiteness of the barriers. The new phenomenon could give experimentalists a tool to decide the phase transition problem.

We have seen above that the "$\pm$ J-model" is governed by 15·J energy barriers. As we have only gaps of 4·J between different energy levels we have only four levels inside a valley. To provide satisfying statistical mechanics it is very obvious that one has to know the number of these states. As seen above the typical shell size is 13 x 13. Therefore I looked at these 13 x 13 lattices where because of the oddness of the system no "non-trivial loops" are found. Therefore the number of valleys for a p.e. 13 x 13 lattice is restricted to 2, only including the inverse spin configurations.

The number of low-lying excited states is now calculated by combining the numerically exact TM results for a configuration with a low-temperature expansion of the partition function Z. We have

$$Z = g_0\, e^{E_0/T} + g_1\, e^{(E_0 - 4\cdot J)/T} + g_2\, e^{(E_0 - 8\cdot J)/T} + \dots \qquad (18)$$

g_i are the number of states and E_0 is the ground-state energy of the configuration. As we deal with a lot of exponential functions the numerical procedure is quite involved. But in principle one only has to solve a linear equation for the g_i's, considering Z at different temperatures T. We are able to calculate the lowest 10 states. The result considering 20 different lattices is:

$$\ln \frac{g_i}{g_0} = 7 \cdot i^{0.8} \qquad (19)$$

A corresponding plot of (19) shows a straight line in fig. 12. Taking this result into account we can provide a low-temperature series for the $\pm$ J-model by inserting (19) in (18). For the free energy we obtain:

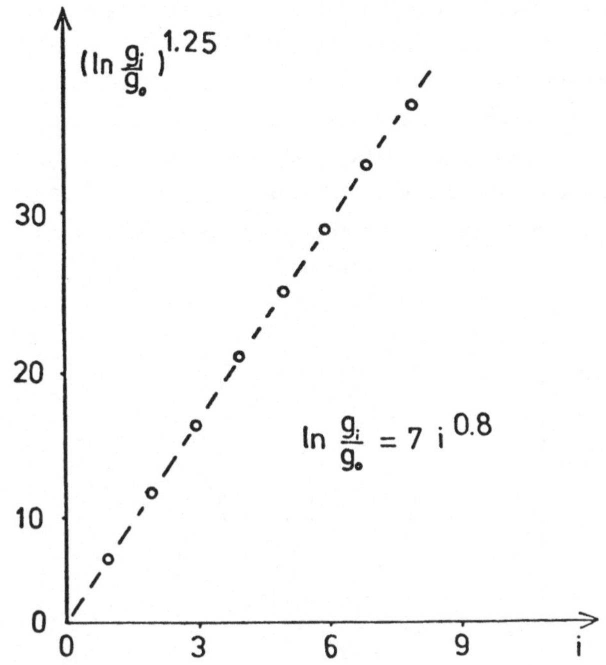

Figure 12

$$F \equiv \frac{1}{N} \cdot T \ln Z = \mathcal{E}_o + s_o/T + \frac{1}{\lambda^2} \ln \sum_{i=0}^{9} \exp \{J \cdot i^{0.8} - 4 \cdot i \cdot J/T\} \tag{20}$$

where $\mathcal{E}_o = -1.41 \cdot J$ the ground-state energy/spin, $s_o = 0.073$ the ground-state entropy/spin and $\lambda = 13 \pm 1$ the average shell size. The series accurately yield up to a temperature of $T = 0.5 \cdot J$ seen from comparison to TM.

But we now return to our main purpose: the specific heat. The result (19) enables us to provide a restricted partition function Z_{rest} for one valley. We have

$$Z_{rest} = \tilde{g}_o\, e^{E_o/T} + \tilde{g}_1\, e^{(E_o - 4J)/T} + \ldots.. \\ + g_4\, e^{(E_o - 16J)/T} + \ldots. \tag{21}$$

where the $\tilde{g}_i$ belong to one valley while the g_i belong to the whole lattice. The whole partition function is calculated by TM and we have to subtract the contribution of the $g_o - g_3$ due to the inverse valley of the considered 13 x 13 configurations. The point is: Considering short observation times, spin glasses occupy low states in only one valley, but above the energy barrier all possible states are included following the Boltzmann weights. From the restricted partition function we now obtain by differentiation the behaviour of the short-time specific heat (related to one valley). The result is shown in fig. 13. We first notice the behaviour of the equilibrium or long-time specific heat. The dotted line is denoted by Z and $t \to \infty$. We have the well-known rounded cusp. But we notice for the short-time specific heat (denoted by $Z_r \to 1$ valley) a break-away from the equilibrium curve at the freezing temperature T_f. A dip of the curve is observed between $T_f = 1.3$ and $T \simeq 0.9$. At higher temperatures the curve follows the equilibrium behaviour. At lower temperatures the new curve lies above the old one.

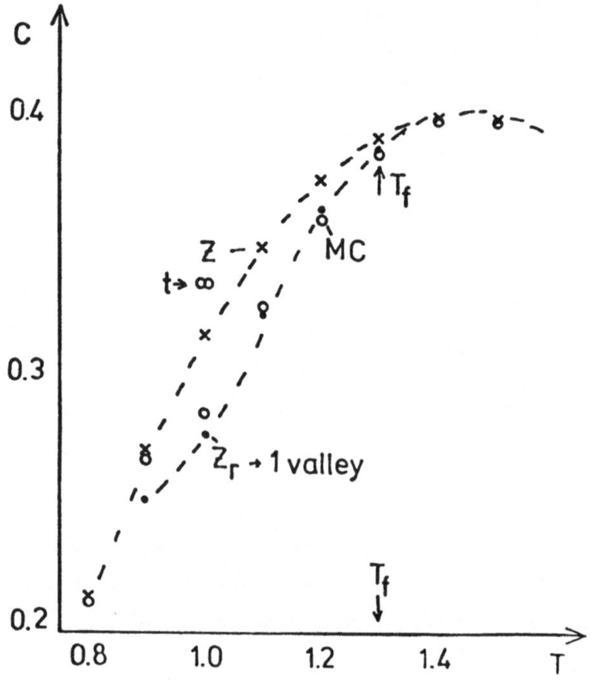

Figure 13

But what is the connection to experiments? Theorist's experiment is Monte Carlo. We have to ensure that the system occupies the states in only <u>one</u> valley and all the states above according to the Boltzmann factor. In MC this is done in the following way: We start at a ground state. Then we turn on the considered temperature ($0.9 < T < 1.3$). The system now occupies the states in the corresponding valley. In short runs (about 2000 MCS/spin) it is ensured that we do not reach the next valley. Thus averaging over about 50 runs per sample and temperature we end up with the result denoted by the open circles in fig. 13. We notice that the MC results for 2000 MCS/spin follow the predicted numerically exact Z_r-result. Waiting longer times we reach the equilibrium curve, i.e. MC reaches the other valleys. As we deal just below T_f, the observation times are relatively short and still accessible by MC. Real experiments now have to be provided in the same way: Starting from a ground state we should obtain in relatively rapidly heated experiments a deviation of the specific heat just below T_f from the long-time curve. The long-time curve on the other hand corresponds to the following procedure: Carefully cooling the sample and then heating up from temperature to temperature already yields equilibrium results at the considered temperatures as theoretically seen in MC. Therefore the time dependence effect of the specific heat has not been observed so far in spin-glass experiments /13,17/. Obviously it is necessary to heat up from a ground state at every measured temperature to avoid the effect that other valleys are already occupied destroying the small effect.

The experiments have to be carried out in the following way:

1.) Cooling slowly to very low temperatures to obtain ground states.
2.) Heating relatively rapidly to the temperature to be considered.
3.) Measurement of the specific heat at different observation times (max. several minutes).
    Temperature range to be considered $0.5 \cdot T_f < T < 1.5 \cdot T_f$.

A dip of the specific heat as qualitatively shown in fig. 13 is related to finite energy and entropy barriers and therefore to the non-existence of a transition.

If there is no time-dependence, the behaviour of real spin glasses is quite different from that of short-range spin-glass models and we should consider models providing a phase transition. The knowledge of the numbers allows some further considerations. It is possible to calculate the time necessary to escape from the bottom of the valley to the summits of the surrounding hills, i.e. to the 16·J niveau. First I provide some MC experiments: The system started at a ground state, then I measured the time to reach the first 16·J state. In the theory of statistical mechanics I suppose that the MC runs were long enough allowing the system to follow the Boltzmann weights. All the states inside one valley are occupied according to their weight, i.e. we have a partition function for the valley:

$$Z_v = \sum_{i=0}^{3} \tilde{g}_i \, e^{(E_o - 4i \cdot J)/T} \tag{22}$$

The time to reach one of the 16 · J states is then given by the ratio of the corresponding probabilities:

$$t = Z_v/g_4 \cdot e^{(E_o - 16 \cdot J)/T} \tag{23}$$

Fig. 14 shows the behaviour of the time according to equation (23).

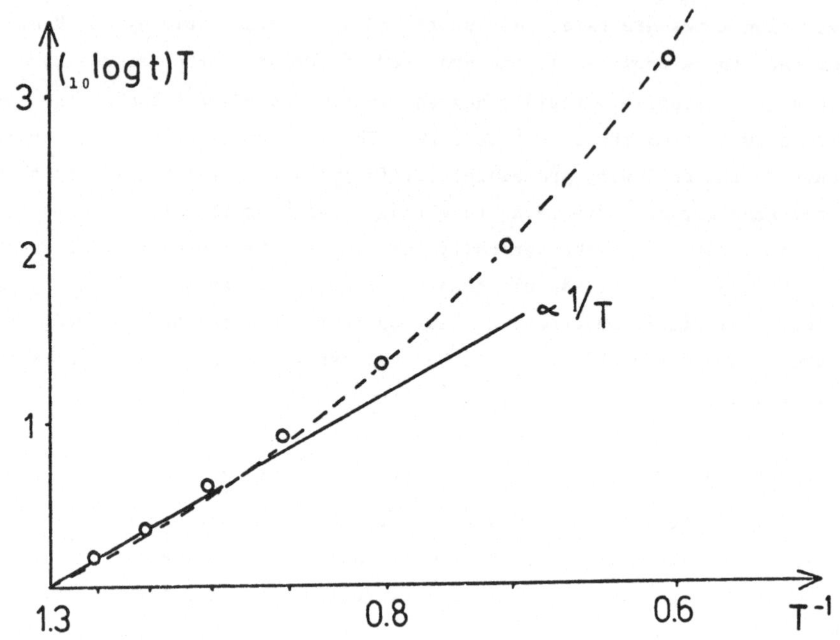

Figure 14

A plot of T log t against the inverse temperature T^{-1} is shown. At temperatures just below $T_f(\approx 1.3)$ we notice a nearly linear increase of T log t. At temperatures lower than 0.9 the gradient of the curve is steeper. The open circles are due to the mentioned MC runs. They are in fair agreement with the theoretical curve. On the other hand the linear increase of $T\ell nt$ proportional to T^{-1} for $1.3 > T > 0.9$ is in agreement with the long-time MC simulations of A. P. Young. He found that $T\ell nt \propto T^{-1}$ in a temperature range down to $T = 0.9$. Lower temperatures were not included in his paper /18/. A description of his simulations is given in his contribution to these proceedings.

So far my considerations carried out looking at the influence of the valley-hill-picture on spin glasses. In conclusion of this chapter I would like to point out that first my contributions are supported by A. P. Young's T^{-1} proportionality for $T\ell nt$. I obtained the same behaviour taking the knowledge of the numbers g_i or $\tilde{g}_i$, respectively, into account. But a far more important result is the time dependence of the specific heat. Experiments carried out as mentioned above should allow a decision in favour or against a transition in real spin glasses. I hope the theoretical result will give rise to a new series of time-dependent measurements.

V. Conclusion

In a short conclusion I would like to point out what is clear or unclear following my numerical results:

1.) No phase transition in short-range spin-glass models in two dimensions.

2.) In three dimensions I expect the same behaviour as in two. But the case is not as clear as my systems may have been too small.

3.) At low temperatures we have huge observation times to reach equilibrium. This time may exceed at low enough temperatures even 100 years. Real spin glasses are therefore non-ergodic at low temperatures.

4.) The existence or nonexistence of finite energy and entropy barriers is related to the time-dependence of the specific heat. Corresponding experiments could answer this vital question in the theory of spin glasses.

APPENDIX

Numerical Methods:

Here I give a short description of the numerical methods to obtain the subsequent results. Mainly I used two major approaches to calculate thermodynamical quantities as specific heat, susceptibilities, magnetizations etc. The first is my own invention, lateron called "Transfermatrix Method", as we will see later, a somewhat misleading denotation. The main purpose is to calculate "numerically exact" the partition function of a given two or three-dimensional lattice with short-range interaction,preferably between Ising spins. The thermodynamic quantities are obtained by numerical differentiation. The second method is the standard Monte-Carlo simulation. Here the spin glass is simulated using a 1-spin-dynamic. It is a dynamical approach and one expects that the dynamic corresponds to that of real spin glasses. The most exciting results were obtained by comparing results of both methods allowing to distinguish between equilibrium and nonequilibrium effects.

a) Transfermatrix Method

As outlined above the denotation is somewhat misleading, but in principle the method works like the well known analytic Transfermatrix approach /19/. We start with the first row and then we add spins row by row. Here I restrict myself to the description of a nearest neighbour Ising lattice in two dimensions of size L x L. Further extensions are straightforward. The partition function Z, or better the logarithm of Z to avoid overflow effects, is calculated recursively. Mainly we have three steps:

1.) We start with the first row: the computer generates all possible configurations of the first row. Then a configuration - following a simple logical scheme - of the second row is fixed and combined with all states of the first row. The statistical weights are calculated. Then we sum over all states of the first row, i.e. we take the trace of the first row.

2.) The next configuration of the second row is now combined with all states of the first. We follow the scheme outlined in 1.). Then the result is stored separately. Furtheron we continue with all possible configurations of the second row. As we have 2^L states we need a corresponding storage capacity of the computer. Therefore the method is restricted to $L \leqslant 18$. Now the program looses the information of the first row. All states of the first row are "summed out".

3.) Now we continue from row to row. At the end we sum all states of the last row to obtain the partition function, i.e. the sum over all possible states. Here I give a short description in a simplified programming language:

We calculate an L x L lattice.

$F = 0$

$C_1 = 1$

Do 1 i = 1, L

Do 2 k = 2, L

$E = J(i, k / i, k + 1)/T$ horizontal bonds

$H = h_{ik}/T$ magnetic (random) field

$C = 0$

Do 3 $j = 1, 2^{k-1}$

Do 4 $j_1 = 1, 2^{L-k}$

$I1 = j_1 + 2(j - 1) \cdot 2^{L-k}$

$IE = I1 + 2^{L-k}$

$A(I1) = A(I1) \cdot \exp(E + H)/C1$

$A(IE) = A(IE) \cdot \exp(-E - H)/C1$

$C = C + \{A(I1) + A(IE)\}/2^L$

4 Continue

$E = -E$

3 Continue

$F = F - \frac{1}{N} \ln C$

2 Continue

So far we calculated the horizontal contributions. The loop 1 Do i = 1, L gives the index of the row.

The loops 2 - 4 generate successively all possible configurations of the i-th row. At the same time the configurations obtain their Boltzmann weight. This is stored in the array A containing 2^L places.

The loops 2 - 4 now simulate step by step blocks of up and down spins multiplying the statistical weights of the bonds and the applied magnetic field. As easily seen, the system calculates all possible configurations of a row.
To avoid numerical problems we make use of the constants C and C_1 keeping the numbers in the machine well below the overflow limit.
The next part contains the "vertical" contributions:

```
C2 = 1
C = 0
Do 5 k = 1, N
E = J(i, k / i + 1, k)/T vertical bonds
Do 6 j = 1, 2^(k-1)
Do 7 j1 = 1, 2^(L-k)
I1 = j1 + 2 · (j - 1) · 2^(L-k)
IE = I1 + 2^(L-k)
x = A(I1)/C2
y = A(IE)/C2
A(I1) = x · exp E + y · exp(-E)
A(IE) = x · exp(-E) + y · exp E
C = C + (A(I1) + A(IE))/2^L
7 Continue
6 Continue
C2 = C
F = F - (1/N) ln C
5 Continue
1 Continue
C3 = Σ_i^(2^L) A(i)
F = F - (1/N) ln C3
```

Finally we obtained $F = \frac{1}{N} \ln Z$. The loops 5 - 6 here generate all the configurations of the (i + 1)th row and couple them to the i-th row - coupling J(i, k ,i + 1, k).

Here an example:
We assume the spins of the first row are generated. The array A(1) contains all spins up: ↑....↑ - correspondingly $A(2^{L-1} + 1)$ has all spins up but the first one down: ↓↑....↑. The loops 5 - 6 now first put the first spin of the second row in place:
We have the spin up:

```
↑....↑                              ↓↑....↑
.                                   .
.               and                 .
↑    weight e^E                     ↑    weight e^-E
```

and then down:

$$\begin{array}{lcl} \uparrow\ldots.\uparrow & & \downarrow\uparrow\ldots.\uparrow \\ \vdots & \text{and} & \vdots \\ \downarrow \quad e^{-E} & & \uparrow \quad e^{E} \end{array}$$

The program now takes the trace over the first spin of the first row, i.e. it sums the configurations:

$$\begin{array}{lcl} \uparrow\ldots.\uparrow & & \downarrow\uparrow\ldots.\uparrow \\ \vdots & \text{and} & \vdots \\ \uparrow & & \uparrow \end{array}$$

and stores the sum in A(1). Analogously the sum with the first spin down is stored at $A(2^{L-1} + 1)$. Now all the other spins are treated in the same way.

I hope the principle of the method is clear now and the reader will be able to write his own program.

b) Monte Carlo Simulations

Here I give only a very simple cooking recipe for very bloody beginners. All the others should make use of K. Binder's profound description of the method /14/. We use a 1-spin-dynamic, i.e. the program picks up a spin at a random site in the lattice. Then it calculates its energy surface and decides - following the Boltzmann weight - whether to flip or not by comparing the weight to a random number simulating the heat bath. The interesting quantities are directly obtained from the lattice - calculating a time average - as it is assumed that the computer visits the states of the system following their corresponding weights, i.e. configuration 1 with weight exp $-E_1/T$ and configuration 2 with exp $-E_2/T$ are visited in a ratio $\exp-(E_1 - E_2)/T$. I now give again a short description of the program.

```
Do 1 ki = 1, MCS  number of MC steps per spin
Do 2 i = 1, L
Do 2 j = 1, L
k = random 1
ℓ = random 2
E = {IS(k - 1,ℓ) · J(k - 1, ℓ/k, ℓ)
+ IS(k + 1, ℓ) · J(k, ℓ/k + 1, ℓ)
+ IS(k, ℓ - 1) · J(k, ℓ - 1/k, ℓ)
+ IS(k, ℓ + 1) · J(k, ℓ/k, ℓ + 1)}
· IS(k1, k2)
```

```
DE = exp(-E/T)
R = random 3
if (R · GT · DE) goto 22
IS(k, ℓ) = -IS(k, ℓ)
22 Continue
Interesting quantities calculated.
2 Continue
1 Continue
```

I should note that one has to use different random number generators random 1, 2, etc. to avoid dangerous correlations between the numbers.

At the end I hope that the reader will be able to write his programs. I wish him a lot of very successful numerical work.

References

1. S. F. Edwards and P. W. Anderson, J. Phys. F5, 965 (1975)
2. G. Toulouse, in: Disordered systems and localization, Springer Lecture Notes in Physics 149 (1981)
3. I. Morgenstern and K. Binder, Phys. Rev. Lett. 43, 1615 (1979)
4. I. Morgenstern and K. Binder, Phys. Rev. B22, 288 (1980)
5. I. Morgenstern and K. Binder, Z. Phys. B39, 227 (1980)
6. I. Morgenstern and H. Horner, Phys. Rev. B25, 504 (1982)
7. D. Sherrington and S. Kirkpatrick, Phys. Rev. Lett. 35, 1792 (1975)
8. P. Hoever, W. F. Wolff and J. Zittartz, Z. Phys. B44, 129 (1981)
9. S. Kirkpatrick, Phys. Rev. B16, 4630 (1977)
10. A. J. Bray and M. A. Moore, J. Phys. F7, L333 (1977)
11. K. Binder, Fundamental Problems in Statistical Mechanics V, North-Holland, Amsterdam (1980)
12. R. E. Walstedt and L. R. Walker, Phys. Rev. Lett. 47, 1624 (1981)
13. J. A. Mydosh, in: Disordered Systems and localization, Springer Lecture Notes in Physics 149 (1981)
14. K. Binder, in: Monte Carlo Methods in Statistical Physics, Springer, Berlin (1979)
15. A. P. Malozemoff and Y. Imry, Phys. Rev. B24, 289 (1981)
16. B. Barbara, A. P. Malozemoff and Y. Imry, Phys. Rev. Lett. 47, 1852 (1981)
17. J. Souletie, private communication
18. A. P. Young, Phys. Rev. Lett. 50, 1509 (1983)
19. See e.g. McCoy and Wu, The Two-Dimensional Ising Model, Harvard (1973)

NUMERICAL STUDIES OF SPIN GLASSES

A.P. Young
Department of Mathematics
Imperial College
London SW7 2BZ, U.K.

Computer simulations have played a very important role in the theory of spin glasses because analytic methods have proved to be so difficult. For example we do not yet have a completely satisfactory mean field theory, nor do we know with certainty what is the lower critical dimension, d_L, defined to be that dimension below which fluctuation effects destroy the transition predicted by mean field theory. Furthermore simulations are attractive because there are simple models for spin glasses which can be treated very efficiently by a computer.

There are several good reviews [1] on the application of computer simulations to spin glasses so I shall not attempt to review the field here. Instead I shall limit myself to describing work that I have been personally involved in. This divides naturally into two parts. The first is to clarify the mean field theory by studying the infinite range model of Sherrington and Kirkpatrick [2]. The second objective has been to investigate short range models with a view to determining d_L and to investigating static and (very importantly) dynamic effects in two and three dimensions. The simplest model to study is the Ising model, where the spins take values ± 1, and, although it is a considerable simplification of a real system, it appears to display all the characteristic spin glass behaviour [3]. I shall only discuss results for Ising models. In a separate lecture at this colloquium Binder [4] will describe his own results on Ising models with short range interactions. His calculations and mine have a somewhat different emphasis but agree where they overlap. Computer simulations on a much more realistic model of Heisenberg spins with RKKY interactions are discussed at this meeting by Walstedt [5].

INFINITE RANGE MODEL

Sherrington and Kirkpatrick, SK [2], suggested it would be useful to study a spin glass model with infinite range interactions because, by analogy with ferromagnetism, the exact solution of this model could be called the mean field theory of the spin glass problem. It was initially hoped that this model would be easy to solve and that the solution would be a reasonable approximation to realistic systems with short range interactions in three dimensions. Unfortunately it appears that neither hope has been realised. As we shall see, the exact solution may have been found, though even this is not clear, and is rather complicated. The model certainly has a phase transition and should therefore be very different in behaviour from real systems if, as is plausible, $d_L = 4$ (see below) so realistic models should have no transition in three dimensions. Nonetheless, as discussed below, experimentalists often compare their data with predictions of the SK model, with fair success. This is one reason for trying to understand the model. In addition the SK model represents a tantalising challenge for theorists which is studied 'because it is there'.

The Hamiltonian of the SK model is

$$H = - \sum_{<i,j>} J_{ij} S_i S_j - h \sum_i S_i \tag{1}$$

where $S_i = \pm 1$ ($i = 1 \ldots N$) is an Ising variable, and the interactions J_{ij} are independent random variables with mean J_o/N and variance J/N, the same for all pairs of spins. I shall only discuss the case of $J_o = 0$. A uniform field, h, has also been included in eq. (1). The original SK solution involves a single order parameter, q, defined by

$$q = \langle\langle S_i \rangle_T^2 \rangle_J \tag{2}$$

where $\langle \ldots \rangle_T$ denotes a statistical mechanics average for a given set of interactions and $\langle \ldots \rangle_J$ is an average over the interactions. For $h = 0$ SK found a transition at $T_c = J$ (setting Boltzmann's constant to unity). Subsequently Almeida and Thouless, AT [6], found that this solution is unstable below a line in the h-T plane terminating at $h = 0$, $T = T_c$; see Fig. 1.

Several attempts were then made to find a stable solution in the low temperature region, of which the most interesting is due to Parisi [7]. This has an infinite number of order parameters which are parametrised by a function $q(x)$ where $0 < x < 1$ (see Fig. 2.). Parisi's theory used the 'replica trick' and at first the significance of the dummy variable x was unclear. In particular it was not obvious what, in Parisi's theory, corresponds to the order parameter, q, in eq. (2). Subsequently, though, it has become clear [8,9] how to relate physical quantities to variables calculated by replicas.

The key physical observation in this, is that below the AT line a system can exist in one of many phases, presumably represented by solutions of the mean field equations of Thouless, Anderson and Palmer, TAP [10], which are known [11] to have an

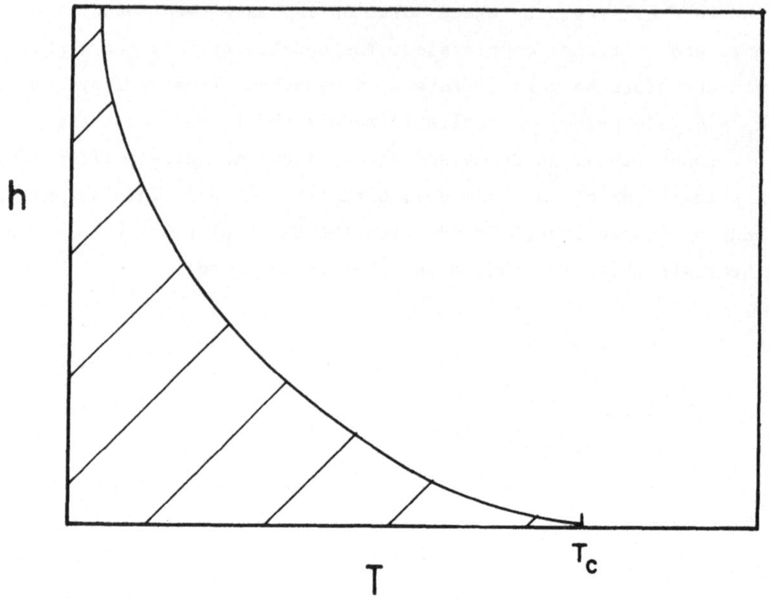

Fig. 1. The shaded region is where the SK solution is unstable. The limit of stability $T_c(h)$ is the AT line and, close to T_c, varies like $T_c(h) - T_c \propto h^{2/3}$

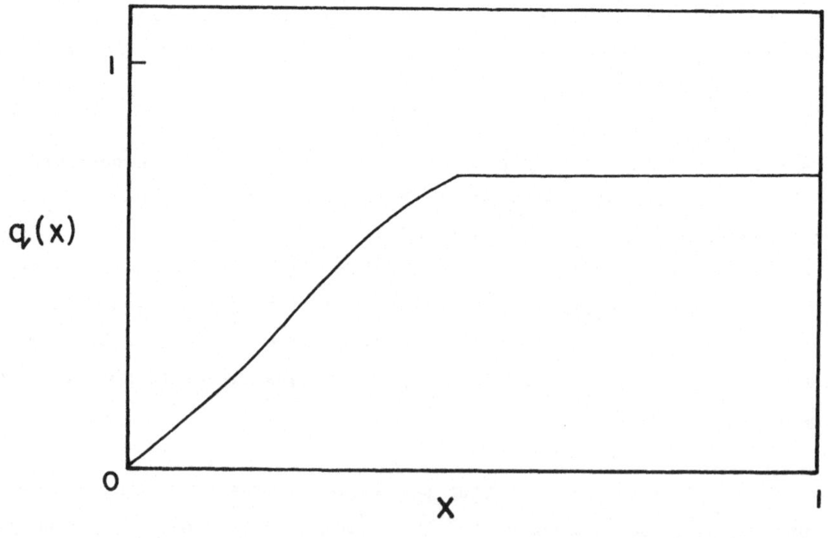

Fig. 2. A sketch of Parisi's order parameter function, q(x), for $T < T_c$, h = 0.

enormous number of solutions. If one follows the dynamics of the system then, for $N \rightarrow \infty$, these phases must be stable and the system will remain forever in the phase in which it was prepared. On the other hand a statistical mechanics average, i.e. $<\ldots>_T$, includes contributions from each phase 's', with a weight P(s) [8]. Hence

$$<S_i>_T = \sum_s P(s)\, m_i^s \tag{3}$$

where m_i^s is the magnetisation of site i when the system is in phase 's'. Hence the statistical mechanics order parameter, q, defined by eq. (2) is given by

$$q = < \sum_{s,s'} P(s)\, P(s')\, q^{ss'} >_J \tag{4}$$

where

$$q^{ss'} = \frac{1}{N} \sum_i m_i^s\, m_i^{s'} \tag{5}$$

is the overlap between the magnetisations in phases 's' and 's''. One can then show [9] that dx/dq, the derivative of the inverse of Parisi's order parameter function, is a probability distribution for solutions to have overlap q. i.e.

$$\frac{dx}{dq} = W(q) = < \sum_{s,s'} P(s)\, P(s')\, \delta(q^{ss'} - q) >_J \tag{6}$$

so that q, defined by eq. (4), is the first moment of the distribution, i.e.

$$q = \int q' W(q') dq' = \int_0^1 q(x) dx \tag{7}$$

In Parisi's theory q has a simple value,

$$q = 1 - T/T_c \tag{8}$$

for $h \rightarrow 0$.

Eq. (4) involves interference between different phases, which is rather unphysical. A more physical order parameter would describe ordering in a single phase. To avoid having to specify which phase we are referring to, we define the Edwards Anderson order parameter, q_{EA}, as a weighted average by

$$q_{EA} = \sum_s P(s)\, q^{ss} \tag{9}$$

If many solutions have significant statistical weight then $q_{EA} > q$. One can also show that [8]

$$q_{EA} = q(x = 1) \tag{10}$$

An alternative interpretation of 'x' has been given by Sompolinsky [12]. Barriers between different phases must diverge for $N \to \infty$ and so there must be relaxation times which also diverge in the thermodynamic limit. Sompolinsky interprets $q(x)$ as a dynamical correlation function on a time scale t_x where the times t_x all diverge when $N \to \infty$. This predicts that q_{EA} is given by eq. (10) but gives $q = q(x = 0)$ (=0 for $h \to 0$) instead of eq. (7).

Monte-Carlo simulations can be thought of as mimicking the dynamics of the system (at least for Ising spins) and can therefore give information on dynamics as well as statics. The purpose of the numerical calculations was therefore twofold. First of all to test the hypothesis that relaxation times diverge in the thermodynamic limit and secondly to run the simulation for longer than the longest relaxation time to get statistical mechanics averages which can be compared against the two different interpretations of Parisi's theory.

To investigate dynamical effects it is useful to calculate the time dependent auto-correlation function

$$q(t) = \frac{1}{N} \sum_{i=1}^{N} \langle\langle S_i(t_o)\, S_i(t_o + t)\rangle_T\rangle_J \tag{11}$$

where t_o is the equilibration time (measured in Monte-Carlo steps per spin). It is important that t_o is longer than the longest relaxation time for the energy so that the system is truly in equilibrium at time t_o. Since relaxation times diverge for $N \to \infty$ it is necessary, for each finite size, to check a posteriori that a large enough t_o was allowed.

In zero field the Hamiltonian is invariant under time reversal, so that for every phase there is a time reversed phase with all the magnetisations changed in sign. Hence for sufficiently long times $q(t)$ must tend to zero. It appears [13] that for each sample $q(t)$ one has very long relaxation time (the ergodic time) where all the spins turn over and beyond which $q(t)$ tends to zero. For $N \to \infty$ this time is much longer than the other timescales (for instance those on which the energy equilibrates). However for the rather small sizes available to computer simulations this separation of times is not clearcut and it is useful to consider another quantity, $q^{(2)}(t)$, which is insensitive to fluctuations on the ergodic time. $q^{(2)}(t)$ is defined by

$$q^{(2)}(t) = \frac{1}{N(N-1)} \sum_{i \neq j} \langle\langle S_i(t_o) S_j(t_o) S_i(t_o + t) S_j(t_o + t)\rangle_T\rangle_J \tag{12}$$

and follows the correlations of pairs of spins. It will be found that there is a spectrum of relaxation times which contribute to $q^{(2)}(t)$ up to a maximum value, τ. For $t>>\tau$ one has

$$q^{(2)}(t) \to q^{(2)} = \langle\langle S_i S_j \rangle_T^2 \rangle_J \tag{13}$$

and, according to ref. (8), $q^{(2)}$ is given in Parisi's theory by

$$q^{(2)} = \int_0^1 q^2(x)\,dx \tag{14}$$

One finds, plotting $q^{(2)}(t)$ against $\ln t$, that $q^{(2)}(t)$ reaches $q^{(2)}$ at a well defined time τ, where there appears to be a change in slope. Data for $q^{(2)}(t)$ against $\ln t/\ln \tau$ for several sizes at $T = 0.4\,T_c$, $h = 0$ is shown in fig. 3. All the data lies pretty well on the same universal curve. Furthermore, as shown in fig. 4, $\ln \tau$ increases with the size of the system roughly as $N^{1/4}$. This is direct evidence for the SK model being non-ergodic. If one represents the decay of $q^{(2)}(t)$ by a spectrum of relaxation times, in turn arising from a spectrum of barrier heights then the barrier height spectrum is roughly uniform up to a value ΔE_{max} where $\Delta E_{max} = T \ln \tau \propto N^{1/4}$. This leads to a spectrum of relaxation times where, for $N \to \infty$, each timescale is much longer than the previous one, as assumed by Sompolinsky [12].

Also shown in fig. 4 is data for $\ln \tau$ at $T = 0.4$, $h = 1.2T_c$ which lies above the AT line (at $T = 0.4$ the critical AT field, is $0.8T_c$). Clearly relaxation times stay finite indicating that there is only one phase for the system to be in and, furthermore, one finds the results for $q^{(2)}$ and q converge very rapidly with increasing N to the SK predictions (in the SK theory $q^{(2)} = q^2$).

We have therefore found direct evidence that the SK solution is correct above the AT line, where relaxation times are finite, but that the model is non-ergodic below this line where presumably there are many phases available to the system.

Having characterised the relaxation averages we can now study statistical mechanics averages and compare them with Parisi's theory. Because of fluctuations on the ergodic time it is difficult to extract q from the long time limit of q(t). We therefore calculate

$$q_{mod}(t) = \frac{1}{N} \langle\langle \left| \sum_{i=1}^{N} S_i(t_o) S_i(t_o + t) \right| \rangle_T \rangle_J \tag{15}$$

which is insensitive to fluctuations on the ergodic time. A small uniform field probably suppresses fluctuations on the ergodic time in which case

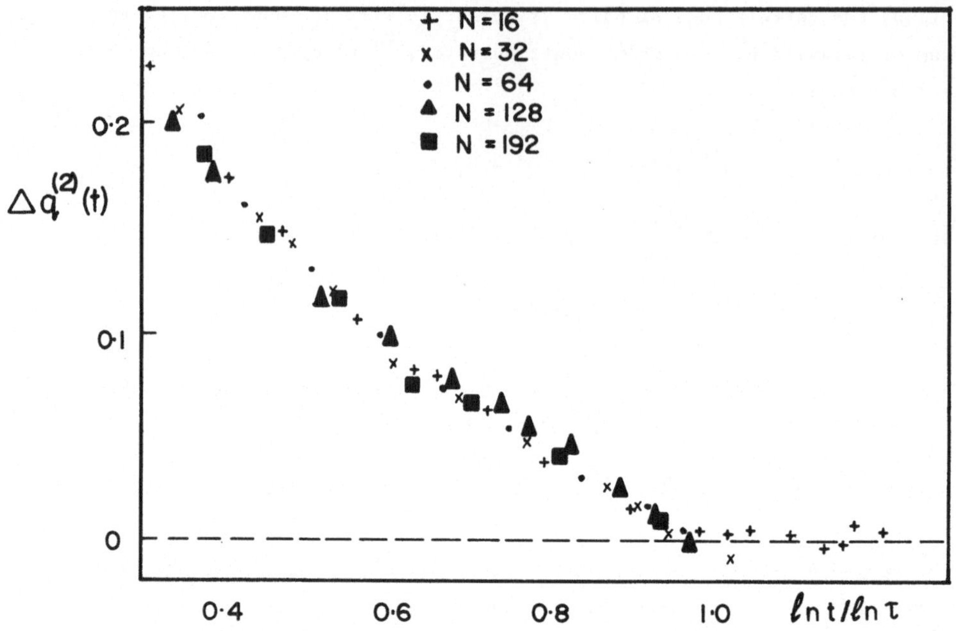

Fig. 3. A plot of $\Delta q^{(2)}(t) = q^{(2)}(t) - q^{(2)}$ against $\ell nt/\ell n\, \tau$ at $T = 0.4\ T_c$, $h = 0$ for several sizes. The data appears to lie on a single universal curve which is roughly a straight line up to $\ell nt/\ell n\, \tau = 1$ where there appears to be a change in slope.

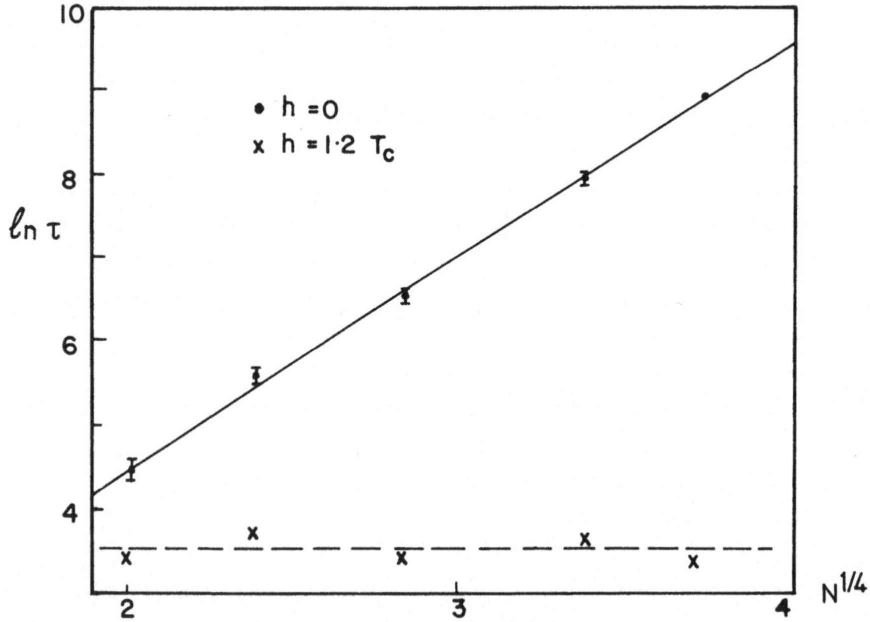

Fig. 4. A plot of $\ell n\, \tau$ against $N^{1/4}$ for $h = 0$ and $h = 1.2\ T_c$ at $T = 0.4\ T_c$. The critical AT field is $0.8\ T_c$ at this temperature.

$$q_{mod} = \lim_{t>>\tau} q_{mod}(t) \tag{16}$$

evaluated in zero field would be equal, in the thermodynamic limit, to q(h) as $h \to 0$. This supposition is hard to test by simulations on rather small sizes.

Fig. 5 shows $\sqrt{q}^{(2)}$ and q_{mod} for several sizes at $T = 0.4\ T_c$, $h = 0$. The data appears to tend to the Parisi values for $[\int q^2(x)dx]^{1/2}$ and $\int q(x)dx$ respectively. This supports Parisi's theory with the interpretation of dx/dq as a probability distribution. It does not appear to be consistent with the dynamical interpretation of x [12].

What fig. 5 demonstrates is that the first 2 moments of the distribution of W(q), eq. (6), are given reasonably well by Parisi's theory. It would be much more convincing, of course, to reproduce the whole distribution. Recent calculations [14] have indeed

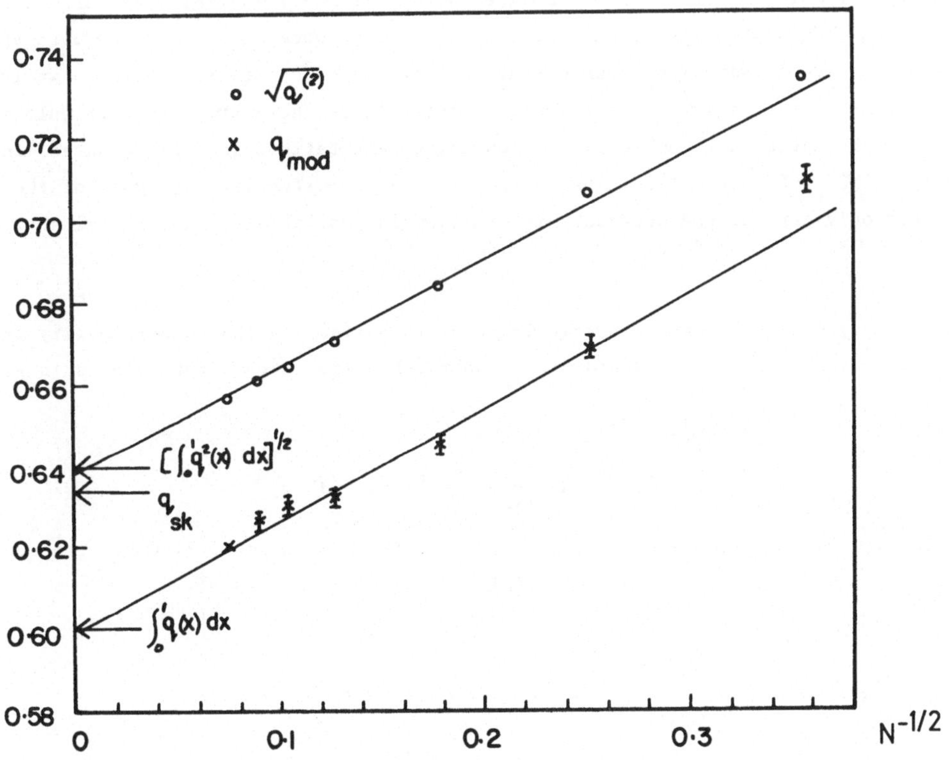

Fig. 5. Data for $\sqrt{q}^{(2)}$ and q_{mod} is plotted against $N^{-1/2}$ for $T = 0.4$, $h = 0$. The arrows mark the Parisi predictions for these quantities and q_{SK}, the value of q in the SK theory.

found a W(q) which is very similar to Parisi's dx/dq except that for finite sizes, the delta function arising from the plateau region in q(x), see Fig. 2, is broadened by finite size effects. From this we conclude that the SK model is certainly understood qualitatively and that Parisi's theory, with dx/dq as a probability distribution is likely to be the exact solution. If it is not exact, Parisi's results must be very close to being so.

SIMULATIONS ON SHORT RANGE MODELS

One result from mean field theory which is not in doubt is that a transition does occur. It is then anticipated that mean field theory should be a reasonably good approximation to systems with finite range interactions provided the space dimension, d, is sufficiently high that fluctuation effects, neglected in mean field theory, do not destroy the transition. The lower critical dimension, d_L, is that dimension below which no transition occurs. For ferromagnetism $d_L = 1$ for Ising spins and $d_L = 2$ for vector spin models with rotational invariance in spin space. Both these results are obtained straightforwardly by considering the low energy excitations; domain walls for Ising spins, and spinwaves for vector models. It is, however, very difficult to determine the nature and number of the low energy excitations for spin glasses. In principle this information would be contained in a full stability analysis of the SK model but only part of the necessary information is available for the Parisi solution [15].

Alternatively one can look for a transition by investigating the susceptibility which diverges as the transition temperature is approached from above. For Ising spin glasses this is

$$\chi_{SG} = \frac{1}{N} \sum_{i,j} \langle\langle S_i S_j \rangle_T^2 \rangle_J \tag{17}$$

One can also investigate how individual terms in the sum in eq. (17) vary with the distance R_{ij} between spins i and j. Defining

$$\Gamma(R_{ij}) = \langle\langle S_i S_j \rangle_T^2 \rangle_J \tag{18}$$

then if

$$\Gamma(R_{ij}) \propto \exp(-R_{ij}/\xi) \tag{19}$$

one can extract a spin glass correlation length ξ, which should also diverge at the transition.

The first calculations of χ_{SG} were high temperature series expansions [16]. An analysis of the series gave a transition for d > 4 but with apparently no transition at lower dimensions. However subsequent analysis of the series in d = 3 [17] showed that situation is not completely clearcut, some methods of analysis of the series giving a transition, others not. In fact there are really not enough terms available for this rather irregular series to say definitely whether or not a transition occurs in d = 3.

Early Monte-Carlo calculations in d = 2 and 3 [18] showed spin glass 'freezing' which was varyingly interpreted as a sharp phase transition or a process of gradual freezing where relaxation times smoothly (but rapidly) increase as T is lowered. For two dimensions Morgenstern and Binder [19] showed very clearly that no sharp transition occurs at the temperature where freezing is observed in Monte-Carlo simulations because the correlation length in eq. (19), calculated by statistical mechanics, is only about 2 lattice spacings at these temperatures. These calculations used exact transfer matrix techniques for finite systems for sizes of L x L where L ≤ 18. Since $\xi \ll L$ at the temperature discussed finite size effects should not be very significant. Morgenstern and Binder could not definiteley rule out a transition at very much lower temperatures but this seems unlikely and their results are consistent with a power law divergence of ψ_{SG} and ξ as $T \to 0$, namely

$$\psi_{SG} \propto T^{-4}$$

$$\xi \propto T^{-2} \tag{20}$$

Morgenstern and Binder applied the same methods in d = 3 [20] but could only handle much smaller sizes L x L x M where L ≤ 4 and M ≤ 10, which is probably too small to make firm predictions. As with the high temperature series expansions the important case of d = 3 remains elusive. There is also an analytic approach [21] based on the dynamical mean field of ref. 12 which predicts $d_L = 4$. This may well be right but it is a pity that the numerical results have so far been unable to confirm it convincingly.

If $d_L = 4$ the experiments on spin-glasses must be explained by 'gradual freezing'. However there are a number of problems with this idea:

(i) There is a dramatic increase [22,23] and apparent divergence in the non-linear susceptibility, $\chi_{n\ell}$, in the same temperature region as the cusp in the linear susceptibility. $\chi_{n\ell}$ is essentially χ_{SG} and so should diverge if there is a spin glass transition but remain finite in the gradual freezing picture.

(ii) There is evidence [24,25] for a transition line in the h-T plane similar to the AT [6] line for the SK model. Even the power law variation $T_c(h) - T_c(0) \propto h^{2/3}$, for small h, predicted by AT seems to be found experimentally.

(iii) For certain materials at least, the temperature of the a.c. susceptibility peak varies little with frequency [26]. A much larger logarithmic variation is predicted by the gradual freezing approach.

It appears that either the prediction that $d_L = 4$ is wrong, or correlation lengths and relaxation times, while not strictly diverging, increase considerably over a narrow temperature range due to <u>cooperative</u> effects between the spins. In order to test out these two alternatives one needs to know quantitatively the range of correlations in space and time as functions of h and T. While Morgenstern and Binder [19,20] have discussed the spatial extent of correlations virtually no precise results on relaxation times have been given, and very little is known about the effect of a magnetic field. It was therefore felt useful to investigate these questions in detail for a two dimensional model [27]. The results are consistent with Morgenstern and Binder's claim of a transition only at T = 0 and, we shall see, do not explain quantitatively most of the experimental results referred to above. Some preliminary data for d = 3 will also be presented which show that relaxation times and χ_{SG} increase somewhat more rapidly than in d = 2 but one cannot really predict whether or not a transition occurs at low temperatures because relaxation times are too long in this region to equilibrate the sample.

The Hamiltonian is the same as eq. (1) but now the J_{ij} are for nearest neighbours only on a square or a simple cubic lattice and have a '±J' distribution,

$$P(J_{ij}) = \frac{1}{2}\left[\delta(J_{ij} - 1) + \delta(J_{ij} + 1)\right] \qquad (21)$$

where J has been set equal to unity. As for the infinite range model, information on dynamics is conveniently extracted from the auto-correlation function q(t), defined by eq. (11). For $t \to \infty$, $q(t) \to q$ which is non zero only if $h \neq 0$. The functional form of q(t) varies with T, and a single parameter, τ, which gives a measure of the relaxation times, is defined by

$$\tau = (1-q)^{-1} \int_0^\infty [q(t) - q]dt \qquad (22)$$

τ can be thought of as an average relaxation time [27]. Since we shall see that τ increases exponentially as T decreases one presumes that relaxation occurs by activation over barrier heights and it is useful to define a characteristic barrier height ΔE by

$$\Delta E = T \ln\tau \qquad (23)$$

Static information is derived from χ_{SG} and ξ defined by eqs. (17-19).

Results for $\chi_{SG}^{\frac{1}{2}}$ and ξ with h = 0 are shown in fig. 6. for square lattices of size L x L with L = 64 and 128. The points are averages over 20 samples for L = 128 and 60 samples for L = 64. No dependence on system size was found, as expected because $\xi << L$. It was not possible to equilibrate the system at lower temperatures because relaxation times are too long. The data is certainly consistent with Morgenstern and Binder's predictions given in eq. (20). Assuming a T^{-4} variation for χ_{SG} below T = 1 then at T = 0.58, the significance of which is explained below, one has $\chi_{SG} \simeq 300$ which can be compared with the experimental data of [22] who find $\chi_{SG} \sim 500$ at $T = 12.5^{\circ}$ K and $\chi_{SG} \sim 3500$ at $T = 11.15^{\circ}$K. Note that the increase observed experimentally is much faster than T^{-4} and is consistent with a finite temperature transition, whereas the simulation data is not.

Next the data for the characteristic energy barrier ΔE is plotted in fig. 7 against T^{-1} for h = 0. The straight line fit is excellent and indicates that the relevant barrier heights diverge like T^{-1} as $T \to 0$. If we want to transform Monte Carlo steps per spin into real time we have to multiply by some microscopic timescale τ_{micro}.

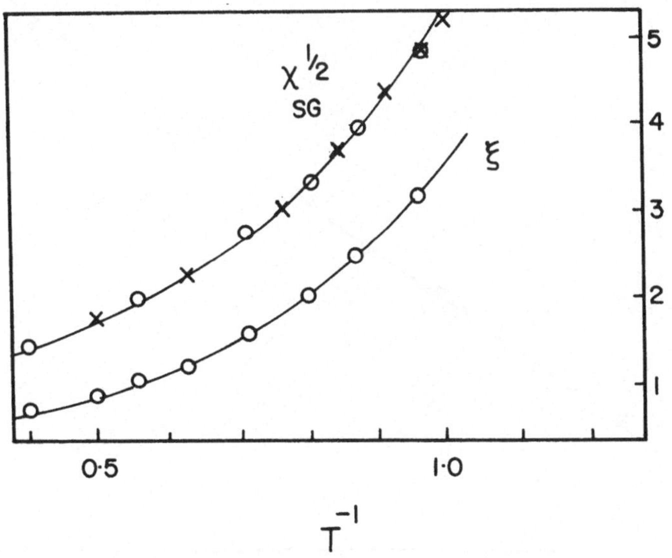

Fig. 6. Results for $\chi_{SG}^{\frac{1}{2}}$ and ξ from Monte Carlo simulations for a nearest neighbour Ising spin glass with $\pm J$ interactions on a square lattice of size L x L where L = 64 (circles) and 128 (crosses).

It is not clear exactly what this should be but for a typical CuMn sample we will guess 10^{-11} seconds. What follows will not be sensitive to this precise value. If ν is the frequency of an a.c. susceptibility measurement and we equate $(\nu\,\tau_{micro})^{-1}$ at the freezing temperature with our value for τ in the simulation then the freezing temperature for $\nu = 10\ H_z$ say corresponds in the simulation to the temperature where $\tau = 10^{10}$. Extrapolating the data in fig. 7 gives T = 0.58. At this temperature $|d\,\ell nT/d\ell n\tau| \simeq 1/50$ smaller than the result of a simple Arrhenius law where $|d\ell nT/d\ell n\tau| = 1/\ell n\tau \simeq 1/23$. This shows that cooperative effects, which give a temperature dependent ΔE, make the freezing temperature vary rather more slowly with ν, than does the Arrhenius law. It is however still much larger than experiments of Mulder et al. [26] on several RKKY systems where $d\ell n\, T_f/d\ell n\, \nu \simeq 1/500$. Once again the two-dimensional model is unable to explain experiments in any sort of quantitative way.

Including a magnetic field one can draw lines in the h-T plane of constant τ and these are shown in fig. 8. Rather surprisingly the data at low temperature satisfies $\delta T \propto h^{2/3}$, the same power as the AT line. Kinzel and Binder [4,28] have obtained

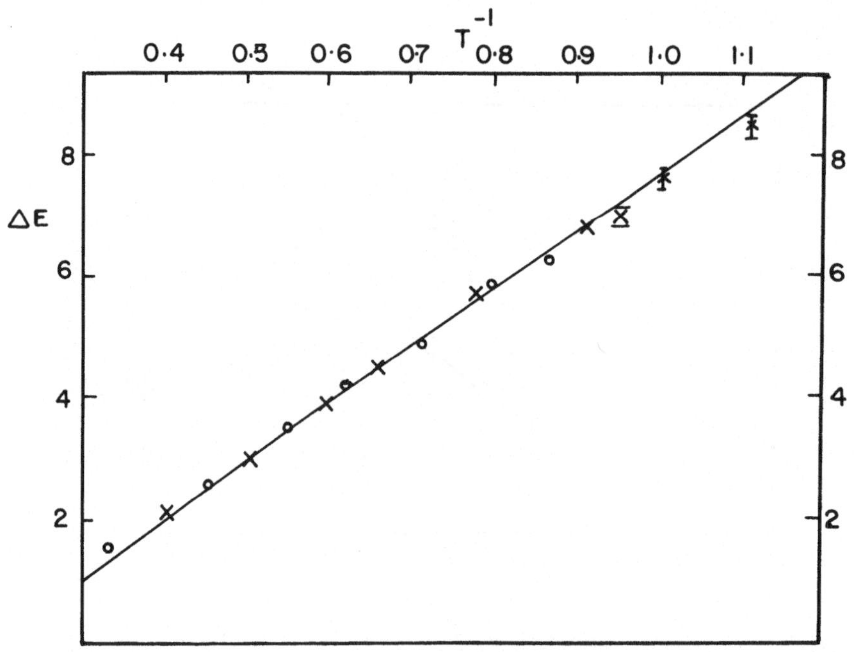

Fig. 7. The characteristic barrier height ΔE for a 2-dimensional Ising spin glass plotted against T^{-1}.

similar results. However, this is not an AT line since there is no transition at finite temperature and what is observed is a purely dynamical effect. Interestingly, just this sort of effect has been seen in the insulating spin glass $Eu_{0.4}Sr_{0.6}S$ [25]. This raises the possibility that insulating systems might be describable by gradual freezing whereas experiments on metallic systems such as CuMn definitely seem to indicate a transition. We shall refer to this again at the end.

It is possible that the discrepancies noted above between experiments on RKKY systems such as CuMn and the d = 2 simulations may be due to the different space dimensions. To look into this question Monte-Carlo simulations have been started for a 64^3 simple cubic lattice. Fig. 9 shows results for χ_{SG}^{-1} against T down to T = 1.55, the lowest temperature at which the system can be equilibrated within available computer time. Also shown is an analysis of the Fisch-Harris [16] high temperature series by

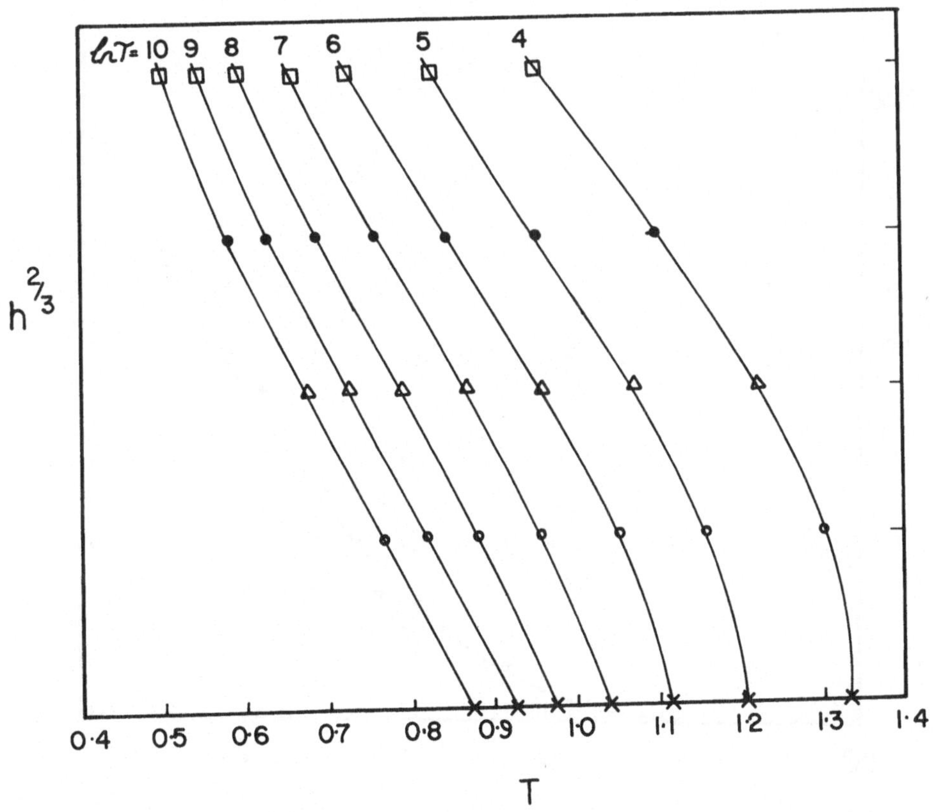

Fig. 8. Lines of constant τ are plotted in the $h^{2/3}$-T plane showing a series of dynamical 'Almeida-Thouless like' lines for different timescales. This is for a 2-dimensional Ising model.

R.G. Palmer [17]. Actually two different methods of analysis are shown, indistinguishable within the accuracy of the figure, one of which has no transition, the other of which has a transition at T = 0.89. These analyses agree well with the Monte-Carlo results and neither can really tell whether or not there is a transition at low temperature. The correlation length is just over 3 lattice spacings at T = 1.55. This is very much smaller than the size of the system so it is not anticipated that finite size effects are important, though this has not been checked explicitly by looking at other sizes.

Results for ΔE^{-1} are shown in fig. 10. Over the temperature range studied the data fits a Vogel-Fulcher law, i.e.

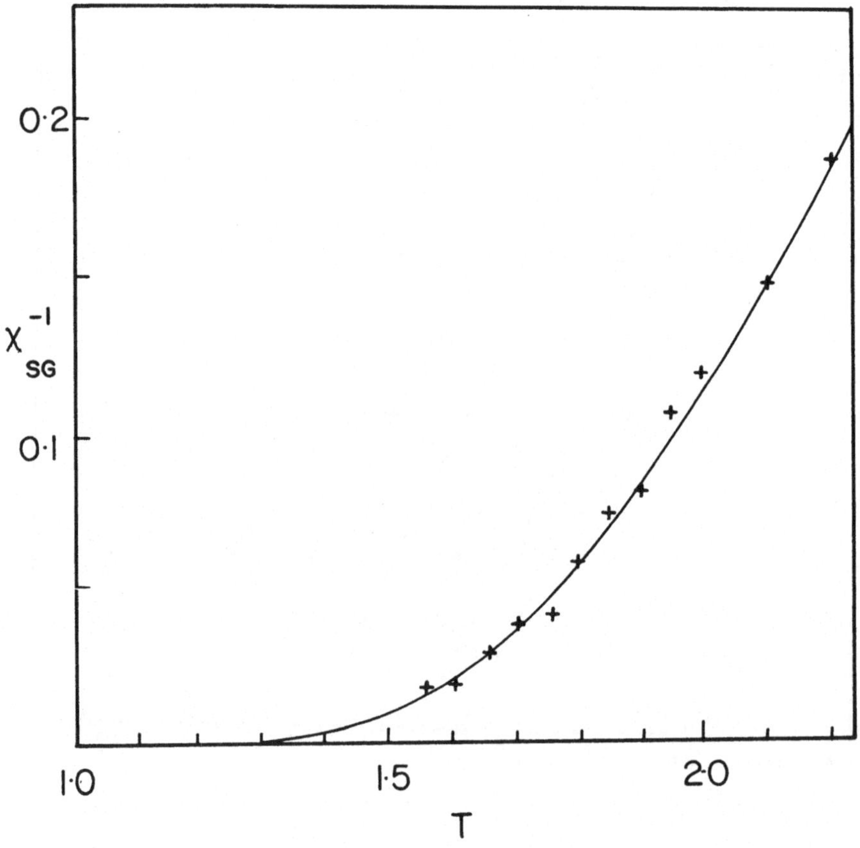

Fig. 9. χ_{SG}^{-1} against T for a 3-dimensional nearest neighbour Ising spin glass model with ±J interactions. The solid curve is an analysis of the high temperature series from ref. [17]; see text.

$$\Delta E \propto (T - T_o)^{-1} \tag{24}$$

with $T_o \simeq 1.1$. Since T_o is well below the lowest temperature studied we cannot say that a divergence occurs there. However, the results cannot be well fitted by the variation observed in two dimensions, where

$$\Delta E = a + b/T, \tag{25}$$

see fig. 7. In fact, the results for d = 3 can be fitted if one adds a T^{-2} term to eq. (25). The two dimensional results, by contrast, fit eq. (25) better than a Vogel-Fulcher law.

We have seen that computer simulations are completely consistent with there being no transition at finite temperature in d = 2, but with a power law divergence of χ_{SG}, ξ and ΔE as $T \to 0$. Unfortunately, the simulations do not provide clear evidence as

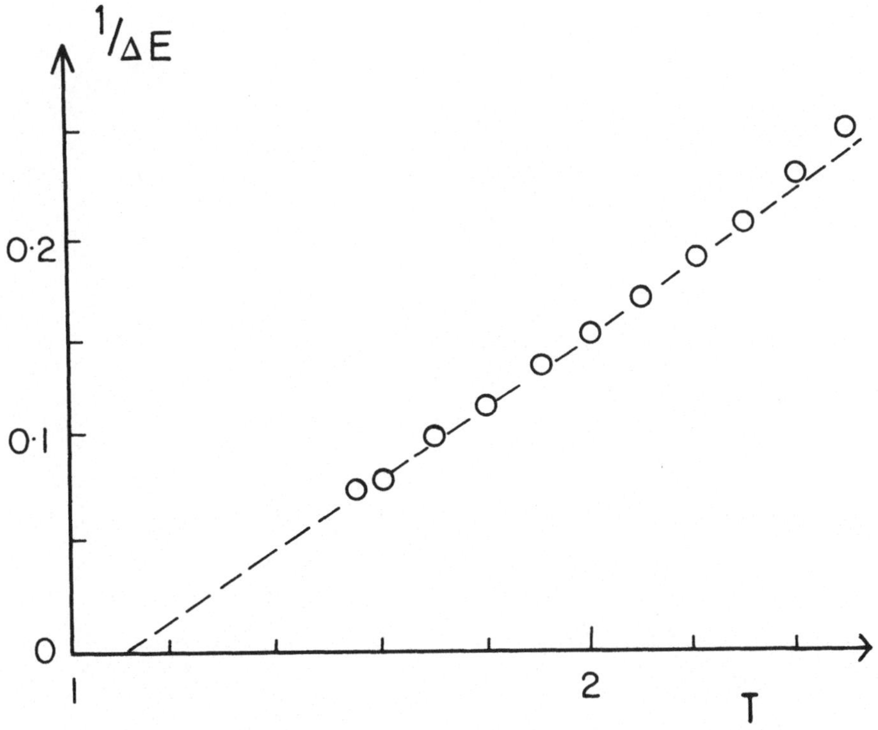

Fig. 10. ΔE^{-1} against T for a 3-dimensional Ising spin glass. Over this range of temperature the data fits a Vogel-Fulcher law.

to whether or not a transition occurs in $d = 3$.

Experiments on the insulating spin glass $Eu_{0.4}\ SR_{0.6}\ S$ [25] which show a series of dynamical 'AT lines' are consistent with gradual freezing. Unfortunately there do not yet appear to have been careful measurements of $\chi_{n\ell}$, which would provide the clearest evidence for a spin glass transition. It is therefore possible that insulators, which have short range interactions apart from dipole-dipole forces, exhibit gradual freezing. I would encourage experimentalists to carry our careful measurements of $\chi_{n\ell}$ on insulating materials to test this hypothesis.

Metallic systems, on the other hand, have RKKY interactions and Dzyaloshinski-Moriya [29] couplings. The folklore is that these are really short range, though this has not been demonstrated completely convincingly in my opinion. RKKY systems such as CuMn really seem to exhibit a transition at finite temperature as shown most clearly by the non-linear susceptibility [22,23]. This raises the possibility that metallic and insulating spin glasses might lie in different universality classes, even perhaps with different values of d_L.

REFERENCES

1. K. Binder and D. Stauffer: In *Monte Carlo Methods*, ed. by K. Binder, Topics in Curr.Phys., Vol. 7 (Springer, Berlin, Heidelberg, New York 1979) p. 301; also in *Monte Carlo Methods* II (to appear).
2. D. Sherrington and S. Kirkpatrick, *Phys.Rev.Lett.* 35, 1792 (1975), referred to as SK.
3. D. Sherrington, in these proceedings.
4. K. Binder, in these proceedings.
5. R. Walstedt, in these proceedings.
6. J.R.L.de Almeida and D.J. Thouless, *J.Phys.* A 11, 983 (1978), referred to as AT.
7. G. Parisi, *J.Phys.* A 13, 101 (1980); ibid *J.Phys.* A 13, L115 (1980); ibid *Phys. Rev.Lett.* 43, 1574 (1979).
8. C.de Dominicis and A.P. Young, *J.Phys.* A 16, 2063 (1983).
9. G. Parisi, *Phys.Rev.Lett.* 50, 1946 (1983); A. Houghton, S. Jain and A.P. Young, *J.Phys.* C 16, L375 (1983).
10. D.J. Thouless, P.W. Anderson and R.G. Palmer, *Phil.Mag.* 35, 593 (1977) referred to as TAP.
11. A.J. Bray and M.A. Moore, *J.Phys.* C 13, L469 (1980); C.de Dominicis, M. Gabay, T. Garel and P. Orland, *J.de Physique* 41, 923 (1980); F. Tanaka and S.F. Edwards, *J.Phys.* F 10, 2769 (1980).
12. H. Sompolinsky, *Phys.Rev.Lett.* 47, 935 (1981) and these proceedings.
13. N.D. Mackenzie and A.P. Young, *Phys.Rev.Lett.* 49, 301 (1982) and to be published.
14. A.P. Young, unpublished.
15. C.de Dominicis and I. Kondor, *Phys.Rev.* B 27, 606 (1983).
16. R. Fisch and A.B. Harris, *Phys.Rev.Lett.* 38, 785 (1977).
17. R.G. Palmer, unpublished.
18. K. Binder and D. Stauffer, *Z.Phys.* B 26, 339 (1977); S. Kirkpatrick, *Phys.Rev.* B 16, 4630 (1977); A.J. Bray, M.A. Moore and P. Reed, *J.Phys.* C 11, 1187 (1978).
19. I. Morgenstern and K. Binder, *Phys.Rev.Lett.* 43, 1615 (1979); ibid *Phys.Rev.* B 22, 288 (1980).
20. I. Morgenstern and K. Binder, *Z.Phys.* B 39, 227 (1980).
21. A. Zippelius, these proceedings.
22. R. Omari, J.J. Préjean and J. Souletie, to be published. J. Souletie, these proceedings.
23. P. Monod and H. Bouchiat, *J.Phys.Lett.* 43, 45 (1982); B. Barbara, A.P. Malozemoff and Y. Imry, *Phys.Rev.Lett.* 47, 1852 (1981).
24. R.V. Chamberlin, M. Hardiman, L.A. Turkevich and R. Orbach, *Phys.Rev.* B 25, 6720 (1982); Y. Yeshurun, L.J.P. Kelesen and M.B. Salamon, *Phys.Rev.* B 26, 1491 (1982).
25. N. Bontemps, J. Rajchenbach and R. Orbach, *J.de Physique Lett.* 44, L47 (1983); J.A. Hamida, C. Paulsen, S.J. Williamson and H. Maletta, preprint; H. Maletta, these proceedings.
26. E.D. Dahlberg, M. Hardiman, R. Orbach and J. Souletie, *Phys.Rev.Lett.* 42, 401 (1979); C.A.M. Mulder, A.J.van Duyneveldt and J.A. Mydosh, *Phys.Rev.* B 25, 515 (1982).
27. A.P. Young, *Phys.Rev.Lett.* 50, 917 (1983); and to be published.
28. W. Kinzel and K. Binder, *Phys.Rev.Lett.* 50, 1509 (1983).
29. A. Fert and P.M. Levy, *Phys.Rev.Lett.* 44, 1438 (1980).

CONCLUSION

THE SPIN GLASS: STILL A PROBLEM?

by David Sherrington
Physics Dept., Imperial College, London SW7 2BZ, UK.

The simple question of my title is not amenable to a single monosyllabic answer. It can be viewed at various levels, qualitative or quantitative, of essentials or also of details, of idealized models or of reality. The answer depends on the interpretation of the question. I believe, however, that it is fair to say that progress in our understanding over the last few years has been considerable and I shall try to review, from a personal viewpoint, what appear to be the points of basic agreement and comprehension and also some of the puzzles remaining. Many of these aspects have been discussed in detail in the specialized talks of this meeting and the reader is referred to them as necessary for supplementation of my comments.

The number of papers on spin glasses published during the last decade is well in excess of 1000 and no attempt will be made to give complete references. I apologize at the outset to those who have contributed to our understanding but who are not explicitly referenced. More complete attributions can be found in the recent reviews of Fischer [1] and of Rammal and Souletie [2].

For several years there was considerable disagreement over the features needed to warrant the name "spin glass", as also over the appropriateness of the expression, compared with, for example, "mictomagnet". Some imprecision remains and a concise universally acceptable rigorous definition is still difficult to devise, but a general qualitative agreement appears to have emerged, with contention mainly only over details. Let me first give a restricted "definition" which I shall subsequently modify in details. That is that a spin glass is a state with

(i) thermodynamically frozen magnetic moments,

(ii) no average periodic long-range magnetic order, and

(iii) experimentally, severe history-dependence, remanence and slow relaxation after magnetic perturbation; theoretically, many (meta) stable states of number $N_m(N)$ increasing monotonically with the number of magnetic atoms or spins, N.

Condition (i) is sometimes relaxed to replace "thermodynamically frozen" by "apparently frozen over long time-scales", but strictly the former is required for a true ordered phase. Furthermore, we wish to exclude any systems with hidden but deterministic ordering, such as Mattis models [3]. Formally, this excludes systems for which local gauge transformations can make the Hamiltonian unfrustrated or periodically frustrated (see below), but operationally it can be viewed as the requirement that the actual order be essentially irreproducible (ie. with measure zero in the thermodynamic limit) in heating and cooling experiments. Additionally, the definition can be generalized to apply after the subtraction of any average periodic order, a procedure necessary to deal with a system in which an applied field induces overall magnetization or where there is coexisting ferromagnetism or other periodic order. Finally, there is imprecision over the meaning of metastability as used in (iii) above. Most theoretical studies examine stability only with respect to single spin flips, but more generally the requirement is expected to apply to stability with respect to any number of spin flips which does not increase as the system size N is increased towards the thermodynamic limit $N \rightarrow \infty$, i.e. intensive flips.

All the above properties are necessary. On the other hand, it is not relevant to the identification of a system as a spin glass whether it is metallic or non-metallic, whether it contains dipolar or quadrupolar moments, magnetic or electric, whether it obeys concentration-scaling laws or not, or the temperature for which the specific heat is a maximum. It should, however, be noted at this stage that theoretical arguments suggest that the lower critical dimension for true thermodynamic spin-glass order in systems with short-range exchange interactions is probably four (although three remains a possibility) so that many experimental "spin glasses" may warrant only the relaxed version of condition (i) above.

What are the ingredients a system needs to be a potential spin glass? These appear to be quenched spatial disorder (of the Hamiltonian) and frustration [4]. The quenched disorder can be associated with lattice sites, involving different spins or magnetic vacancies, different anisotropies or, perhaps, local fields; or it can be associated with the exchange interactions between the spins (bond disorder); or it can be topological as on an amorphous structure. Frustration is the inability of a system to satisfy all its ordering instructions and in a magnetic context it may occur due to mixed ferromagnetic and antiferromagnetic interactions, or to competing exchange and anisotropies or fields, or due to purely antiferromagnetic interactions in systems with odd-numbered exchange-bond rings. We know examples where disorder alone can give randomly frozen moments with no apparent long-range order, and also others where periodic frustration gives rise to metastable states but it appears that <u>both disorder and frustration are necessary ingredients for a true spin glass</u>, although they are not sufficient.

I have already indicated that in a general qualitative sense there is a "universality" among spin glasses - I shall not discuss the critical phenomenal interpretation of this term. There are, however, several secondary differences. Let us consider a subset: (i) Ising or vector spins. In the absence of a magnetic field these are expected to be qualitatively similar but it has been predicted [5], and probably observed [6], that a vector system in a field will exhibit a phase with spin-glass order transverse to the field, reminiscent of a spin-flop phase of a pure antiferromagnet. In Fig. 1 are shown qualitatively the predicted mean field phase diagrams in the presence of ferromagnetic ordering forces.

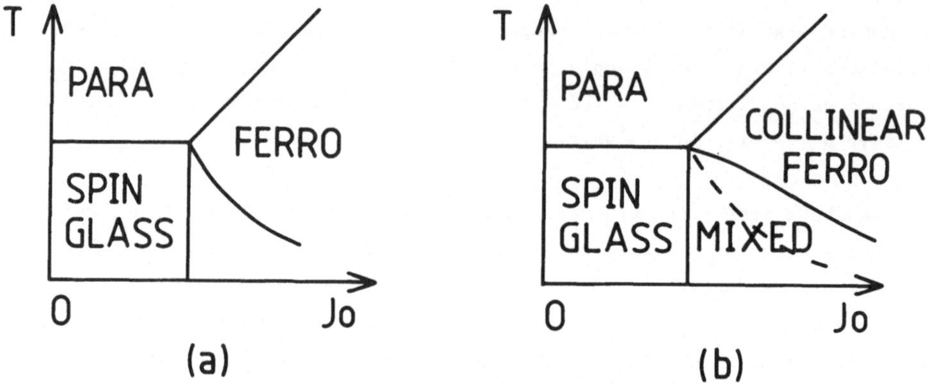

Fig. 1. Mean-field phase diagrams for (a) Ising, (b) classical m>0 vector Edwards-Anderson models. Full lines denote phase boundaries, the dashed line indicating a crossover. Hatching indicates regions exhibiting (quasi) non-ergodicity (condition (iii)). MIXED refers to phase with both ferromagnetic and transverse spin glass order.

(ii) Dipolar as compared with quadrupolar or Potts spins. Quadratic exchange between conventional (dipolar) spins leads to a symmetry between ferromagnetic (parallel) and antiferromagnetic (anti-parallel) ordering which is absent in many more complicated systems; for example quadrupolar or Potts spins are precisely ordered by ferromagnetic interactions but antiferromagnetic interactions merely favour orthogonality, a less precise state than the antiparallel ordering of the dipolar example. This leads to several unusual properties compared with dipolar spins [7]. Bi-quadratic exchange is asymmetric even for dipolar vector spins. (iii) Competing ferro-antiferromagnetic exchange as compared with purely antiferromagnetic exchange and topological disorder. Although in the former case many practical spin glass examples are known, in the latter the situation is less clear - certainly antiferromagnetism is weakened by disordering an unfrustrated lattice and there is some indication of the onset of history dependence in amorphous FeF_3 at a temperature much lower than the Néel temperature of the crystalline form [8] but, by contrast, no evidence of order has been seen to the lowest temperatures studied in homogeneously random antiferromagnets such as In in CdS [9]. (iv) Short-range interaction (nearest and next nearest neighbour) systems, such as semiconducting $Eu_xSr_{1-x}S$, as compared with metallic systems, such as AuFe, having longer range RKKY interactions. Experimentally the metallic systems seem to exhibit greater similarity to mean field spin glass predictions than do the semiconductors.

For theoretical considerations it is useful to make two separate categorizations, first into mean field theory and beyond, secondly into phase transition-like and low-temperature aspects.

Let us start with mean field theory. At a qualitative, handwaving level [10] it is easy to see the possibility of a non-periodically ordered phase. A complete mean field theory, as characterized by an appropriate infinite-range model [11] is much more complicated than for conventional pure problems, but equally has proven very instructive. It has taken eight years to achieve an essentially complete understanding of the thermodynamics of this model but now that appears to be the case. I shall describe briefly some important consequences of the model and novel features of its analysis.

The model is the infinite-range version of the classic Edwards-Anderson model [12].

$$H=-\sum_{(ij)} J_{ij}\, \underline{S}_i \cdot \underline{S}_j - \sum_i \underline{H} \cdot \underline{S}_i \quad ; i, j, = 1, \ldots N, \lim_{N \to \infty} \tag{1}$$

where the J_{ij} are distributed randomly with mean J_o/N, variance J^2/N. The important consequences are (i) a phase transition to a spin glass state exists in the thermodynamic limit, (ii) metastable states are important - for a system of N spins the number of states metastable against single spin flips scales as $N_s(N) \sim \exp(\alpha(T)N)$ where $\alpha(T)$ is non-zero for $T < T_g(H)$ [13], (iii) in dynamics, Langevin or Glauber, there enter many lifetimes $\tau_i(N)$ which increase exponentially with N, typically as [14, 15]

$$\ln \tau_i(N) \sim E_i/J \sim \lambda_i N^a \quad ; 0 < a < 1 \tag{2}$$

where the E_i can be identified as energy barriers between metastable states. From the monotonic increase of the lifetimes with N it follows that in the thermodynamic limit the model is non-ergodic; the metastable states become stable in this limit.

There are two complementary mean field approaches to the SK model of eqn.(1). The first, known popularly as TAP [16], looks at the self-consistent mean field solutions without recourse to prior averaging. It is very instructive conceptually in that it demonstrates clearly the relevance of the (meta)stable states and their multiplicity. The second, known as the replica method [12], uses a trick to replicate the system n times, average over J_{ij}, evaluate the resultant effective system in the large N limit, and let $n \to 0$. Interchange of limits $N \to \infty$ and $n \to 0$ is necessary for progress but appears to be without serious consequence. The relevant order parameter for spin glass behaviour is $q^{\alpha\beta} = \langle S_i^\alpha S_i^\beta \rangle_n$; $\alpha \neq \beta$, where α, β are replica labels while i labels a site and $\langle\ \rangle_n$ a thermodynamic average over the n-replicated system. In the simplest approximation one makes the "obvious" ansatz that $q^{\alpha\beta}$ is independent of α, β and chooses it by requiring that the free energy be extremal (in fact maximal since the number of ($\alpha\beta$) combinations becomes negative in the limit $n \to 0$). In view of our present knowledge of metastable state multiplicity it is not surprising that this ansatz is inadequate, as inter-replica fluctuation analysis shows explicitly [17]. Much effort has gone into attempts to overcome these instabilities and a remarkable initially-mysterious suggestion for replica-symmetry breaking was proposed by Parisi [18] and has survived all later stability tests [19]. In retrospect it seems to me that a scheme such as Parisi's is inevitable. The reason is that it is clear from the state structure, as well as from explicit analysis, that one needs to be able to break replica-symmetry continually and hence requires a procedure which can be repeated ad infinitum. Thus one is driven necessarily to consider fractal decomposition. The simplest possibility is indicated schematically in Fig. 2. It corresponds essentially to Parisi's scheme

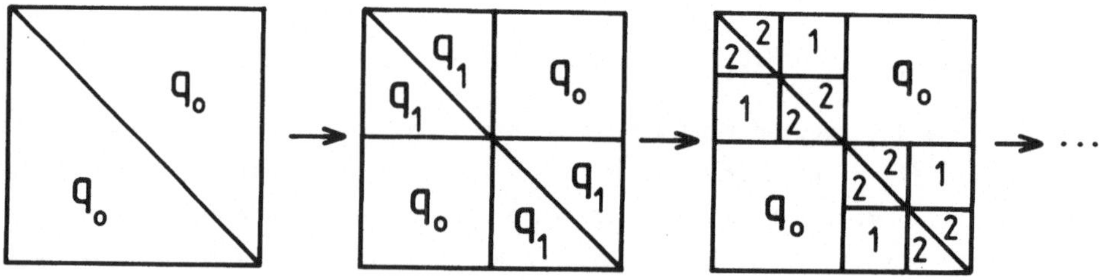

Fig. 2. Schematic indication of Parisi scheme. The block represents the $q^{\alpha\beta}$ matrix and the elements the stages of fractal decomposition. The end-stage provides a continuous function q (x) after an appropriate limit is taken.

and leads to an order function $q(x)$, where conventionally x is restricted to the interval (0, 1). A more complicated fractal decomposition starting from the Sommers ansatz [20] has been shown [21] to lead to the thermodynamic equation of Sompolinsky [14] which involves two non-independent functions $q(x)$, $\Delta(x)$. It is physically satisfying that both formulations yield the same thermodynamics; both for Gibbs averages, which involve integrals over the whole x interval, and for averages restricted to single free energy valleys, which are given by $x = 1$. On the other hand, each of the two formulations has specific advantages - Parisi's for direct relation to the TAP state analysis, Sompolinsky's for mathematically convenient closed-form expression, as well as interpretation of $\Delta(x)$ as anomalous response. In fact, until very recently [22, 23] , the physical significance of $q(x)$ was obscure but now it is realised that the inverse $x(q)$ is the integrated probability overlap of the various

TAP solutions - the overlap of TAP solutions S and S' is defined as $N^{-1} \sum_i M_i^{S} M_i^{S'}$ and x(q) is the fraction of states with overlap less than q; a related but more complicated interpretation is available for the Sompolinsky formulation. An alternative but related qualitative interpretation due to Sompolinsky is that x labels timescales with $\tau(x) \gg \tau(x + \delta x)$; $\delta x > 0$, N large. Of course this interpretation is strictly valid only for N finite, since as $N \to \infty$ the barriers become infinite.

The infinite-range model has been the subject of several dynamical studies. For $T > T_g$ it is relatively straightforward with the local magnetization decay distinguished from a conventional system principally by the much greater temperature-range over which slowing-down occurs. For $T < T_g$ power law decay is predicted [24] in place of the conventional exponential, again a reflection of the continuation to T=0 of massless modes and transitions. Dynamics has been most frequently considered in a Langevin context and analysed using a functional integral version of the Martin-Siggia-Rose formulation [25]. Initially this was proposed as an alternative to replicas [26] and was used to derive a convenient closed form expression for the free energy [14]. However, the non-ergodic nature of the model for $N \to \infty$ appears to require the reintroduction of replicas [23] to obtain the Gibbs limit.

The situation for finite-range systems is much more of a problem and many uncertainties remain. Whenever one considers real systems one is interested in two special space dimensions, the upper critical dimension above which the critical exponents are those of mean field theory and the lower critical dimension beneath which no transition occurs. For real spin glasses the upper critical dimension is generally believed, on the basis of dimensional analysis within a renormalization group formulation, to be six [27], although it has recently been proposed [28] that the onset temperature in a magnetic field will be modified for $d < 8$. Due to the absence of adequate knowledge of the ground state(s) and low-lying excitations a rigorous determination of the lower critical dimension is not possible and one is driven to rely on computer studies, of numerically exact correlation ranges [29] and of Monte Carlo simulations [30] , and high temperature series analyses [31]. These demonstrate fairly convincingly that the lower critical dimension is greater than two but are insufficiently accurate to rule unequivocally on dimension three. However, the generally favoured lower critical dimension estimate/guess is four. A second conventional analysis of normal phase transition systems is to use renormalization-group theory to study the modification of critical exponents just below the upper critical dimension. In the absence of a field or nearby transition to spontaneous magnetization this has been performed on the paramagnetic side of the transition without obvious problems [27], but when $H \neq 0$ [32] or near the paramagnet-ferromagnet-spin glass multicritical point [33] no physically sensible solution has been found.

On the other hand, it appears that, irrespective of the true lower critical dimension, systems with sufficient and appropriate frustration and disorder can have many metastable states at low temperatures. In consequence one can reasonably expect that over short times at low temperatures such systems will behave qualitatively analagously to spin-glasses above their lower critical dimension. Thus one expects the observed slow relaxation and response behaviour, spin-wave normal modes [34] should be quite well-defined although the absence of translational invariance may explain the difficulty of their observation in neutron scattering experiments, and in metallic spin glasses containing impurities with strong spin-orbit coupling Dzyaloshinskii-Moriya interactions will lead to significant anisotropy [35] , with interesting consequences for magnetic resonance [36, 37] and perhaps also for tipping the balance towards the spin-glass ordering in three dimensions [38].

Experimentally, the existence or absence of a phase transition in three dimensions in real systems is less clear. Certainly there are many examples of quite sharp changes in slope of zero field susceptibilities, of very stable plateaux in field-cooled susceptibilities, of apparent divergence at $T=T_g$ of non-linear paramagnetic susceptibility, and of Almeida-Thouless like scaling with field of the reduction in the temperature at which slow response onsets. Many of these effects are, however, believed to be non-equilibrium, determined by the restricted ability of a system to explore in a finite time all the phase space available to it when metastable states are important - in a time t effectively only barriers of height less than order $kT \ln (t/\tau_0)$, where τ_0 is an attempt time, will be surmounted. Experimental evidence for such a hypothesis is provided by the existence of an approximate scaling "law" [39]

$$\partial\chi(\omega) / \partial \ln (1/\omega) = \text{constant}$$

and by the fact that, although the field-induced reduction in the "irreversibility"-onset temperature scales in a manner reminiscent of mean field theory, <u>all</u> the onset temperatures decrease with the measuring time-scale used to characterize reversibility or its absence [40]; it is particularly of note in the latter connection that analogous behaviour is found in computer simulations of the two-dimensional Ising Edwards-Anderson model for which it is known there is no phase transition [30, 41]. Thus, one can probably "explain" the observation of spin-glass like features beneath some characteristic temperature $T_g(t)$ even for dimension less than the lower critical dimension. On the other hand, it is harder to explain the apparent divergence of the <u>paramagnetic</u> non-linear susceptibility. It has been pointed out that some computer-generated non-linear susceptibilities can be fitted to a T^{-b} law as well as to $(T-T_g)^{-c}$ [42] but there exist real experiments [43] showing a much stronger divergence at T_g than could be fitted by T^{-b} with a sensible value of b, so a puzzle still remains as to how such strong paramagnetic indicators of a spin-glass transition can occur in real three-dimensional systems. It may be relevant that the systems in which they have been observed most strongly are metallic.

With respect to the experimental analogues of the phase diagram of Fig. 1 there is still uncertainty about which, if any, of the phase lines exist in a strict thermodynamic sense, about whether the line drawn separating spin-glass and ferromagnetic phases is re-entrant, even if the phases are defined only loosely, and whether there is a fundamental difference between semiconductors with short-range exchange and metals with longer-range RKKY interactions.

Thus, it appears that while the infinite-range spin-glasses are reasonably well-understood several open questions remain for real and short-range systems, although our general conceptualization has advanced there too, with Gibbs thermodynamics recognized as of far less practical relevance in frustrated and disordered systems than in conventional unfrustrated pure ones.

There have also been several spin-offs of note, such as the development of simulated annealling [44] as a technique for optimization of cost functions for design problems exhibiting frustration, albeit of a more complex nature than that in a simple spin glass, and of a spin-glass like model of the brain [45] where the multiple and quasi-fractal nature of the metastable state structure provides a tunable but non-local memory and the modification of that structure by variation of the bond strengths provides the basis of a mechanism for learning. It seems reasonable to assume that the modification to our thinking which the spin-glass problem has engendered will have further ramifications beyond its immediate borders.

REFERENCES

1. K.H. Fischer; Phys. Stat. Sol. (b) 116, 357 (1983)
2. R. Rammal and J. Souletie; in "Magnetism of Metals and Alloys" (ed. M. Cyrot); North Holland (1982)
3. D.C. Mattis; Phys. Lett. A56, 421 (1976)
   R. Medina, J.F. Fernandez and D. Sherrington; Phys. Rev. B21, 2915 (1980)
4. G. Toulouse; Commun. Phys. 2, 115 (1977)
5. M. Gabay and G. Toulouse; Phys. Rev. Lett. 47, 201 (1981)
   See also D.M. Cragg, D. Sherrington and M. Gabay, Phys. Rev. Lett. 49, 158 (1982)
6. J. Lauer and W. Keune; Phys. Rev. Lett. 48, 1850 (1982)
7. D. Elderfield and D. Sherrington; J. Phys. C16, L497 (1983)
8. G. Ferey, F. Varret and J.M.D. Coey; J. Phys. C12, L531 (1979)
9. R.E. Walstedt, R.B. Kummer, S. Geschwind, V. Narayanamurti,and G.E. Devlin; J. Appl. Phys. 50, 1700 (1979)
10. D. Sherrington; AIP Conf. Proc. 29, 224 (1975)
11. D. Sherrington and S. Kirkpatrick; Phys. Rev. Lett. 35, 1792 (1975)
12. S.F. Edwards and P.W. Anderson; J. Phys. F5, 965 (1975)
13. A.J. Bray and M.A. Moore; J. Phys. C13, L469 (1980)
    C. de Dominicis, M. Gabay, T. Garel and P. Orland; J. Physique 41, 923 (1980)
    F. Tanaka and S.F. Edwards; J. Phys. F10, 2769 (1980)
14. H. Sompolinsky; Phys. Rev. Lett. 47, 935 (1981)
15. N.D. Mackenzie and A.P. Young; Phys. Rev. Lett. 49, 301 (1982)
16. D.J. Thouless, P.W. Anderson and R.G. Palmer; Phil. Mag. 35, 593 (1977)
17. J.R. de Almeida and D.J. Thouless; J. Phys. A11, 983 (1978)
18. G. Parisi; Phys. Rev. Lett. 43, 1754 (1979)
19. C. de Dominicis and I. Kondor; Phys. Rev. B27, 606 (1983)
20. H.J. Sommers; Z. Phys. B31, 301 (1978)
    C. de Dominicis and T. Garel; J. Physique Lettr. 40, L574 (1979)
21. C. de Dominicis, M. Gabay and H. Orland; J. Physique Lettr. 42, L523 (1981)
22. G. Parisi; Phys. Rev. Lett. 50, 1946 (1983)
23. A. Houghton, S. Jain and A.P. Young; J. Phys. C16, L375 (1983)
24. H. Sompolinsky and A. Zippelius; Phys. Rev. Lett. 47, 359 (1981), Phys. Rev. B25, 6860 (1982)
25. P.C. Martin, E.D. Siggia and H.A. Rose; Phys. Rev. A8, 423 (1973)
    C. de Dominicis, J. Phys. C1, 247 (1976)
    H.K. Janssen, Z. Phys. 23, 377 (1976)
26. C. de Dominicis; Phys. Rev. B18, 4913 (1978)
27. A.B. Harris, T.C. Lubensky and J.H. Chen; Phys. Rev. Lett. 36, 415 (1976)
28. J.E. Green, M.A. Moore and A.J. Bray; preprint (1983)
29. I. Morgenstern and K. Binder; Phys. Rev. B22, 288 (1980)
30. A.P. Young; Phys. Rev. Lett. 50, 917 (1983), and unpublished.
31. R. Fisch and A.B. Harris; Phys. Rev. Lett. 38, 785 (1977)
    R.V. Ditzian and L.P. Kadanoff; Phys. Rev. B19, 4631 (1979)
    R.G. Palmer, unpublished
32. A.J. Bray and S.A. Roberts; J. Phys. C13, 5405 (1980)
33. J.H. Chen and T.C. Lubensky; Phys. Rev. B16, 2106 (1976)
34. L.R. Walker and R.E. Walstedt; Phys. Rev. Lett. 38, 514 (1977)
35. A. Fert and P.M. Levy; Phys. Rev. Lett. 44, 1438 (1980)
36. C.L. Henley, H. Sompolinsky and B.I. Halperin; Phys. Rev. B25, 5849 (1982)
37. E.M. Gullikson, D.R. Fredkin and S. Schultz; Phys. Rev. Lett. 50, 537 (1983)
38. R.E. Walstedt and L.R. Walker; Phys. Rev. Lett. 47, 1624 (1981)
39. L. Lundgren, P. Svedlinh and O. Beckman; J. Magn. Mag. Mater. 25, 33 (1981)
40. N. Bontemps, J. Rajchenbach and R. Orbach; J. Physique Lettr. 44, L47 (1983)
    J.A. Hamida, C. Paulsen, S.J. Williamson and H. Maletta; preprint (1983)
41. W. Kinzel and K. Binder, Phys. Rev. Lett. 50, 1509 (1983)
42. K. Binder and W. Kinzel; J. Phys. Soc. Japan Suppl. 52, 209 (1983)
43. R. Omari, J.J. Préjean and J. Souletie; J. Physique (to be published 1983)
44. S. Kirkpatrick, C.D. Gelatt Jr. and M.P. Vecchi; Science 220, 671 (1983)
45. J.J. Hopfield; Proc. Natl. Acad. Sci. USA 79, 2554 (1982)

Lecture Notes in Physics

Vol. 152: Physics of Narrow Gap Semiconductors. Proceedings, 1981. Edited by E. Gornik, H. Heinrich and L. Palmetshofer. XIII, 485 pages. 1982.

Vol. 153: Mathematical Problems in Theoretical Physics. Proceedings, 1981. Edited by R. Schrader, R. Seiler, and D.A. Uhlenbrock. XII, 429 pages. 1982.

Vol. 154: Macroscopic Properties of Disordered Media. Proceedings, 1981. Edited by R. Burridge, S. Childress, and G. Papanicolaou. VII, 307 pages. 1982.

Vol. 155: Quantum Optics. Proceedings, 1981. Edited by C.A. Engelbrecht. VIII, 329 pages. 1982.

Vol. 156: Resonances in Heavy Ion Reactions. Proceedings, 1981. Edited by K.A. Eberhard. XII, 448 pages. 1982.

Vol. 157: P. Niyogi, Integral Equation Method in Transonic Flow. XI, 189 pages. 1982.

Vol. 158: Dynamics of Nuclear Fission and Related Collective Phenomena. Proceedings, 1981. Edited by P. David, T. Mayer-Kuckuk, and A. van der Woude. X, 462 pages. 1982.

Vol. 159: E. Seiler, Gauge Theories as a Problem of Constructive Quantum Field Theory and Statistical Mechanics. V, 192 pages. 1982.

Vol. 160: Unified Theories of Elementary Particles. Critical Assessment and Prospects. Proceedings, 1981. Edited by P. Breitenlohner and H.P. Dürr. VI, 217 pages. 1982.

Vol. 161: Interacting Bosons in Nuclei. Proceedings, 1981. Edited by J.S. Dehesa, J.M.G. Gomez, and J. Ros. V, 209 pages. 1982.

Vol. 162: Relativistic Action at a Distance: Classical and Quantum Aspects. Proceedings, 1981. Edited by J. Llosa. X, 263 pages. 1982.

Vol. 163: J. S. Darrozes, C. Francois, Mécanique des Fluides Incompressibles. XIX, 459 pages. 1982.

Vol. 164: Stability of Thermodynamic Systems. Proceedings, 1981. Edited by J. Casas-Vázquez and G. Lebon. VII, 321 pages. 1982.

Vol. 165: N. Mukunda, H. van Dam, L.C. Biedenharn, Relativistic Models of Extended Hadrons Obeying a Mass-Spin Trajectory Constraint. Edited by A. Böhm and J.D. Dollard. VI, 163 pages. 1982.

Vol. 166: Computer Simulation of Solids. Edited by C.R.A. Catlow and W.C. Mackrodt. XII, 320 pages. 1982.

Vol. 167: G. Fieck, Symmetry of Polycentric Systems. VI, 137 pages, 1982.

Vol. 168: Heavy-Ion Collisions. Proceedings, 1982. Edited by G. Madurga and M. Lozano. VI, 429 pages. 1982.

Vol. 169: K. Sundermeyer, Constrained Dynamics. IV, 318 pages. 1982.

Vol. 170: Eighth International Conference on Numerical Methods in Fluid Dynamics. Proceedings, 1982. Edited by E. Krause. X, 569 pages. 1982.

Vol. 171: Time-Dependent Hartree-Fock and Beyond. Proceedings, 1982. Edited by K. Goeke and P.-G. Reinhard. VIII, 426 pages. 1982.

Vol. 172: Ionic Liquids, Molten Salts and Polyelectrolytes. Proceedings, 1982. Edited by K.-H. Bennemann, F. Brouers, and D. Quitmann. VII, 253 pages. 1982.

Vol. 173: Stochastic Processes in Quantum Theory and Statistical Physics. Proceedings, 1981. Edited by S. Albeverio, Ph. Combe, and M. Sirugue-Collin. VIII, 337 pages. 1982.

Vol. 174: A. Kadić, D.G.B. Edelen, A Gauge Theory of Dislocations and Disclinations. VII, 290 pages. 1983.

Vol. 175: Defect Complexes in Semiconductor Structures. Proceedings, 1982. Edited by J. Giber, F. Beleznay, J. C. Szép, and J. László. VI, 308 pages. 1983.

Vol. 176: Gauge Theory and Gravitation. Proceedings, 1982. Edited by K. Kikkawa, N. Nakanishi, and H. Nariai. X, 316 pages. 1983.

Vol. 177: Application of High Magnetic Fields in Semiconductor Physics. Proceedings, 1982. Edited by G. Landwehr. XII, 552 pages. 1983.

Vol. 178: Detectors in Heavy-Ion Reactions. Proceedings, 1982. Edited by W. von Oertzen. VIII, 258 pages. 1983.

Vol. 179: Dynamical Systems and Chaos. Proceedings, 1982. Edited by L. Garrido. XIV, 298 pages. 1983.

Vol. 180: Group Theoretical Methods in Physics. Proceedings, 1982. Edited by M. Serdaroğlu and E. İnönü. XI, 569 pages. 1983.

Vol. 181: Gauge Theories of the Eighties. Proceedings, 1982. Edited by R. Raitio and J. Lindfors. V, 644 pages. 1983.

Vol. 182: Laser Physics. Proceedings, 1983. Edited by J. D. Harvey and D. F. Walls. V, 263 pages. 1983.

Vol. 183: J. D. Gunton, M. Droz, Introduction to the Theory of Metastable and Unstable States. VI, 140 pages. 1983.

Vol. 184: Stochastic Processes – Formalism and Applications. Proceedings, 1982. Edited by G. S. Agarwal and S. Dattagupta. VI, 324 pages. 1983.

Vol. 185: H. N. Shirer, R. Wells, Mathematical Structure of the Singularities at the Transitions between Steady States in Hydrodynamic Systems. XI, 276 pages. 1983.

Vol. 186: Critical Phenomena. Proceedings, 1982. Edited by F.J.W. Hahne. VII, 353 pages. 1983.

Vol. 187: Density Functional Theory. Edited by J. Keller and J.L. Gázquez. V, 301 pages. 1983.

Vol. 188: A. P. Balachandran, G. Marmo, B.-S. Skagerstam, A. Stern, Gauge Symmetries and Fibre Bundles. IV, 140 pages. 1983.

Vol. 189: Nonlinear Phenomena. Proceedings, 1982. Edited by K. B. Wolf. XII, 453 pages. 1983.

Vol. 190: K. Kraus, States, Effects, and Operations. Edited by A. Böhm, J. W. Dollard and W. H. Wootters. IX, 151 pages. 1983.

Vol. 191: Photon Photon Collisions. Proceedings, 1983. Edited by Ch. Berger. V, 417 pages. 1983.

Vol. 192: Heidelberg Colloquium on Spin Glasses. Proceedings, 1983. Edited by J. L. van Hemmen and I. Morgenstern. VII, 356 pages. 1983.

Selected Issues from

Lecture Notes in Mathematics